Assessment of the Department of Veterans Affairs Airborne Hazards and Open Burn Pit Registry

David A. Savitz, Anne N. Styka, and David A. Butler, *Editors*

Committee on the Assessment of the Department of Veterans Affairs
Airborne Hazards and Open Burn Pit Registry

Board on the Health of Select Populations

Board on Population Health and Public Health Practice

Health and Medicine Division

A Report of
The National Academies of
SCIENCES · ENGINEERING · MEDICINE

THE NATIONAL ACADEMIES PRESS
Washington, DC
www.nap.edu

THE NATIONAL ACADEMIES PRESS 500 Fifth Street, NW Washington, DC 20001

This study was supported by Contract No. VA241-P-2024 between the National Academy of Sciences and the Department of Veterans Affairs. Any opinions, findings, conclusions, or recommendations expressed in this publication do not necessarily reflect the views of any organization or agency that provided support for the project.

International Standard Book Number-13: 978-0-309-45117-8
International Standard Book Number-10: 0-309-45117-5
Digital Object Identifier: 10.17226/23677

Additional copies of this publication are available for sale from the National Academies Press, 500 Fifth Street, NW, Keck 360, Washington, DC 20001; (800) 624-6242 or (202) 334-3313; http://www.nap.edu.

Printed in the United States of America

Cover photo by E. B. Boyd. Used with permission.

Suggested citation: National Academies of Sciences, Engineering, and Medicine. 2017. *Assessment of the Department of Veterans Affairs Airborne Hazards and Open Burn Pit Registry*. Washington, DC: The National Academies Press. doi: 10.17226/23677.

The National Academies of

SCIENCES • ENGINEERING • MEDICINE

The **National Academy of Sciences** was established in 1863 by an Act of Congress, signed by President Lincoln, as a private, nongovernmental institution to advise the nation on issues related to science and technology. Members are elected by their peers for outstanding contributions to research. Dr. Marcia McNutt is president.

The **National Academy of Engineering** was established in 1964 under the charter of the National Academy of Sciences to bring the practices of engineering to advising the nation. Members are elected by their peers for extraordinary contributions to engineering. Dr. C. D. Mote, Jr., is president.

The **National Academy of Medicine** (formerly the Institute of Medicine) was established in 1970 under the charter of the National Academy of Sciences to advise the nation on medical and health issues. Members are elected by their peers for distinguished contributions to medicine and health. Dr. Victor J. Dzau is president.

The three Academies work together as the **National Academies of Sciences, Engineering, and Medicine** to provide independent, objective analysis and advice to the nation and conduct other activities to solve complex problems and inform public policy decisions. The National Academies also encourage education and research, recognize outstanding contributions to knowledge, and increase public understanding in matters of science, engineering, and medicine.

Learn more about the National Academies of Sciences, Engineering, and Medicine at **www.national-academies.org**.

COMMITTEE ON THE ASSESSMENT OF THE DEPARTMENT OF VETERANS AFFAIRS AIRBORNE HAZARDS AND OPEN BURN PIT REGISTRY

DAVID A. SAVITZ (*Chair*), Vice President for Research, Professor of Epidemiology, and Professor of Obstetrics and Gynecology, Brown University, Providence, RI
VINÍCIUS C. ANTÃO, Director of Patient Registries, Healthcare Research Institute, Hospital for Special Surgery, New York, NY
JANE E. CLOUGHERTY, Associate Professor and Director of Exposure Science, University of Pittsburgh, PA
MONTSERRAT FUENTES, Dean of College of Humanities and Sciences, Virginia Commonwealth University, Richmond, VA
RICHARD A. KULKA, Consultant, Richard A. Kulka Consulting, Raleigh, NC
FRANCES MURPHY, President and CEO of Sigma Health Consulting, LLC, Silver Spring, MD
CECILE S. ROSE, Professor of Medicine in the Division of Environmental and Occupational Health Sciences, National Jewish Health, Denver, CO
ARMISTEAD G. RUSSELL, Howard T. Tellepsen Chair and Regents' Professor, Georgia Institute of Technology, Atlanta, GA
DAVID H. TRUMP, Chief Deputy Commissioner for Public Health and Preparedness, Virginia Department of Health (retired), Richmond, VA
JOYCE S. TSUJI, Principal Scientist, Exponent, Bellevue, WA
MARK J. UTELL, Professor of Medicine and Environmental Medicine, Director of Occupational and Environmental Medicine, University of Rochester Medical Center, Rochester, NY

Study Staff

DAVID A. BUTLER, Study Director. Scholar and Director, Office of Military and Veterans Health
ANNE N. STYKA, Program Officer
CARY HAVER, Program Officer (from July 2016)
PAMELA RAMEY-McCRAY, Administrative Assistant (from October 2016)
SULVIA DOJA, Research Associate (through August 2016)
NICOLE FREID, Senior Program Assistant (August 2015–September 2016)

Reviewers

This report was reviewed in draft form by individuals chosen for their diverse perspectives and technical expertise. The purpose of this independent review is to provide candid and critical comments that will assist the institution in making its published report as sound as possible and to ensure that the report meets institutional standards for objectivity, evidence, and responsiveness to the study charge. The review comments and draft manuscript remain confidential to protect the integrity of the deliberative process. We wish to thank the following individuals for their review of this report:

Ronald Blanck, Martin, Blanck & Associates
Babette Brumback, University of Florida
Don A. Dillman, Washington State University
Herman Gibb, Gibb Epidemiology Consulting, LLC
Grace LeMasters, University of Cincinnati
Shari Beth Libicki, Ramboll Environ
Dylan Small, University of Pennsylvania
Tyler C. Smith, National University School of Health and Human Services

Although the reviewers listed above have provided many constructive comments and suggestions, they were not asked to endorse the conclusions or recommendations nor did they see the final draft of the report before its release. The review of this report was overseen by **Frank E. Speizer,** Harvard Medical School; **Ellen Wright Clayton,** Vanderbilt University; and **Chris Whipple,** Environ (retired). They were responsible for making certain that an independent examination of this report was carried out in accordance with institutional procedures and that all review comments were carefully considered. Responsibility for the final content of this report rests entirely with the authoring committee and the institution.

Preface

Women and men who were deployed to Iraq, Afghanistan, and other locations in Southwest Asia were subjected to an extensive array of health threats beyond those directly resulting from combat. Environmental exposures, in particular, have been of great concern to many of those who served, and understandably so, and there is perhaps no more worrisome issue in this area than the potential health consequences from exposure to emissions from open burn pits. Undoubtedly, for some period of time, the disposal of all waste materials through uncontrolled incineration led to the exposure of large numbers of service personnel to particulate matter and other health hazards, which in turn created a high probability of both acute and chronic health consequences in these individuals. Congress mandated that the Department of Veterans Affairs (VA) create a registry that would acquire exposure and health information on service members and veterans exposed to burn pits and other airborne hazards. Despite a number of daunting challenges in doing so in a short time frame, VA was fully responsive and developed an ambitious program to enroll volunteer participants. With a substantial volume of information now in hand, our committee was asked to evaluate the effectiveness of this program along a number of dimensions. As the report indicates, we have a number of major concerns and suggestions for improvement, but we acknowledge that we are addressing these issues with the luxury of hindsight, sufficient time to examine and evaluate the work, and the range of expertise needed to make such an assessment. VA faced a much more daunting task and deserves the gratitude of the Congress, service members, and veterans for its efforts, which we applaud. We also want to be clear that as we critically examine the nature of the registry and the data VA has produced, we fully appreciate and support the need to be responsive to the concerns of those who served, and we acknowledge that some of those who were exposed to burn pits and airborne hazards undoubtedly have suffered and continue to suffer adverse health consequences. Our assessment is focused on the registry itself as a means of answering questions about the health consequences of that exposure in a scientifically informative and constructive manner.

The committee wishes to acknowledge the VA staff who responded to our many requests for information related to the registry: Drs. Paul Ciminera, Nicholas G. Lezama, and Michael A. Montopoli; and Mr. Vincent Mitchell. The data analyses contained in this report were performed by the research corporation Westat under the direction of the committee. The committee greatly benefited from the work performed by Dr. Joseph Gasper, Mr. Jason Liu, and Ms. Jennifer Kawata and very much appreciates their rigor, their willing and able response to repeated requests, and the clarity of their presentation. The committee is grateful to the many veterans and experts who attended and provided input or materials during and after the committee's May 2015 workshop. Finally, we need to give a great deal of credit to the Health and Medicine Division staff who contributed profoundly to the committee's report. Dr. David Butler, Ms. Anne Styka, and Ms. Cary Haver provided a critical understanding of the past work on other reports related to burn pits as well as evincing a nuanced ability to inform and guide the committee's work without constraining its conclusions. We also thank Ms. Sulvia Doja, Ms. Pam McCray, and Ms. Nicole Fried for generously and capably providing logistical support to the committee. A thank you is also extended to Mr. Daniel Bearss, who conducted database and literature searches, and Ms. Ellen Kimmel, who assisted the committee with fact checking the report.

David A. Savitz, *Chair*
Committee on the Assessment of the Department of Veterans Affairs
Airborne Hazards and Open Burn Pit Registry

Contents

APPENDIXES

Acronyms and Abbreviations

AFHSC	Armed Forces Health Surveillance Center
AFIOH	Air Force Institute for Occupational Health
AH&OBP	Airborne Hazards and Open Burn Pit
AOR	adjusted odds ratio
BMI	body mass index
CAD	coronary artery disease
CHPPM	Center for Health Promotion and Preventive Medicine
CI	confidence interval
CMI	chronic multisymptom illness
COPD	chronic obstructive pulmonary disease
CTS	Contingency Tracking System
DEERS	Defense Enrollment Eligibility Reporting System
DMDC	Defense Manpower and Data Center
DoD	Department of Defense
DOEHRS	Defense Occupational Environmental Health Readiness System
EPMSP	Enhanced Particulate Matter Surveillance Program
GAO	Government Accountability Office
ICD-9	*International Classification of Diseases, Ninth Revision*
IED	improvised explosive device
IOM	Institute of Medicine
IRR	incidence rate ratio
MI	myocardial infarction

NHIS	National Health Interview Survey
OEF	Operation Enduring Freedom
OIF	Operation Iraqi Freedom
OMB	Office of Management and Budget
OND	Operation New Dawn
OR	odds ratio
PAH	polycyclic aromatic hydrocarbons
PCDD/F	polychlorinated dibenzo-*p*-dioxins and dibenzo-*p*-furans
PM	particulate matter
SSC	Sergeant Thomas Joseph Sullivan Center
VA	Department of Veterans Affairs
VHA	Veterans Health Administration
VOC	volatile organic compound

Summary

Military operations produce a great deal of trash in an environment where standard waste management practices may be subordinated to more pressing concerns. As a result, ground forces have long relied on incineration in open-air pits as a means of getting rid of refuse.

The consequences of this expediency have come to the fore in the most recent conflicts in which the U.S. military have participated. The Department of Defense (DoD) cataloged 63 open burn pit sites in Iraq as of November 2009 and another 197 in Afghanistan as of 2011 (DoD, 2011). DoD estimated that collectively the sites at the large bases alone produced approximately 60,000–85,000 pounds of solid waste per day, including plastics, wood, metal, and—according to some sources—such materials as chemicals (paints, solvents), petroleum, medical waste, munitions, and human waste (IOM, 2011). No inventories of these materials were kept, and little information besides anecdote exists regarding the burning that took place at the many small and often temporary sites used by forces.

Concerns over possible adverse effects of exposure to smoke from trash burning in the theater were first expressed in the wake of the 1990–1991 Gulf War and stimulated a series of studies that indicated that exposures to smoke from oil-well fires and from other combustion sources, including waste burning, were stressors for troops (IOM, 2005). These studies grew in number in the years that followed and resulted in a provision in the National Defense Authorization Act for Fiscal Year 2010 (Public Law 111-84, § 317; enacted October 28, 2009) that prohibited the disposal of waste in open-air burn pits by DoD and called for the department to issue appropriate regulations concerning them. At that same time, the Department of Veterans Affairs (VA) asked the National Academies of Sciences, Engineering, and Medicine (the National Academies) to convene a committee to examine the long-term health consequences of exposure to burn pits in Iraq and Afghanistan. That committee's report (IOM, 2011) included a recommendation that an epidemiologic study be conducted to evaluate the health status of deployed service members.

Congress again took action in 2013 (Public Law 112-260, § 201; enacted January 10, 2013) when it directed VA to establish and maintain a registry for service members who may have been exposed to toxic airborne chemicals and fumes generated by open burn pits. The law also called for an independent scientific organization to prepare a report addressing issues related to the establishment and conduct of the registry and use of its data.

In late 2014 VA asked the National Academies to take on this responsibility. The full statement of task is contained in Box 1-1. In brief, it calls on the committee to analyze the initial months of data collected by the registry and offer recommendations on ways to improve the instrument and best use the information it collects. This report, prepared by the Committee on the Assessment of the Department of Veterans Affairs Airborne Hazards and Open Burn Pit Registry, fulfills the congressional mandate and provides responses to other questions posed by VA.

FRAMEWORK AND ORGANIZATION

The committee organized its response to its statement of task into six chapters addressing the following topics:

- The motivation for and the conduct of the study, including an overview of the issues related to the health effects of exposure to emissions from open burn pits and other airborne hazards present in the Southwest Asia theater of operations, the committee's approach to carrying out its work, and information learned from its public workshop and recent epidemiologic studies of military personnel exposed to burn pits (Chapter 1).
- The use of registries in environmental health research, with a focus on VA and DoD exposure registries; the limitations of the data they gather and of the inferences that can be drawn from them; and an overview of potential comparison populations that might be used in studies of health outcomes in Airborne Hazards and Open Burn Pit (AH&OBP) Registry respondents (Chapter 2).
- An examination of the development and implementation of the AH&OBP Registry and of the content of its key element, the self-assessment questionnaire (Chapter 3).
- The methods used and the results of the committee's analyses of the initial months of data collected by the registry, including descriptive statistics and the demographic and military characteristics of respondents (Chapter 4); and interpretation of the exposure (Chapter 5) and the health outcomes (Chapter 6) data collected by the questionnaire that the committee believes offer the most information value.

Those chapters contain the details that build the foundation for the committee's findings, conclusions, and recommendations presented in Chapter 7.

THE COMMITTEE'S INFORMATION-GATHERING AND ANALYSIS EFFORTS

The scientific foundation for the report's findings, conclusions, and recommendations was developed through a number of information-gathering and data analysis activities. These included conducting a public workshop, directing the work of a consulting firm that performed analyses of registry data, and carrying out a review of relevant research literature, including previous National Academies studies on such topics as the use of military burn pits, efforts to monitor air quality in the Persian Gulf region, the state of the literature regarding the health effects of exposure to combustion products in general and burn pit emissions, and related topics.

The committee's workshop—an element of the statement of task—was conducted in May 2015 and included presentations by professionals responsible for carrying out air emissions testing in-theater, researchers studying health outcomes in repatriated military personnel and veterans, a physician providing health care to these individuals, and veterans and veteran service organization representatives sharing their personal experiences and knowledge. Their input gave the committee great insight into the circumstances faced by service members during and after their deployment.

The analyses of the registry data were directed by the committee but carried out by a contractor that obtained the information directly from VA, and neither committee members nor National Academies staff had any access to the data. This was due to VA protocols regarding the management of data they considered to be personally identifiable information and the protection of the privacy of registrants and the security of the information they were providing. As a result of these protocols, some data were not available for analysis, and the committee was not able to fully address its charge in this area.

The committee's literature review consisted of a targeted examination of epidemiologic studies of long-term health outcomes in military and veteran populations potentially exposed to burn pit emissions. It was limited to studies that had been published since the last National Academies review of such work in 2011 (IOM, 2011). A total of eight studies were identified and summarized.

Separately, the committee considered whether there were any population groups that might be used in comparisons of demographic or other characteristics with AH&OBP registrants. Active duty military, veteran, and general population groups were examined and their strengths and weaknesses in this application assessed, but the committee ultimately concluded that, given the nature and quality of the registry data, such comparisons were inappropriate.

ENVIRONMENTAL HEALTH REGISTRIES AS AN INFORMATION-GATHERING INSTRUMENT

Registries are structured systems for collecting and maintaining data on a group of people characterized by a specific disease, condition, exposure, or event as a means to facilitate research, monitor health, or provide information to registrants. While they are generally quicker and less expensive to establish than alternative means—such as an epidemiologic study—and while they allow for the ascertainment of several exposures and health outcomes on a defined population, they also have several inherent limitations. Registries that rely on voluntary involvement and self-reported information such as exposures and health outcomes are subject to data biases resulting from such circumstances as selective participation, faulty recall, inaccurate information, or inadvertent or intentional underestimation or exaggeration. They are thus an intrinsically poor source of information on exposures, health outcomes, and possible associations among these events. Even under the best of circumstances, there are substantial limits to the accuracy of the data and—when the respondents make up only a small, unrepresentative fraction of the eligible population—to the generalizability of analyses made with them.

THE AIRBORNE HAZARDS AND OPEN BURN PIT REGISTRY

VA was presented with a challenge when it was directed by the Congress to design, test, and implement an environmental health registry for "individuals who may have been exposed to toxic airborne chemicals and fumes caused by open burn pits" in 12 months. The questionnaire that it developed to collect information from registrants exhibits a number of problems that stem in part from the inherent weaknesses of voluntary, self-report registries but that are exacerbated by a series of flaws in the registry's structure and operation as well as in the questions that are asked and the way they are asked. The committee's review of the AH&OBP Registry questionnaire found that it

- inappropriately uses questions that were validated for and meant to be administered by other survey means such as a face-to-face or computer-assisted phone interview;
- asks questions that may be confusing for respondents because they are ambiguous or otherwise poorly written;
- elicits information on topics such as hobbies and places of childhood residence that do not yield information that could be productively used in any analysis that would be appropriate to undertake using registry data;
- fails to ask questions (regarding non-burn-pit trash burning, for example) that could yield information related to relevant exposures;
- does not take full advantage of its Web-based format to streamline and focus questions based on previous responses;
- does not permit answers to be supplemented or updated later in time; and
- requires respondents to complete a sometimes lengthy set of repetitive questions regarding deployments before addressing core issues like health, increasing the possibility of response fatigue.

The cumulative effect of these flaws is evidenced by the high percentage of respondents who initiated but did not complete the instrument and the number of questions that had large nonresponse rates.

The issue of how to improve the questionnaire depends critically on the registry's intended purpose(s) going forward. Even a well-designed and executed registry would have little value as a scientific tool for health-effects research compared to a well-designed epidemiologic study. The committee concludes that, given the inherent weaknesses of the instrument, the best ways to make use of the AH&OBP Registry are to

1. make it a means for the eligible population to document their concerns over health problems that may have resulted from their service, bring those concerns to the attention of VA and their health care providers, and supply VA with a list of persons who are interested in burn pit exposure issues; and
2. collect self-reported data on exposures and health problems in the respondents that might possibly be used to stimulate research using more sophisticated analysis means.

If VA chooses to use the registry for these purposes, then the questionnaire may be simplified as follows:

The committee recommends that VA eliminate the questionnaire sections addressing locations of previous residences (Section 4), non-military work history (5) and home environment, community, and hobbies (6), which collect data that might only be useful in epidemiologic studies of the population.

Eliminating these categories would make the questionnaire easier and faster to complete and would better focus it on the needs of the eligible population.

More generally, the AH&OBP Registry's data collection, administration, and management efforts would be improved by taking these steps:

The committee recommends that once VA clarifies the intent and purpose of the registry, it develop a specific plan for more seamlessly integrating relevant VA and DoD data sources with the registry's data with the goals of reducing future participant burden, increasing data quality by restructuring questions to minimize recall and other biases, and improving the usefulness of the registry database as an information source for health care professionals and researchers.

The committee recommends that alternative means of completing the questionnaire, such as a mail-in form or via a computer-assisted phone interview, be offered in order to ensure that the subset of eligible persons who do not use or are not facile with the Internet have the opportunity to participate in the registry.

The committee recommends that VA involve external survey experts experienced in Web-based instruments in any restructuring of the registry questionnaire.

ANALYSIS METHODS AND DESCRIPTIVE STATISTICS FOR THE REGISTRY DATA

VA made data from the first 13 months of the operation on the AH&OBP (June 2014–July 2015) available to the committee's contractor. However, VA data security and participant privacy protocols precluded these data from including any items that would allow for a description of the association between respondents' self-reported exposures and their Veterans Health Administration (VHA) health care experience as called for in the statement of task. They also circumscribed the type and level of detail of other requested analyses. The restrictions thus affect the confidence with which the committee can draw conclusions regarding the process of data acquisition and the validity of the information reported on exposure and health outcomes.

Another major limitation is that questionnaire and other data were made available only for those who finished and submitted the questionnaire. A VA report (2015) indicated that nearly 40% of those who began filling out an AH&OBP Registry questionnaire did not complete it; this is an outcome that should be followed up.

The committee recommends that VA evaluate whether and how registrants who did not complete the questionnaire differ from those who did, analyze the determinants of non-completion, and use this information to formulate strategies to encourage registrants to finish and submit their responses and improve the completion rate for future participants.

Approximately 1.0% of eligible Gulf War veterans and 1.7% of eligible post-9/11 veterans (46,404 respondents in total) are represented in the data made available for the committee's analysis. Nearly all respondents report having encountered one or more airborne hazards in theater: 96% of all respondents reported being exposed to a burn pit on at least one deployment, and 85.6% of Gulf War era respondents reported exposure to smoke from oil-well fires, while 85.2% of all respondents reported being exposed to dust storms. The lack of data on those who were deployed and who do not believe they were exposed to burn pits precludes using the registry to compare exposed individuals with unexposed individuals. Therefore, the only means available for evaluating burn pit exposure is to examine gradations of exposure among the respondents.

Analyses of demographic data indicate that neither the Gulf War nor post-9/11 era registry respondents can be considered representative of their respective eligible non-respondent populations, although the differences are more pronounced for post-9/11 respondents and non-respondents. Thus, findings from these data—which represent the experience of a small, non-random, self-selected sample—are not generalizable to the broader, eligible population and cannot be used for making inferences concerning them.

ANALYSIS METHODS AND INTERPRETATION OF REGISTRY EXPOSURE DATA

The committee identified several problems with the way that the registry's exposure data were collected, which were compounded by the inherent limitations of self-reported information. One issue is that response fatigue resulting from the way that the exposure questions are structured may affect the accuracy of information provided by respondents who were deployed multiple times or to multiple locations. Another is that the questions do not provide information on the intensity of exposure and are in any case limited by the great variety of chemical constituents and particulate matter characteristics that made up that exposure and the lack of the information concerning them. A high fraction of registry participants reported potential exposures to both burn pit emissions and dust, and there was a tendency for individuals who reported exposures to one type of source to report exposures to other sources as well. These issues again highlight the lack of representativeness of the data and undermine its usefulness in evaluating associations between exposures and health outcomes.

Given the charge—and a concern for over interpreting the data at hand—the committee developed a reduced set of metrics to categorize exposure potential for the purpose of analysis. Because there were many sources of airborne emissions that contributed to a service member's exposures to particulate matter and chemical exposures and insufficient data by which to determine which sources contributed the most or pose the most harm, the committee chose to weigh each potential exposure equally and to focus on the totality of exposures.

On the basis of its evaluation, the committee concludes that the exposure data are of insufficient quality or reliability to make them useful in anything other than the most general assessments of exposure potential. Given this limitation, the committee believes that there may be some circumstances where supplementing these data with information from on-site environmental monitoring, meteorological, satellite, or other relevant measurements or observations might yield results that would suggest that some individuals or groups experienced greater or lesser exposures to specific constituents; these results might in turn stimulate more detailed assessments of health outcomes in particular populations.

ANALYSIS METHODS AND INTERPRETATION OF REGISTRY HEALTH OUTCOMES DATA

The committee took an approach analogous to that used for exposure data to characterize the health outcomes data for analysis purposes—specifically, generating variables using multiple grouped indicators of these outcomes. Health outcomes related to the symptoms, conditions, and diseases associated with the respiratory and the cardiovascular systems were identified as the best candidates for study because these are the most plausible and well-documented potential health effects of the exposures of concern. However, the limitations of the AH&OBP questionnaire and the data collected by it are still too great to allow any firm conclusions to be drawn from this analysis.

Generally speaking, the committee found that the observed prevalences of respiratory and cardiovascular outcomes appear consistent with what would be expected in a population that is predominantly male, aged 25–60, and for whom about one-third report a current or former history of smoking. It concluded that the health data may be of sufficient quality to justify internal comparisons in which data from subsets of registrants with varying levels of potential exposure are compared with one another. An examination of multiple indices of exposure to burn pit emissions and other hazards associated with deployment showed that registrants who reported more exposures of all types also tended to report more health problems of all types.

The committee's exposure potential variables had strong and consistent associations with self-reported asthma; any respiratory symptom; emphysema, chronic bronchitis, or chronic obstructive pulmonary disease (COPD); functional limitations due to lung or breathing problems; cardiovascular disease; and hypertension. Importantly, though, the analyses also uncovered some unexpected findings that are not consistent with currently understood scientific mechanisms of exposure and outcome, such as a statistically significant association between higher self-reported levels of asbestos exposure and a higher prevalence of neurologic, immune, or liver conditions. Such outcomes strongly suggest that the results of analyses of registry data cannot be taken at face value and that the identified associations may be an artifact of the population's selection and the limitations of the self-reported exposure and disease data.

Again, the bottom line is that registry analyses are not generalizable and can only describe what exposures and conditions the population of registry respondents are reporting; registry data cannot be used to determine cause or to estimate prevalence in the total eligible population of service members or veterans. The committee wishes to emphasize that it would have reached this same determination had the analyses found no associations or weak associations between the exposures and health outcomes.

The strong conclusion that can be drawn is that a more rigorous and appropriate study design is needed to examine the relationship between the exposures encountered during deployment to the Southwest Asia theater of operations and health outcomes. While the registry provides a forum for collecting and recording information on those who were deployed and are motivated to participate, it cannot answer such questions.

The committee recommends that other means for evaluating the potential health effects associated with airborne hazards and open burn pit exposures be developed, such as a well designed epidemiologic study.

The 2011 report *Long-Term Health Consequences of Exposure to Burn Pits in Iraq and Afghanistan* (IOM, 2011) contains advice and recommendations on how such a study might be conducted.

The committee concludes that, while medically verified health outcomes information only exists for the subset of the population that uses VA health care—data that were not available to the committee but that are contained in VHA records—there is potential value in linking the registry data to health care use data and conducting analyses. Comparisons between self-reported information collected by the questionnaire and diagnoses in VA medical records for respondents who use VA health care would provide further information concerning the level of validity of self-reported health outcomes in the population of respondents.

Given this and the committee's other findings regarding the registry,

The committee recommends that VA's messaging be explicit about the limitations on the ability of the AH&OBP Registry to generate valid information that can be used to improve VA health and benefits programs or to inform treatment of individuals potentially exposed to burn pits or other airborne hazards in theater in order to ensure that participants and others do not form unrealistic expectations about the value of participation or the capabilities of the registry.

OTHER FINDINGS, CONCLUSIONS, AND RECOMMENDATIONS

The committee was also asked to offer observations on some additional issues surrounding the registry and the actions being taken by DoD and VA to address airborne hazards and open burn pit questions. Specifically, the legislation that directed VA to establish the registry called for

- An assessment of the effectiveness of actions taken by the [Department of Veterans Affairs and the Department of Defense] to collect and maintain information on the health effects of exposure to toxic airborne chemicals and fumes caused by open burn pits.
- Recommendations to improve the collection and maintenance of such information.
- Using established and previously published epidemiological studies, recommendations regarding the most effective and prudent means of addressing the medical needs of eligible individuals with respect to conditions that are likely to result from exposure to open burn pits (Public Law 112-260 § 201(b)(1)(A)(i–iii)).

To date, other than the AH&OBP Registry and the airborne exposures and health information collected as part of such efforts as the Gulf War Registry and Millennium Cohort Study, there are no systematic data collection or maintenance efforts focused on the effects of burn pit emissions.[1] Very limited in-theater air pollution data gathering efforts have generated information that would aid in studies of those who served in the same place and at the same time as measurements were made. Two previous National Academies reports have offered recommendations on how more rigorous and useful data could be collected: *Review of the Department of Defense Enhanced Particulate Matter Surveillance Program Report* (NRC, 2010) and *Long-Term Health Consequences of Exposure to Burn Pits in Iraq and Afghanistan* (IOM, 2011). This committee concludes that the recommendations these reports offer regarding, respectively, environmental sampling in a combat theater and the conduct of a prospective study of the long-term health effects of exposure to burn-pit emissions are still salient and, if implemented, would materially improve the knowledge base concerning the health effects of past, present, and future in-theater exposures.

The committee's review of the literature found there have been very few epidemiologic studies of service members and veterans exposed to open burn pits. The available information does not suggest any general course of action for addressing the medical needs of this population beyond the steps that health care providers should already be carrying out: taking a thorough history, including all occupational exposures, listening carefully to the patient, and structuring the clinical response accordingly. The health care provider instructions for AH&OBP Registry clinical examinations produced by VA (2016) are sound guidance on this.

The information developed by the registry has limited value for improving individual patient care. However, while these data may be inappropriate for evaluating the association between exposures and health outcomes, there are other ways in which they may be useful. As has already been mentioned, the committee believes that the registry's primary utility is that it provides a means for veterans and service members to document their concerns about wartime exposures and the health problems that might have resulted from them and to bring these to the attention of both VA and their health care providers. The self-reported signs, symptoms, and diseases identified by registrants constitute a record that can alert providers to concerns and problems that may be forgotten about or missed during clinical encounters.

The registry questionnaire collects a number of pieces of information that would facilitate conversations between a patient and a health care provider, without regard to whether the information might be relevant to AH&OBP exposures. For example, someone who reported difficulty walking long distances or climbing stairs might be experiencing joint pain, respiratory problems, atherosclerotic vascular disease with congestive heart failure, obesity, or even anxiety. Similarly, a complaint of chest pain can have multiple causes in addition to angina and coronary artery disease, including gastroesophageal reflux disease, chest wall pain or costochondritis, and anxiety. And often these symptoms can be multi-factorial in origin. Registry questionnaire responses are already accessible to VA health care providers as part of a veteran's electronic health record, and a complete set of responses may be downloaded and printed for a respondent to take to a clinical visit with a provider in or outside of the VA system.

The committee recommends that VA enhance the utility of the AH&OBP Registry by developing a concise version of participant's questionnaire responses focused on information that would be most useful in a routine clinical encounter and make it available for download.

[1] DoD and VA collect and analyze data on all medical conditions in the populations that participate in their health care programs, but these are not specific to airborne hazards or burn pit emissions.

Providers often have little time to get histories and patients do not always do a good job of raising concerns, so a succinct summary would greatly benefit both.

The data the registry provides on the number of respondents who report particular health problems may also be useful to VA. For example, several thousand individuals have indicated that they have diagnosed or self-reported cardiopulmonary symptoms. If these persons subsequently present for an evaluation or treatment at rates that would not otherwise have been anticipated by VA, it would indicate that the registry could be used as tool for anticipating future demand for particular provider services. However, it remains to be seen whether this would be the case, and the number of individuals who have thus far completed the questionnaire is only a tiny fraction of the overall population eligible for VA care.

Given the demonstrated concerns of respondents regarding the health effects of exposure to airborne hazards and open burn pit emissions, it is unclear why so few have yet to arrange for the optional in-person clinical evaluation by a VA provider that is made available as part of the registry.

The committee recommends that VA continue its efforts to make it easier for participants to schedule and get the optional health examination offered as part of the AH&OBP Registry—such as through targeted follow-up of respondents who indicate interest—and that it investigate the reasons why such a small percentage of respondents who indicate interest in an exam (~2.5%, to date) request one.

Adding a means of scheduling an exam as part of the questionnaire—a capability that the committee understands is being implemented—is a useful first step.

CLOSING OBSERVATIONS

The committee recognizes the great interest that active duty military personnel and veterans who served in Iraq, Afghanistan, and the greater Southwest Asia theater of operations have in understanding potential threats to their health from airborne hazards and open burn pit exposures. As its analysis has made clear, though, there are inherent features of registries that rely on voluntary participation and self-reported information that make them fundamentally unsuitable for addressing the question of whether these exposures have, in fact, caused health problems. Addressing the issues identified by the committee would, though, improve the AH&OBP Registry's utility as a means of

- generating a roster of concerned individuals that VA can use for targeted outreach, surveillance, and health-risk communication;
- creating, via the completed questionnaire, a record of self-reported exposures and health concerns that is recorded in the participant's VA electronic health record; and
- allowing VA users and nonusers who take part in the optional clinical exam to articulate concerns they may have to a health care provider and, if warranted, undergo appropriate diagnostic testing or referral, and begin treatment to improve symptoms.

All parties—service members, veterans, and their families; VA; Congress; and other concerned people—would benefit from having a realistic understanding of the strengths and limitations of registry data so that they can make best use of them and, if desired, conduct the kind of investigations that might yield salient health information and improve health care for those affected.

REFERENCES

DoD (Department of Defense). 2011. Exposure to toxins produced by burn pits: Congressional data request and studies. In Memorandum for the Assistant Secretary of Defense for Health Affairs. Washington, DC: Department of Defense.

IOM (Institute of Medicine). 2005. *Gulf War and health, volume 3: Fuels, combustion products, and propellants*. Washington, DC: The National Academies Press.

IOM. 2011. *Long-term health consequences of exposure to burn pits in Iraq and Afghanistan.* Washington, DC: The National Academies Press.

NRC (National Research Council). 2010. *Review of the Department of Defense Enhanced Particulate Matter Surveillance Program report.* Washington, DC: The National Academies Press.

VA (Department of Veterans Affairs). 2015. *Report on data from the Airborne Hazards and Open Burn Pit (AH&OBP) Registry, June 2015.* http://www.publichealth.va.gov/docs/exposures/va-ahobp-registry-data-report-june2015.pdf (accessed September 10, 2016).

VA. 2016. War Related Illness and Injury Study Center. Airborne Hazards & Open Burn Pit Registry: Introduction to airborne hazards for providers. http://www.warrelatedillness.va.gov/education/factsheets/airborne-hazards-and-open-burn-pit-registry-for-providers.asp (accessed October 18, 2016).

1

Introduction

The involvement of the National Academies of Sciences, Engineering, and Medicine (the National Academies) with issues regarding the health impacts associated with burn pit emissions resulting from military operations began with a 2009 request from the Department of Veterans Affairs (VA) to examine the evidence regarding long-term effects from exposure to burn pits in the Iraq and Afghanistan theaters and to evaluate the feasibility and design issues surrounding an epidemiologic study of exposed veterans (IOM, 2011). While the statement of task for that study specified using Joint Base Balad in Iraq as an example of a location with high burn pit exposure, the committee responsible for conducting it extended the task to include all military personnel deployed to Iraq and Afghanistan who may have experienced exposure. Using peer-reviewed literature and data provided by the Department of Defense (DoD) and VA to examine the long-term health effects of exposure to burn pits, the 2011 report committee concluded that there was limited but suggestive evidence of an association between exposure to combustion products and reduced pulmonary function. However, it determined that there was inadequate or insufficient evidence of an association between exposure to combustion products and cancer, respiratory diseases, circulatory diseases, neurologic diseases, and adverse reproductive and developmental outcomes in the populations studied (service members, firefighters, and incinerator workers).

That Institute of Medicine (IOM) committee recommended that a tiered approach be applied to conducting an epidemiologic study that would characterize exposures and account for other sources of air pollutants in the ambient environment. The three tiers of the recommended study were characterized by the decreasing specificity of exposure and would answer different research questions. That committee suggested that the first tier be focused on determining whether proximity to burn pit operations at Joint Base Balad increased the risk of adverse health outcomes, using data such as dates of deployment, duties on base, and location of housing relative to the burn pit, and taking into account wind-dispersion models to assess individual exposure to Joint Base Balad burn pit emissions. The second tier would address whether the installation of incinerators at Joint Base Balad between 2008 and 2010 had reduced the incidence of disease or intermediate outcomes by evaluating chronic health outcomes in those deployed before and those deployed after the burn pit was shut down, factoring in the increased use of incinerators over those 2 years. The third tier would be an examination of whether deployment at Joint Base Balad during full burn pit operation increased the risk of adverse health outcomes compared with deployment elsewhere in Iraq or Afghanistan or with no deployment, taking into account the fact that the burn pit emissions occurred in the presence of particulate matter and other pollutants from other sources. With this approach, that committee concluded that a study of health effects resulting from exposure to burn pit emissions would be feasible, but the committee also realized that a study's ability to produce useful and actionable results would depend on it having a well thought-out design, thorough exposure assessment, and careful follow-up.

VA published preliminary plans formulated in response to the conclusions and recommendations contained in the 2011 report in a February 4, 2013, posting in the *Federal Register* (*Federal Register*, 2013). To address the need for further study of the long-term health effects of exposure to airborne hazards in Iraq and Afghanistan, VA proposed to (1) design appropriate studies that would incorporate objective measures such as clinical exams, (2) establish an independent oversight mechanism, and (3) conduct a cohort study to assess potential long-term effects related to burn pit emissions in the context of other ambient exposures. The cohort study would involve a population-based prospective study that would include baseline and repeated clinical examinations with sufficient follow-up to address the potential long-term health effects of deployment to Iraq and Afghanistan as well as potential burn pit exposure.

At around the same time (on January 10, 2013), Congress passed and the President signed Public Law 112-260,[1] § 201 of which directed VA to establish an open burn pit registry within 1 year after enactment. Appendix A of this report provides an excerpt of the law calling for the current study. The law directed the Secretary of VA to coordinate with DoD to establish and maintain an open burn pit registry for eligible individuals who may have been exposed to toxic airborne chemicals and fumes caused by open burn pits. It specified that the registry should include information that would be necessary to ascertain and monitor the health effects of members of the armed forces exposed to toxic airborne chemicals and fumes caused by open burn pits. The law instructed VA to develop a public information campaign to inform eligible individuals about the registry and to periodically notify eligible individuals of significant developments in the study and treatment of conditions associated with exposure to toxic airborne chemicals.

Section 201(b)(1) of the law directed the Secretary to enter into an agreement with an independent scientific organization to prepare a report that addressed several issues related to the establishment and conduct of the registry and use of its data.

THE COMMITTEE'S CHARGE AND APPROACH TO ITS TASK

In accordance with the directive contained in § 201(b)(1) of Public Law 112-260, VA entered into an agreement with the National Academies to form an expert committee to perform an assessment of the registry created by the law, which it named the Airborne Hazards and Open Burn Pit (AH&OBP) Registry. The committee's statement of task is shown in Box 1-1.[2] In brief, it directs the committee to conduct an analysis of data collected by the registry in the initial months of its operation and to offer observations and recommendations regarding its operation and best use of the data it collects.

The committee convened by the National Academies included experts in epidemiology, environmental and occupational health, exposure assessment, military and veteran's health, statistics, survey methodology, and toxicology. It initially comprised eight members, who held three in-person meetings between March and August 2015. These meetings included a presentation by VA staff to present and elucidate their charge to the committee and an information-gathering workshop that is detailed in the section below. Three more members were added in late 2015 to supplement the committee's expertise. That expanded committee met three additional times in person and twice by phone over 2016 to consider evidence and write its report.

INFORMATION GATHERING

Several activities were undertaken to develop the scientific foundation for the report's findings, conclusions, and recommendations. These included gathering material from the peer-reviewed literature; requesting informa-

---

[1] *Title II–Health Care*, Public Law 112-260, Establishment of Open Burn Pit Registry, § 201 (January 10, 2013).

[2] Note that the Statement of Task refers to the Institute of Medicine (IOM). As of March 2016, the Health and Medicine Division of the National Academies of Sciences, Engineering, and Medicine continues the consensus studies and convening activities previously undertaken by the IOM.

BOX 1-1
Committee's Statement of Task

In response to a mandate contained in Section 201 of Public Law 112-260, the Institute of Medicine (IOM) will convene an ad hoc committee to provide recommendations on collecting, maintaining, and monitoring information collected by the Department of Veterans Affairs' (VA's) Airborne Hazards and Open Burn Pit Registry (AH&OBPR). The committee will assess the effectiveness of the VA's information gathering efforts and provide recommendations for addressing the future medical needs of the affected groups. The study will review data collection methods and outcomes, as well as analyze the self-reported veteran experience data gathered in the registry. It will also focus on the assessment of the effectiveness of the actions taken by the VA and Department of Defense and provide recommendations for improving the methods enacted.

The committee will supervise the conduct of an analysis of the first ~6 months of data collected by the AH&OBPR using de-identified datasets provided by VA, summarize information on the cohort, and compare its demographic characteristics with the characteristics of all those who were deployed during the Gulf conflicts. The committee will also solicit veteran input and provide methodological recommendations on how to best ascertain and monitor the health effects of the exposure of members of the Armed Forces to toxic airborne chemicals caused by open burn pits and other potential airborne hazards during deployment. IOM will offer recommendations on improving the collection and maintenance of information in the AH&OBPR.

These recommendations shall include but not be limited to

1. how to categorize the self-reported exposures in the AH&OBPR,
2. suggested changes to the current information collection instrument, and
3. methodological approaches to the analysis of these data.

As part of the process, the committee will plan and conduct a workshop to receive suggestions and input from veterans about their experiences so that the recommendations can be informed by insights from this group. Following the completion of the analysis, the committee will prepare a report summarizing the approach and findings of this analysis and their recommendations.

The report's description of the initial data gathered in the AH&OBPR will address

1. associations of self-reported exposures with self-reported health conditions,
2. associations of self-reported exposures with Veterans Health Administration (VHA) health care experience, and
3. an analysis of how registry participants differ in demographic or exposure status (to the extent available data allows) from non-participant groups, such as all deployers or appropriate U.S. comparison populations.

The committee will use this information and additional research in order to evaluate how best to use the registry's self-reported data to benefit active-duty military personnel and veterans who were exposed to burn pit emissions. Public Law 112-260, § 201(b)(1)(A) requests that this report include

1. an independent scientific assessment of "the effectiveness of actions taken by the Secretaries [of Veterans Affairs and Defense] to collect and maintain information on the health effects of exposure to toxic airborne chemicals and fumes caused by open burn pits,
2. recommendations to improve the collection and maintenance of such information, and
3. using established and previously published epidemiologic studies, recommendations regarding the most effective and prudent means of addressing the medical needs of eligible individuals with respect to conditions that are likely to result from exposure to open burn pits."

tion directly from VA and from experts in the field; examining other pertinent published literature, government documents and reports, testimony presented to Congress, surveys, and exposure and diseases registries; attending professional meetings and educational events; and consulting relevant National Academies reports. In particular, the committee found a great deal of relevant background information on such topics as the use of military burn pits and the state of the literature regarding the health effects of exposure to combustion products in genera—and to burn pit emissions in particular—in the 2011 IOM report *Long-Term Health Consequences of Exposure to Burn Pits in Iraq and Afghanistan*, some results of which are touched on in the following section. Details concerning these topics may be found in that report.

Prior National Academies Reports Regarding Burn Pit Exposures and Health Outcomes

Three previous National Academies reports examined health outcomes related to specific exposures that occurred during deployment to Southwest Asia,[3] including exposures from or related to burn pit emissions. They are briefly summarized below.

The third volume of the report series *Gulf War and Health* contained a comprehensive review of the literature addressing the association between exposure to fuels, combustion products, and propellants present in the 1990–1991 Gulf War theater and health outcomes (IOM, 2005). Combustion products were defined as "smoke from fires, exhaust from burning fuels, and products of other combustion sources" (p. 39), and it was noted that these are also constituents of air pollution in general. The committee responsible for that report classified its findings into a set of categories adapted from the scheme used by the International Agency for Research on Cancer (IARC, 2006). It stated that

> The strongest finding was that there is sufficient evidence of an association between combustion products and lung cancer. The committee also found limited or suggestive evidence of an association between combustion-product exposure and cancers at several other sites (oral, nasal, laryngeal, and bladder), incident asthma, and two reproductive outcomes after exposure during pregnancy: preterm birth and low birthweight or intrauterine growth retardation. (p. 6)

Responding to a request from the U.S. Army, the National Academies formed an expert committee to review a report (Engelbrecht et al., 2008) that summarized the results of DoD's Enhanced Particulate Matter Surveillance Program (EPMSP) (NRC, 2010). EPMSP was an effort to characterize and quantify particulate matter in the ambient environment at 15 sites[4] in the Persian Gulf Region over 12 months in 2006–2007. This committee was also asked, among other tasks, to consider whether and how such data and other information collected by the U.S. Army Center for Health Promotion and Preventive Medicine (CHPPM) might be put to use in assessing the health outcomes of deployed personnel. It commended the effort but found that the design and conduct of the EPMSP limited its usefulness in health studies. The committee concluded that it was plausible that exposure to ambient pollution was associated with adverse health outcomes but that the interpretation of the information collected in theater was encumbered by uncertainties regarding the actual exposures, the small number of study subjects, and the limited amount of exposure data. It recommended that "[a] more complete inventory of all major sources of ambient pollutants and potential emissions in the theater should be constructed before assessment of health effects to ensure that all relevant pollutants are monitored" (p. 9).

Of particular relevance for the current report, the previously mentioned report *Long-Term Health Consequences of Exposure to Burn Pits in Iraq and Afghanistan* summarized the health effects associated with exposures to 51 pollutants that were detected in air samples taken at Joint Base Balad in Iraq in 2007 and 2009 (IOM, 2011). There were a number of potential sources of these, and that committee concluded that burn pits were not a major source

[3] VA defines the Southwest Asia Theater of operations to include the following locations: Iraq, Kuwait, Saudi Arabia, Bahrain, Gulf of Aden, Gulf of Oman, Oman, Qatar, and the United Arab Emirates; the waters of the Persian Gulf, Arabian Sea, and Red Sea; and the airspace above these regions.

[4] The 15 sites were in the following countries: Djibouti (one), Afghanistan (two, in Bagram and Khowst), Qatar (one), United Arab Emirates (one), Iraq (six, in Balad, Baghdad, Tallil, Tikrit, Taji, and Al Asad), and Kuwait (four, in northern, central, coastal, and southern Kuwait).

of many of them. Furthermore, some key air pollutants were not assessed, including ozone, carbon monoxide, nitrogen dioxide, and sulfur dioxide or other chemicals associated with combustion such as hydrogen cyanide.

Because there were few studies on exposures of military populations to burn pits, that committee expanded its literature review to include toxicologic studies. It also looked at studies of surrogate populations whose members had occupational or residential exposures to combustion products, such as firefighters, incinerator workers, and those living near an incinerator. The committee used a weight-of-evidence approach to determine the strength of the association between exposure to combustion products and each health outcome. Three major classes of chemicals detected at Joint Base Balad—polychlorinated dibenzo-*p*-dioxins and dibenzo-*p*-furans, polyaromatic hydrocarbons, and volatile organic compounds—and particulate matter have been associated with a wide array of long-term health effects in many organs and organ systems, including the adrenal glands, blood, lungs, liver, kidneys, stomach, spleen, and cardiovascular, respiratory, reproductive, and central nervous system (IOM, 2011).

However, the 2011 report concluded that there was "inadequate/insufficient evidence of an association between exposure to combustion products and cancer, respiratory disease, circulatory disease, neurologic disease, and adverse reproductive and developmental outcomes in the populations studied" (IOM, 2011, p. 7). It did find limited or suggestive evidence of an association between exposure to combustion products and reduced pulmonary function in the study populations, but it was unable to determine whether the long-term health effects are likely to result from service members exposed to emissions from burn pits—specifically the one in operation at Joint Base Balad—because high ambient concentrations of particulate matter from both natural and anthropogenic sources likely modified the effects, but could not be accounted for or adjusted in the analyses. Therefore, that committee concluded that the long-term health risk of airborne, burn pit, and other related exposures was not clearly defined. The report stated that none of the individual chemical constituents of the combustion products emitted from the burn pit appeared to have been present at concentrations high enough to be responsible for any of the adverse health outcomes. However, it was also noted that "the possibility of exposure to mixtures of those chemicals raises the potential for health outcomes associated with cumulative exposure to combinations of the constituents in burn pit emissions" (p. 8). Moreover, given the limitations of the literature, the information "might not provide a comprehensive picture of the risks posed to military personnel from burn pit emissions" (IOM, 2011, p. 103).

Registry Data Analysis Efforts

In order to conduct the analysis of the initial months of data collected by the AH&OBP Registry, the National Academies contracted with Westat, Inc., a statistical survey research corporation. The data made available by VA—which did not contain any personally identifiable information—were delivered to Westat in January 2016, and its staff conducted work under the direction of the committee. Neither committee members nor National Academies staff had any access to the data—they received summary statistics and analysis results only.

Separately, VA contracted directly with Westat to conduct a series of analyses of AH&OBP Registry data under VA's direction which resulted in three reports (Gasper and Kawata, 2015; VA, 2015a,b) and a publication in the *Journal of Occupational and Environmental Medicine* (Liu et al., 2016). The committee used these as references but emphasizes that the VA–Westat efforts were entirely independent of the committee's work, covered time periods that were different than the data made available for the committee's analyses, and used additional data sources that were not made available to the committee.

Some information requested by the committee was not released because of VA protocols regarding the management of data that VA considered to be personally identifiable information and the protection of the privacy of registrants and the security of the information VA was providing. These protocols limited the committee's ability to fully address its charge.

Summary of the Committee's Workshop

As part of fulfilling its statement of task, the committee held a workshop to assist in information gathering which informed discussions throughout this report. It heard from presenters with knowledge of veterans health issues related to airborne hazards and open burn pits, in-theater burn pit exposures characterization, military environmental health registries, and also from veterans and veteran service organizations with perspectives on burn pit exposure effects. A roundtable of veterans—including those who had been deployed to bases with burn pits—and family members of veterans offered their views and engaged in dialogue with the committee members.

The workshop agenda is reproduced in Appendix B. Copies of speakers' written presentations are posted to the National Academies website,[5] and highlights of this information are summarized below.

John Kolivosky, PE, provided an overview of an in-theater ambient air monitoring program. Mr. Kolivosky was the project lead on a burn pit assessment conducted in the Deployment Environmental Surveillance Program at the Army Institute of Public Health, and he provided the committee with details on how the effort was conducted and on the data that were collected. The operation of air sampling devices was the responsibility of Army Preventive Medicine personnel. They would conduct site surveys to identify potential exposure to airborne hazards and collect 24-hour time-composite samples using Environmental Protection Agency methods or the equivalent. The samples were sent out of theater to a laboratory for analysis and archived in the Defense Occupational Environmental Health Readiness System. Some 20,000 samples are in the database; most of these were taken within the boundaries of bases. Particulate matter (PM)—both PM_{10} and $PM_{2.5}$[6] —as well as heavy metals and volatile organic compound levels were measured at various times. Mr. Kolivosky indicated that there were a number of challenges associated with operating air sampling equipment in theater that limited the ability to collect good data. He and his colleagues have since published two papers on the results of the air monitoring effort (Blasch et al., 2016a,b).

Maj Charlie Toth, U.S. Air Force, PE, spoke about environmental sampling at the Balad Air Base.[7] Maj Toth focused on his knowledge of the design of the burn pit study sampling plan for the site and on personal observations from the actual collection of samples and subsequent report. The sampling plan was a joint effort with the Army CHPPM and Air Force Institute for Occupational Health. The goal was to quantify worst-case effects of exposures to the open burn pits. The intent of investigators was to gather this information while waste incinerators were still in place and the exposure was still comparable to what the exposed population had experienced. Maj Toth described his experience as an environmental engineer tasked with collecting samples near Balad Air Base and the resulting report published from the data gathered (CHPPM and AFIOH, 2009). Between January and May 2007, samples were collected to assess the levels of dioxins, furans, polycyclic aromatic hydrocarbons, volatile organic compounds, and PM_{10} particulates. Maj Toth suggested that having a dedicated sample team or providing more training for the use of the instruments could have improved the collection of samples. During his time in theater, Maj Toth said he recalled seeing items including plastics, metal/aluminum cans, rubber, chemicals such as paints and solvents, petroleum, oil, lubricant products, munitions, unexploded ordnance, wood waste, and incomplete combustion by-products in burn pits with jet fuel (JP-8) being used as the accelerant. The report summarizing this work noted that the actual number of final samples collected was relatively small and that 15% of the samples were rejected because of damage from shipping or failed pumps. The report's findings did not include input or verification from members of the team involved in the collection of data, which led, in Maj Toth's view, to flaws and misinterpretation of information.

John Rinker, CIH, discussed information developed in his master's thesis, "Retrospective Geospatial Modeling of PM10 Exposures from Open Burning at Joint Base Balad, Iraq" (Rinker, 2011), and provided the committee with an understanding of what data were gathered with respect to burn pit emissions and exposures. The intent of his study was to create an exposure-zone map using spatially interpolated particulate air sampling point data from Joint Base Balad and exposure contours from dispersion model outputs. He concluded that source-specific individual exposures could be estimated with his dispersion model and information on subjects' time-activity patterns.

Gary Gackstetter, DVM, PhD, and Tomoko I. Hooper, MD, briefed the committee on lessons learned from other health registries and the Millennium Cohort Study. Drs. Gackstetter and Hooper are part of the research team conducting the Millennium Cohort Study, a prospective longitudinal epidemiological research effort intended to evaluate the impact of military exposures, including deployment, on long-term health outcomes. While it was not

[5] See http://www.nationalacademies.org/hmd/Activities/Veterans/VABurnPitRegistry/BurnPitRegistryWorkshop.aspx (accessed February 9, 2017).

[6] $PM_{2.5}$ refers to airborne particles 2.5 microns in diameter and smaller, while PM_{10} refers to particles between 2.5 and 10 microns in diameter.

[7] Balad Air Base was renamed Joint Base Balad in June 2008. It reverted to Balad Air Base in December 2011 when the Iraq Air Force assumed responsibility for it.

specifically established for the purpose of examining exposures and health outcomes related to service in Iraq and Afghanistan, its original mandate envisioned the possibility of such studies, and 61% of the more than 200,000 participants in the study have deployed there, and the information obtained from them has been used to investigate burn pit-related issues (Gray et al., 2002; Smith et al., 2009, 2012). They presented background information on the Millennium Cohort and offered their perspectives on the utility of registries versus prospective epidemiologic studies in research on health outcomes in military populations. (Chapter 2 addresses the Millennium Cohort Study in a discussion of populations whose health outcomes could be compared with results generated by the AH&OBP Registry.)

Anthony Szema, MD, an adjunct professor at Stony Brook University and clinical assistant professor at Hofstra University, provided a perspective on the AH&OBP Registry and on registries in general as a tool for gathering information on the health of veterans. He presented his views on the limitations of the AH&OBP Registry and on topics the questionnaire does not address such as spirometry, allergy skin testing, contact dermatitis to metals and chemicals, gastroesophageal reflux disease, and lung biopsies. Dr. Szema suggested that some exposure questions were limited or absent, for example questions about dust in containerized housing units, JP-8 jet fuel, grinding metals, and asbestos. He contrasted the AH&OBP questionnaire with information gathering on the BurnPits360.org website.

A number of points were raised in the roundtable discussion that concluded the workshop. CPT (ret.) LeRoy Torres of the U.S. Army Reserve spoke to the committee via a Web link from his home. He related his experience serving at Camp Anaconda,[8] his exposure to the burn pit that operated there, and his subsequent health problems. Rosie Torres, the executive director of Burnpits 360°, which was established by her and CPT Torres, presented information on the registry that the organization maintains on its website. Ms. Torres suggested that the AH&OBP Registry include an open field that would allow the self-reporting of symptoms and allow family members of deceased eligible persons to submit information. Peter Sullivan of The Sergeant Thomas Joseph Sullivan Center echoed the suggestion that the registry be open to submission of information on deceased service members and veterans and added that mortality data should be collected for those already enrolled in the registry. He indicated that it was important for registry data to be used not only for research but also for improvements in clinical diagnosis, treatment, and prevention. Daniel Sullivan of The Sergeant Thomas Joseph Sullivan Center suggested that it would be useful for VA to involve veterans, their families, and advocacy organizations concerned with burn pit exposure issues in their outreach and education activities surrounding the registry. Representatives of three veterans service organizations—Adrian Atizado of Disabled American Veterans (via teleconference), Thomas Berger, PhD, and Rick Weidman of Vietnam Veterans of America, and Carlos Fuentes of Veterans of Foreign Wars—highlighted reports of difficulties encountered by some eligible persons in registering for and completing the questionnaire. They said that participants they had heard from had found that the online system would sometimes freeze and that some had encountered problems with correcting and supplementing the list of locations where they had served. The multiple formal and informal names used for some bases and difficulties in spelling location names were identified as particular problems. Furthermore, comments were offered that VA had not done a good job in encouraging people who did not have exposure to burn pits or were not experiencing health problems to fill out the registry questionnaire.

The committee is deeply grateful to the participants in the workshop. All of the information presented was factored into the committee's considerations whether or not any particular piece of information is explicitly mentioned here.

Recent Epidemiologic Studies of Military Personnel Exposed to Burn Pits

The committee's statement of task directed it to use established and previously published epidemiologic studies to provide recommendations regarding the most effective and prudent means of addressing the medical needs of eligible individuals with respect to conditions that are likely to result from exposure to burn pit emissions. To accomplish this, a targeted review was conducted of epidemiologic studies published since the last National Acad-

[8] Later called Joint Base Balad.

emies review of such work in *Long Term Health Consequences of Exposure to Burn Pits in Iraq and Afghanistan* (IOM, 2011).

The literature search was limited to epidemiologic studies of long-term health outcomes (not acute[9] or short-term) experienced by service members and veterans of the 1990–1991 Gulf War, the stabilization period (1992–September 2001), and post-9/11[10] (Operation Enduring Freedom [OEF], Operation Iraqi Freedom [OIF], and Operation New Dawn [OND]), which were published in or after 2010 through January 2016 and were not cited in the 2011 IOM report. Search terms included "open burn pit," "military or veterans," and "environment or occupational or war exposure." The committee did not include or review studies of populations exposed to other sources of burning materials, such as firefighters and incinerator workers. Multiple databases including Embase, Medline, Scopus, Web of Science, Cochrane Database of Systematic Reviews, PubMed Systematic Reviews, PubMed, and ProQuest were searched. The committee also searched other sources such as the Defense Technical Information Center database, and reports released by VA, RAND Corporation, DoD, the Government Accountability Office, and the Congressional Research Service.

The literature search identified 23 studies of potential relevance regarding the long-term health consequences of exposure to burn pit emissions in Southeast Asia, specifically Iraq and Afghanistan. The committee reviewed the titles and abstracts of each study, and omitted those publications that described case reports, acute outcomes, or populations of civilians or refugees in the Persian Gulf region, as well as studies that focused on treatments and meeting abstracts. For each of the remaining studies, the full text was obtained, reviewed, and discussed by the committee. Since the literature search was conducted, the committee became aware of one additional publication that presented an analysis of AH&OBP Registry questionnaire data (Liu et al 2016); it is described elsewhere in the report. After exclusions, eight publications remained for review. All studies were conducted using U.S. service members. Five publications reported on new epidemiologic studies (Abraham et al., 2014; Jones et al., 2012; Powell et al., 2012; Sharkey et al., 2015; Smith et al., 2012) and three publications were reviews (Abraham et al., 2015; Falvo et al., 2015; Morris et al., 2011). The five publications of new epidemiologic studies are discussed below and summarized in Table 1-1. Although literature reviews are cited in the discussion below, they were not of primary concern to the committee because they offer no original data.

Four publications looked specifically at proxies of burn pit exposure (Abraham et al., 2014; Jones et al., 2012; Powell et al., 2012; Sharkey et al., 2015; Smith et al., 2012). Two publications (Abraham et al., 2014; Sharkey et al., 2015) reported updated results from an earlier study conducted by the Armed Forces Health Surveillance Center (AFHSC et al., 2010) and three published additional analyses (Jones et al., 2012; Powell et al., 2012; Smith et al., 2012) of a study by the Naval Health Research Center using data from the Millennium Cohort Study (AFHSC et al., 2010). Three publications focused on respiratory outcomes (Abraham et al., 2014; Sharkey et al., 2015; Smith et al., 2012), and one each focused on autoimmune diseases (Jones et al., 2012) and chronic multisymptom illness (Powell et al., 2012). No new studies examined outcomes of cardiovascular conditions.

Respiratory Outcomes

Two publications reported on additional follow-up data for a study of respiratory conditions conducted by the Armed Forces Health Surveillance Center (AFHSC et al., 2010). The original report was central to the IOM's review of the health effects associated with burn pit exposure at Joint Base Balad (IOM, 2011). The AFHSC et al. (2010) study examined medical encounters of Army and Air Force personnel 36 months after deployment to Joint Base Balad or Camp Taji (with burn pits), Camp Beuhring or Camp Arifjan (without burn pits), or the Republic of Korea (urban air pollution and PM exposure) from 2005 to 2007. Personnel who served within 3 miles of burn pits were considered exposed (15,908 at Joint Base Balad and 2,522 at Camp Taji) and were compared with 51,299 personnel at bases without burn pits and 237,714 personnel in the United States who had not deployed as of April 2006.

[9] The committee defined acute outcomes as those that manifested within 6 months of exposure.

[10] Operation Enduring Freedom (OEF; October 7, 2001–December 28, 2014); Operation Iraqi Freedom (OIF; March 19, 2003–August 31, 2010); Operation New Dawn (OND; September 1, 2010–December 15, 2011); Operation Freedom's Sentinel (January 1, 2015–present)—the most recent U.S. mission in Afghanistan—the purpose of which is to focus on training, advising, and assisting Afghan security forces (Torreon, 2015).

TABLE 1-1 Epidemiologic Studies of Military Personnel Exposed to Burn Pits Published Since *Long-Term Health Consequences of Exposure to Burn Pits in Iraq and Afghanistan* (IOM, 2011)

Study	Study Population	Exposure	Health Outcomes	Results	Comments
Abraham et al., 2014 Retrospective cohort study Follow up to AFHSC et al., 2010	Military personnel deployed between January 2005 and June 2007: 18,430 deployed to locations in Iraq with burn pits (15,908 at Joint Base Balad and 2,522 at Camp Taji); 6,337 deployed to locations without burn pits in Kuwait (1,906 at Camp Buehring, and 4,431 at Camp Arifjan); 44,962 deployed to Korea 112,091 stationed in the United States (referent group).	Exposure limited to deployment to bases with burn pits, without burn pits, or not deployed to the Southeast Asia theater of operations.	Medical encounters for respiratory symptoms (ICD-9: 786), asthma (ICD-9: 493), and COPD and allied conditions (ICD-9 codes 490–492 and 494–496) that occurred within 48 months of service (2006–2011).	No statistically significant IRRs comparing bases with burn pits to bases without burn pits. OIF deployment (both bases with and without burn pits) associated with respiratory symptoms (IRR = 1.25; 95% CI 1.20–1.30); asthma (IRR = 1.54; 95% CI 1.33–1.78); and COPD (IRR = 1.12, 95% CI 0.96–1.31) relative to personnel stationed in the United States. OIF deployment (both bases with and without burn pits) associated with respiratory symptoms (IRR = 1.04; 95% CI 1.00–1.09); asthma (IRR = 1.05; 95% CI 0.90–1.23); and COPD (IRR = 1.24, 95% CI 1.03–1.48) relative to personnel deployed to Korea.	Analyses controlled for age, rank, sex, race. Camp Arifjan was located in an industrial setting. Same population as Sharkey et al., different ICD-9 codes define the outcomes and a smaller U.S.-stationed referent group. Outcome ascertainment limited to encounters that occurred at a military hospital or care center.

continued

TABLE 1-1 Continued

Study	Study Population	Exposure	Health Outcomes	Results	Comments
Jones et al., 2012 Prospective cohort study	Army and Air Force personnel who completed both the 2004–2006 and 2007–2008 Millennium Cohort Study questionnaire cycles: 19,157 service members in the lupus study population (3,201 within 3 miles). 18,848 service members in the rheumatoid arthritis (RA) study population (3,145 within 3 miles).	Proximity to bases with burn pits (2, 3, or 5 miles from Joint Base Balad, Camp Taji, or Camp Speicher) between 2003 and 2008 based on DoD deployment data.	New self-reported provider-diagnosed lupus and RA.	Lupus: Restricted to study population within 3 miles of a burn pit (N = 21 cases; 15 unexposed and 6 at Joint Base Balad). Cumulative days of exposure were not related to lupus. Deployment to Joint Base Balad was associated with lupus (OR = 3.65, 95% CI 1.56–8.51). RA: Restricted to study population within 3 miles of a burn pit (N = 234 cases; 193 unexposed, 26 at Joint Base Balad, 9 at Camp Taji, 6 at Camp Speicher). Neither cumulative days of exposure nor camp were related to RA. Analyses conducted using a 5-mile radius for both lupus and RA showed similar results to the 3-mile radius.	Only 4 cases were confirmed by medical records. Lupus analyses adjusted for sex, birth year, race/ethnicity, and routine skin contact with paints, solvents or substances. RA analyses adjusted for sex, birth year, marital status, service component, pay grade, service branch, smoking status, mental and physical component scores, and exposure to microwaves.
Powell et al., 2012 Prospective cohort study	21,400 military personnel who completed the 2004–2006 and 2007–2008 survey cycles of the Millennium Cohort Study.	Location within 3 miles of Joint Base Balad, Camp Taji, or Camp Speicher based on DoD deployment data (N = 3,578).	Chronic multisymptom illness (CMI) defined as self-reporting at least one symptom in at least two of the following symptom constructs: general fatigue; mood and cognitive problems; and musculoskeletal discomfort assessed at baseline and at follow-up.	At baseline, 18.4% of the unexposed and 15.8% of the exposed groups reported CMI; at follow-up 25.9% of the unexposed and 26.7% of the exposed groups reported CMI. Deployment within a 3-mile radius of a documented burn pit was not significantly associated with CMI ($p = 0.23$).	Analyses adjusted for sex, birth year, education, service component, service branch, pay grade, smoking status, alcohol-related problems, mental health symptoms, and baseline CMI status.

				Risk of CMI was elevated for >209 cumulative days of exposure (OR = 1.19, 95% CI 1.02–1.40) but not for lesser durations. No specific location was associated with increased reports of CMI (p = 0.36). Analyses using 2 and 5 miles did not show a difference.	
Sharkey et al., 2015 Retrospective cohort study Follow up to AFHSC et al., 2010	Military personnel deployed between January 2005 and June 2007: Locations in Iraq with burn pits: 15,908 at Joint Base Balad, 2,522 at Camp Taji; Locations without burn pits: 1,906 at Camp Buehring, 4,431 at Camp Arifjan; 44,962 deployed to Korea; 237,714 U.S.-deployed (referent group).	Exposure limited to deployment to bases with burn pits, without burn pits, or not deployed to the Southeast Asia theater of operations.	Medical encounters for respiratory symptoms (ICD-9: 460–519), acute respiratory infections (ICD-9: 460–466); asthma (ICD-9: 493), and COPD (ICD-9 codes 490–492 and 494–496); respiratory and other chest symptoms (ICD-9: 786) that occurred within 48 months of service (2006–2011).	No camps with statistically significantly increased IRRs compared with U.S.-stationed service members in adjusted analyses; several were significantly decreased.	This study adds 12 months of follow-up to the study conducted by AFHSC et al. (2010) that was reported in IOM (2011). Outcome ascertainment limited to encounters that occurred at a military hospital or care center. Analyses adjusted for age, grade, sex, race, and service branch.
Smith et al., 2012 Prospective cohort study	22,844 Army and Air Force personnel deployed to Iraq or Afghanistan after 2003 and who participated in the Millennium Cohort Study and completed surveys in 2004–2006 and 2007–2008.	Location within 3 miles of Joint Base Balad, Camp Taji, or Camp Speicher based on DoD deployment data (N = 3,585).	New conditions self-reported on the 2007–2008 survey but not on the 2004–2006 survey: (1) newly reported chronic bronchitis or emphysema; (2) newly reported asthma; and (3) respiratory symptoms of persistent or recurring cough or shortness of breath.	No increased risks if deployed within 3 miles of burn pit, with cumulative days of being within 3 miles of a burn pit, or with any particular base. No differences at 5 miles. Increased symptom reporting was observed among deployed Air Force personnel located within 2 miles of Joint Base Balad (OR = 1.24, 95% CI 1.01–1.52).	Analyses adjusted for sex, birth year, marital status, race/ethnicity, education, smoking status, aerobic activity, service branch, service component, rank, and occupation.

NOTE: CI = confidence interval; COPD = chronic obstructive pulmonary disease; ICD-9 = *International Classification of Diseases, Ninth Revision*; IRR = incident rate ratios; OIF = Operation Iraqi Freedom; OR = odds ratio; RA = rheumatoid arthritis.

Incidence rates for medical encounters (defined by *International Classification of Diseases, Ninth Revision* [ICD-9] codes) that occurred within 36 months of April 2006 at military medical facilities were reported. Overall, the incidence rates of health outcomes were similar among personnel deployed to bases with burn pits, those deployed to bases without burn pits, and those deployed to Korea (AFHSC et al., 2010). Both Abraham et al. (2014) and Sharkey et al. (2015) reported on analyses with an additional 12 months of follow-up (48 months total, starting April 2006) for respiratory outcomes. While both studies used the same cohort and data sources, there were slight methodologic differences, including the size of the nondeployed referent group (112,091 in Abraham et al. and 237,714 in Sharkey et al.) and the covariates used. None of the analyses were able to control for smoking or other important exposures related to respiratory disease, and both were limited by their exclusive use of military medical records (medical encounters outside of the military system were not included).

Sharkey et al. (2015) conducted the same analysis using the same cohort as that reported by AFHSC et al. in 2010 with 48 months of follow-up instead of 36 months. The analyses adjusted for age, grade, sex, race, and service branch. As was found in the earlier analysis, the risks of respiratory illnesses at the four Southwest Asia bases and Korea were all similar to, or statistically significantly lower than, the risks for personnel who remained in the United States. Incidence rate ratios (IRRs) were reported for the bases, but no comparisons between bases with and bases without burn pits were made. For respiratory diseases as a whole (ICD-9 codes 460–519), IRRs for medical encounters ranged from 0.70 to 1.01. IRRs for acute respiratory infection medical encounters (ICD-9 codes 460–466) ranged from 0.64 to 0.95; chronic obstructive pulmonary disease (COPD) medical encounters (ICD codes 490–492 and 494–496) ranged from 0.66 to 1.00; asthma (ICD-9 code 493) ranged from 0.76 to 0.95; and medical encounters for respiratory symptoms and other chest symptoms (ICD-9 code 786) ranged from 0.80 to 1.04. In all but one instance, the lowest IRR was for Camp Buehring, and the highest was for Camp Arifjan (for asthma, the highest IRR was for Camp Taji)—the two camps without burn pits.

Abraham et al. (2014) conducted another analysis of the same population and data with a few differences. First, the U.S.-based reference group was much smaller than that used previously and by Sharkey et al. The reference group for Abraham et al. was defined as service members stationed in the United States as of April 2006 and for whom deployment occurred after June 2007, resulting in a reference group of 112,091 U.S.-based personnel (the number of personnel at the four bases and Korea remained the same between studies). Second, the authors made additional comparisons between the four bases and Korea and between bases with and without burn pits.

Compared with the rates for all U.S. personnel, the rates of medical encounters for respiratory diseases among personnel deployed to the four bases were elevated (respiratory symptoms: IRR = 1.25, 95% CI 1.20–1.30; asthma: IRR = 1.54, 95% CI 1.33–1.78; COPD and allied conditions: IRR = 1.12, 95% CI 0.96–1.31). Rates for bases with burn pits and without burn pits were also significantly elevated for both asthma and respiratory symptoms (as well as for Joint Base Balad, Camp Taji, and Camp Arifjan, individually). Compared with the rates for personnel stationed in Korea, the rates of medical encounters for respiratory symptoms and asthma were only slightly elevated and not statistically significant. However, rates of medical encounters for COPD and allied symptoms were elevated compared with personnel stationed in Korea, although the difference was statistically significant only for Camp Arifjan (IRR = 1.43, 95% CI 1.05–1.94) and the pooled estimates that included it. Comparing the bases with burn pits with the bases without burn pits (Balad and Taji to Arifjan and Buehring) showed no statistically significant associations (respiratory symptoms: IRR = 0.95, 95% CI 0.88–1.03; asthma: IRR = 0.93, 95% CI 0.69–1.25; COPD and allied conditions: IRR = 0.90, 95% CI 0.65–1.23). The results of these studies indicate that post-deployment health risk is elevated but that the risk is not associated with burn pits (Abraham et al., 2014).

Three publications (Jones et al., 2012; Powell et al., 2010; Smith et al., 2012) reported on the health of participants of the Millennium Cohort Study who had deployed within 3 miles of a burn pit between 2003 and 2008, and were additional analyses of investigations performed by the Naval Health Research Center (AFHSC et al., 2010). The earlier results by the Naval Health Research Center were based on a 5-mile radius and were reviewed in the 2011 IOM report (AFHSC et al., 2010). The Millennium Cohort consisted of more than 27,000 personnel deployed in support of OEF/OIF and included more than 3,000 participants considered exposed, with at least one deployment within a 3-mile radius of a documented burn pit (at Joint Base Balad, Camp Taji, or Camp Speicher). Burn pit exposure proxies were defined as the cumulative number of days within the specified radius and the cumulative number of days within the radius at specific bases. Exposed participants were compared with participants who were deployed to other locations in Iraq and Afghanistan. The Millennium Cohort is considered representative of U.S. military personnel, with reliable self-reported information obtained prior to enrollment which serves as a baseline for subsequent health status. The self-reported health of respondents who participated in the 2004–2006 and 2007–2008 survey cycles was examined (AFHSC et al., 2010).

New cases of self-reported asthma, chronic bronchitis or emphysema, and persistent cough or shortness of breath were reported at follow-up (2007–2008) but not at the baseline assessment (2004–2006). There were no significant differences in newly diagnosed asthma, bronchitis, emphysema, or self-reported respiratory symptoms between those deployed to areas within 5 miles of burn pits and those not exposed, nor were there differences by base site. Analyses were adjusted for smoking status, physical activity, and other covariates measured at baseline (AFHSC et al., 2010). The findings using a 3-mile radius were similar to those using a 5-mile radius (Smith et al., 2012). After adjusting for all covariates in the multivariable logistic regression, the authors found that deployment within 3 miles of the burn pits did not significantly increase the risk for newly reported chronic bronchitis or emphysema (adjusted odds ratio [AOR] = 0.91; 95% CI 0.67–1.24), newly reported asthma (AOR = 0.94; 95% CI 0.70–1.27), or self-reported respiratory symptoms (AOR = 1.03; 95% CI 0.94–1.13) when compared with deployments to other regions of Iraq or Afghanistan with no documented burn pits. Risk did not increase with cumulative days of exposure or by base. Deployment location within 3 miles of the burn pits was not statistically associated with an increase in odds of newly reported symptoms when compared with those who were deployed to other locations ($p = 0.71$). The study found no significant associations with cumulative days exposed ($p = 0.63$) nor with deployment to a specific burn pit site ($p = 0.97$). Within the 2-mile radius, results were similar with the exception of U.S. Air Force personnel deployed to Joint Base Balad, who were found to be at an increased risk for respiratory symptoms (AOR = 1.24; 95% CI 1.01–1.52) when compared with those deployed to other locations.

Autoimmune Diseases

Similar to Smith et al. (2012), Jones et al. (2012) reported on the risks of disease among service personnel deployed within a 3-mile radius of a burn pit. Newly reported lupus and rheumatoid arthritis were identified at baseline using the question, "Has your doctor or other health professional ever told you that you have any of the following conditions?" At follow-up, participants were asked the same question but in the context of "in the last 3 years." Case confirmation of newly reported lupus and rheumatoid arthritis was performed through a review of electronic medical records. Analyses were restricted to U.S. Army and Air Force personnel.

For lupus, the study population consisted of 19,157 Millennium Cohort members, with 3,201 having served within 3 miles of a burn pit; 21 new cases of lupus were identified (Jones et al., 2012). At the 5-mile radius of a burn pit, there was no association with a new diagnosis of lupus, cumulative exposure, or being deployed to Camp Taji or Camp Speicher. However, a significant increase was observed in the likelihood of a lupus diagnosis for those deployed to Joint Base Balad (OR = 3.52, 95% CI 1.59–7.79) compared with those deployed to locations without burn pits (AFHSC et al., 2010; IOM, 2011). At the 3-mile radius, there was no association between cumulative days of burn pit exposure and lupus. However, all six of the lupus cases within 3 miles were stationed at Joint Base Balad, which resulted in a statistically significant risk (OR = 3.65, 95% CI 1.56–8.51).

For rheumatoid arthritis, the study population consisted of 18,848 Millennium Cohort Study members, with 3,145 having served within 3 miles of a burn pit, and 234 new cases of rheumatoid arthritis (Jones et al., 2012). Within 5 miles, there was no association with deployment to a burn pit location, cumulative days exposed, or specific site, with the exception of an increase in rheumatoid arthritis diagnoses for those exposed to burn pits for 132–211 days (OR = 2.03, 95% CI 1.18–3.49), although exposure for more than 211 days was not statistically significant (AFHSC et al., 2010; IOM, 2011). Within 3 miles of a burn pit, there was no association of rheumatoid arthritis with either cumulative days of exposure or site (Jones et al., 2012).

Electronic medical records were used to confirm 2 lupus cases and 10 rheumatoid arthritis cases among active-duty personnel diagnosed while in the military. Among the verified cases, no association between lupus or rheumatoid arthritis and exposure to burn pits was found (AFHSC et al., 2010; IOM, 2011; Jones et al., 2012). The investigators found that deployed personnel exposed to documented burn pits in the combined Joint Base Balad, Camp Taji, and Camp Speicher sites were not at an elevated risk of lupus or rheumatoid arthritis (Jones et al., 2012).

Chronic Multisymptom Illness

Following Smith et al. (2012) and Jones et al. (2012), Powell et al. (2012) also investigated the health effects of exposure to burn pits among Millennium Cohort Study members deployed within 3 miles of a burn pit. In this study, chronic multisymptom illness (CMI) was defined as having reported at least one symptom in at least two of the following symptom constructs: general fatigue, mood and cognitive problems, and musculoskeletal discomfort.

Among the 21,400 study participants, 3,578 had served within 3 miles of a burn pit. There were 956 personnel deployed within 3 miles of a burn pit and 4,608 unexposed personnel with CMI (Powell et al., 2012). CMI was not statistically significantly associated ($p = 0.16$) with being deployed within a 5-mile radius of a burn pit, cumulative exposure to a burn pit overall, or being deployed to Joint Base Balad, Camp Taji, or Camp Speicher when adjusted for baseline covariates (sex, birth year, education, service component, service branch, pay grade, smoking status, alcohol-related problems, mental health symptoms, and baseline CMI status). However, cumulative exposure to a burn pit for more than 210 days was associated with a slight increase in risk for CMI (OR = 1.22, 95% CI 1.04–1.44) (AFHSC et al., 2010; IOM, 2011).

After adjusting for baseline covariates in the models, deployment within a 3-mile radius of a documented burn pit was not found to be significantly associated with CMI ($p = 0.23$), nor were cumulative days of burn pit exposure or individual sites. There was a small significant increase in risk among persons exposed for greater than 209 days (OR = 1.19, 95% CI 1.02–1.40) as there was for those deployed within the 5-mile radius. The authors found no increased risk of reporting CMI symptoms when they compared personnel deployed within 2, 3, and 5 miles of a documented burn pit and when comparing lengths of time at the various bases (Powell et al., 2012).

Discussion

The committee reviewed five publications reporting new epidemiologic studies of long-term health effects in military personnel associated with burn pits. Three examined respiratory outcomes, one investigated rates of lupus and rheumatoid arthritis, and one examined CMI. All suffer from various limitations, such as a reliance on self-reported data (Jones et al., 2012; Powell et al., 2012; Smith et al., 2012) and two publications described nearly the same study but with slight methodological differences, so they cannot be interpreted independently (Abraham et al., 2014; Sharkey et al., 2015).

None of these studies provide a thorough evaluation of the health effects associated with burn pit exposures, even though they all contribute pieces to the overall picture of the evidence base. There are still many gaps in that picture, and there is still work to be done. Furthermore, no single epidemiologic study is ever definitive.

Overall, little new evidence linking respiratory symptoms or diseases with exposure to burn pits has been published, although some studies suggest that deployment to the Southwest Asia theater of operations may play a role (Abraham et al., 2014). Recently published reviews further support this observation and indicate a predominance of airway obstruction and hyperreactivity among deployed personnel (Abraham et al., 2015; Falco et al., 2015; Morris et al., 2011). Furthermore, these studies do not provide new evidence linking burn pit exposure to autoimmune disease or CMI.

The 2011 IOM report indicated that there was "limited/suggestive evidence of an association between exposure to combustion products and reduced pulmonary function," but evidence was inadequate/insufficient for other health outcomes mostly based on studies in firefighters and incinerator workers or communities around incinerators. While the new evidence is less than sufficient, it does not show that service members are at an increased risk of health effects associated with burn pits in particular, although other hazards may be important contributors to respiratory symptoms and disease.

ORGANIZATION OF THE REPORT

The remainder of this report is divided into six chapters plus supporting appendixes. Chapter 2 presents a broad overview of the use of registries for health research and a discussion of some of the inherent limitations of using registry data for assessing associations and drawing conclusions concerning the relationship between exposures and health outcomes.

Chapter 3 centers on the origin, development, and implementation of the AH&OBP Registry and its key element, the self-assessment questionnaire. It includes assessments of the design of the questionnaire and the recruitment and enrollment of registry participants.

Chapter 4 is the first of three chapters that describe the methods and results of the committee's analysis of the initial months of AH&OBP Registry data. An overview of the data requested and received is presented along with descriptive statistics regarding the demographic and military service characteristics of registry respondents.

Chapter 5 summarizes the information provided to the committee on respondents' exposure to burn pit emissions and other important airborne exposures. The limitations of these data are discussed, and suggestions are offered for how the data may best be put to use in evaluating the magnitude, duration, and frequency of respondents' exposure.

Chapter 6 turns to the AH&OBP Registry's health outcomes data, focusing on those conditions and diseases that the committee believes are most relevant to assessing the potential health effects of exposures to burn pits and other airborne hazards. It presents descriptive statistics for these outcomes, describes the approaches used to analyze the data, and concludes with a discussion of how this information may be used to assess associations between health outcome and exposures.

The final chapter of this report, Chapter 7, draws together the committee's primary findings, conclusions, and recommendations and offers perspectives on them.

Appendix A contains an excerpt from Public Law 112-260, § 201 regarding the origination of the AH&OBP Registry and this study. The agenda for a public workshop held by the committee in May 2015 is presented in Appendix B. The most recent version of the AH&OBP Registry questionnaire that is publicly available (dated December 15, 2014) is reproduced in Appendix C. Appendix D contains a list of each data variable requested by the committee, the source of each data variable, and whether the variable was made available. Appendix E presents tables of the multivariate model results and estimates that were used to generate the figures in Chapter 6. Appendix F provides biographical sketches of the committee members and the National Academies staff who worked on this report.

REFERENCES

Abraham, J. H., A. Eick-Cost, L. L. Clark, Z. Hu, C. P. Baird, R. DeFraites, S. K. Tobler, E. E. Richards, J. M. Sharkey, R. J. Lipnick, and S. L. Ludwig. 2014. A retrospective cohort study of military deployment and postdeployment medical encounters for respiratory conditions. *Military Medicine* 179(5):540–546.

Abraham, J. H., L. L. Clark, and A. I. Schneiderman. 2015. Epidemiology of airborne hazards in the deployed environment. In C. P. Baird and D.K. Harkins (eds.), *Airborne hazards related to deployment*. Fort Sam Houston, TX: Borden Institute. Pp. 61–72.

AFHSC (U.S. Armed Forces Health Surveillance Center), Naval Health Research Center, and U.S. Army Public Health Command. 2010. *Epidemiological studies of health outcomes among troops deployed to burn pit sites*. Silver Spring, MD: Armed Forces Health Surveillance Center.

Blasch, K. W., J. E. Kolivosky, and J. M. Heller. 2016a. Environmental air sampling near burn pit and incinerator operations at Bagram Airfield, Afghanistan. *Journal of Occupational & Environmental Medicine* 58(8 Suppl 1):S38–S43.

Blasch, K., J. Kolivosky, and B. Hill. 2016b. Occupational exposures among personnel working near combined burn pit and incinerator operations at Bagram Airfield, Afghanistan. *Inhalation Toxicology* 28(5):216–225.

CHPPM and AFIOH (U.S. Army Center for Health Promotion and Preventive Medicine and U.S. Air Force Institute for Operational Health). 2009. *Addendum 2. Screening health risk assessment burn pit exposures, Balad Air Base, Iraq, May 2008.* USACHPPM Report No. 47-MA-08PV-08/AFIOH Report No. IOH-RS-BR-TR-2008-0001. Aberdeen Proving Ground, MD: U.S. Army Center for Health Promotion and Preventive Medicine.

Engelbrecht, J. P., E. V. McDonald, J. A. Gillies, and A. W. Gertler. 2008. *Department of Defense Enhanced Particulate Matter Surveillance Program (EPMSP)*. Final report. Reno, NV: Desert Research Institute. https://phc.amedd.army.mil/PHC%20Resource%20Library/Final%20EPMSP%20Report%20without%20appx%20Feb08.pdf (accessed September 4, 2016).

Falvo, M. J., O. Y. Osinubi, A. M. Sotolongo, and D. A. Helmer. 2015. Airborne hazards exposure and respiratory health of Iraq and Afghanistan veterans. *Epidemiologic Reviews* 37(1):116–130.

Federal Register. 2013. Notice: Initial research on the long-term health consequences of exposure to burn pits in Iraq and Afghanistan. *Federal Register* 78(23):7860–7861.

Gasper, J., and J. Kawata. 2015. An evaluation of selection bias and data quality: *Open Burn Pit and Airborne Hazards Registry*. Prepared for the Department of Veterans Affairs Office of Public Health, Post-Deployment Health Group by Westat. September.

Gray, G. C., K. B. Chesbrough, M. A. K. Ryan, P. Amoroso, E. J. Boyko, G. D. Gackstetter, T. I. Hooper, and J. R. Riddle for the Millennium Cohort Study Group. 2002. The Millennium Cohort Study: A 21-year prospective cohort study of 140,000 military personnel. *Military Medicine* 167(6):483–488.

IARC (International Agency for Research on Cancer). 2006. *IARC monographs on the evaluation of carcinogenic risks in humans: Preamble*. Lyon: IARC. http://monographs.iarc.fr/ENG/Preamble/CurrentPreamble.pdf (accessed September 4, 2016).

IOM (Institute of Medicine). 2005. *Gulf War and health, volume 3: Fuels, combustion products, and propellants*. Washington, DC: The National Academies Press.

IOM. 2011. *Long-term health consequences of exposure to burn pits in Iraq and Afghanistan*. Washington, DC: The National Academies Press.

Jones, K. A., B. Smith, N. S. Granado, E. J. Boyko, G. D. Gackstetter, M. A. K. Ryan, C. J. Phillips, and T. C. Smith for the Millennium Cohort Study Team. 2012. Newly reported lupus and rheumatoid arthritis in relation to deployment within proximity to a documented open-air burn pit in Iraq. *Journal of Occupational and Environmental Medicine* 54(6):698–707.

Liu, J., N. Lezama, J. Gasper, J. Kawata, S. Morley, D. Helmer, and P. Ciminera. 2016. Burn pit emissions exposure and respiratory and cardiovascular conditions among airborne hazards and open burn pit registry participants. *Journal of Occupational and Environmental Medicine* 58(7):e249–e255.

Morris, M. J., L. L. Zacher, and D. A. Jackson DA. 2011. Investigating the respiratory health of deployed military personnel. *Military Medicine* 176(10):1157–1161.

NRC (National Research Council). 2010. *Review of the Department of Defense Enhanced Particulate Matter Surveillance Program report*. Washington, DC: The National Academies Press.

Powell, T. M., T. C. Smith, I. G. Jacobson, E. J. Boyko, T. I. Hooper, G. D. Gackstetter, C. J. Phillips, and B. Smith for the Millennium Cohort Study Team. 2012. Prospective assessment of chronic multisymptom illness reporting possibly associated with open-air burn pit smoke exposure in Iraq. *Journal of Occupational and Environmental Medicine* 54(6):682–688.

Rinker, J. P. 2011. Retrospective geospatial modeling of PM_{10} exposures from open burning at Joint Base Balad, Iraq. Thesis presented to the faculty, Department of Systems and Engineering Management, Graduate School of Engineering and Management, Air Force Institute of Technology, Air University, Air Education and Training Command, in partial fulfillment of the requirements for the Degree of Master of Science in Industrial Hygiene. http://www.dtic.mil/dtic/tr/fulltext/u2/a539684.pdf (accessed September 5, 2016).

Sharkey, J. M., C. Baird, A. Eick-Cost, L. Clark, Z. Hu, S. Ludwig, and J. Abraham. 2015. Review of epidemiological analyses of respiratory health outcomes after military deployment to burn pit locations with respect to feasibility and design issues highlighted by the Institute of Medicine. In C. P. Baird and D. K. Harkins (eds.), *Airborne hazards related to deployment*. Fort Sam, TX: Borden Institute, U.S. Army Medical Department Center and School. Pp. 291–304.

Smith, B., C. A. Wong, T. C. Smith, E. J. Boyko, G. D. Gackstetter, and M. A. K. Ryan for the Millennium Cohort Study Team. 2009. Newly reported respiratory symptoms and conditions among military personnel deployed to Iraq and Afghanistan: A prospective population-based study. *American Journal of Epidemiology* 170(11):1433–1442.

Smith, B., C. A. Wong, E. J. Boyko, C. J. Phillips, G. D. Gackstetter, M. A. K. Ryan, and T. C. Smith for the Millennium Cohort Study Team. 2012. The effects of exposure to documented open-air burn pits on respiratory health among deployers of the Millennium Cohort Study. *Journal of Occupational and Environmental Medicine* 54(6):708–716.

Torreon, B. 2015. U.S. Periods of War and Dates of Recent Conflicts. Congressional Research Service. Report RS21405. https://www.fas.org/sgp/crs/natsec/RS21405.pdf (accessed October 3, 2016).

VA (Department of Veterans Affairs). 2015a. *Report on data from the Airborne Hazards and Open Burn Pit (AH&OBP) Registry*. http://www.publichealth.va.gov/docs/exposures/va-ahobp-registry-data-report-april2015.pdf (accessed September 10, 2016).

VA. 2015b. *Report on data from the Airborne Hazards and Open Burn Pit (AH&OBP) Registry*. http://www.publichealth.va.gov/docs/exposures/va-ahobp-registry-data-report-june2015.pdf (accessed September 10, 2016).

2

Use of Registries in Environmental Health Research

A registry is a structured system for collecting and maintaining data on a group of people characterized by a specific disease, condition, exposure, or event. Registries may be used to facilitate research, monitor health, or to provide information to registrants. This chapter begins with a broad overview of the use of registries for health research. Because registries are one of the primary methods used by the Department of Veterans Affairs (VA) to assemble information on the effects of military-service related exposures, examples of VA and Department of Defense (DoD) registries that address specific exposures, health outcomes, or groups are summarized. The limitations of registry data, specifically the common biases encountered, are then discussed. This is followed by an examination of the characteristics of some potential comparison populations that might be used in studies of health outcomes in the participants in the Airborne Hazards and Open Burn Pit (AH&OBP) Registry. The chapter concludes with a discussion of some of the inherent limitations of using registry data for assessing associations and drawing conclusions concerning the relationship between exposures and health outcomes.

USE OF REGISTRIES IN HEALTH RESEARCH

In an ideal study of the relationship between an exposure and an outcome, the characteristics of the population of interest, the agent(s) and level(s) of exposure they were subjected to, and the outcome or outcomes that were observed would all be carefully and objectively measured. Well-designed epidemiologic studies are the best means of measuring such associations, but large epidemiologic studies take years of planning to design and are expensive to conduct. In addition, the excessive control of study parameters that is often required in these studies may limit the generalizability of conclusions to a broader population. As a result, epidemiologists have developed several means to gather and use the information that is feasible to obtain.

Registries are one means of accomplishing this. They are generally quicker to establish, cost-efficient, and allow for the ascertainment of several exposures and health outcomes on a defined population. In some registries, this information is supplemented or updated over time. The motivating factors for participation and retention vary according to the registry but may include the apparent relevance, importance, and credibility of the registry or its sponsoring organization (Gliklich et al., 2014). Participants may be directly solicited from lists of persons believed to have the desired characteristics (active recruitment) or via broad appeals to populations thought to include potentially qualified subjects (passive recruitment). In registries that depend on passive recruitment, selective participation is a concern because those persons experiencing adverse health outcomes—whatever their cause—may be more motivated to enroll.

Registry data are collected through active or passive means, or a combination of the two. Active data collection consists of collecting data specifically for the purpose of the registry (such as through a clinical examination or questionnaire). Passive data collection relies on sources that were collected for purposes other than the registry (such as administrative health records, claims databases, or pharmacy records).

An epidemiologic registry may be established to serve one or more predetermined scientific, clinical, or policy purposes such as addressing a public concern (Gliklich et al., 2014). Such registries are most useful when designed for a specific goal or research question that then directs how and which data are to be collected. Antão and colleagues (2015) proposed that the considerations for establishing a registry should include the public health significance of the exposure or condition; the scientific contribution that the registry would provide; the ultimate purpose, duration, scope of data collection, and outcomes of the registry; and whether the registry will be a useful mechanism to achieve the stated goal.

There are several possible uses for registries as data-collection instruments. They can be used to systematically document the experiences of people who choose to participate and to more methodically record otherwise anecdotal reports of exposure and disease. A second use of registries is to define or calculate the minimum number of affected individuals even if the universe of potential participants is undefined (subject to the accuracy of self-reported exposure and disease reports). Likewise, the data can be used to determine the minimum number of exposures and diseases reported by participants for descriptive purposes that may serve as a basis for informed speculation about the magnitude of the problem. Other uses of registry data include generating hypotheses about specific exposures and health effects that may not have been considered otherwise and discovering previously unidentified associations between exposures and health problems, particularly if the health event of concern is rare or distinctive. The data may motivate more focused investigations of health outcomes and give researchers information they can use to develop better designed studies.

Whereas many registries are established with the intention of providing quantitative data, more often the data that they can provide are qualitative. For example, registries may serve as an expression of good faith on the part of the sponsor to be responsive to concerns that have been raised and to provide a forum for receiving testimony of those who wish to provide it. In this vein, registries can be used to facilitate communication with and outreach to specific populations, which may include updates on new scientific and medical developments or new programs or policies relevant to the participants. However, the committee notes that although several registries have been established, particularly by government agencies, to be responsive to concerned constituents, that motivation should not be the primary reason for establishing a registry.

The World Trade Center Health Registry is one example of a registry that was appropriately designed and established to track the health of participants after a unique exposure (the World Trade Center terrorist attacks on September 11, 2001) and that resulted in many scientific publications on the health conditions of participants over time. Its data were used to demonstrate increased reporting of newly diagnosed respiratory symptoms, asthma, posttraumatic stress disorder, and serious psychological distress (for example, Bowler et al., 2012; Brackbill et al., 2013; Farfel et al., 2008). Although the studies based on this registry data were not without limitations, such as by selective participation over time, there were significant findings; for example, rescue and recovery workers who wore respirators when responding were less likely to report respiratory problems 5 to 6 years after the day of the attacks than those who went without adequate respiratory protection (Antão et al., 2011). In addition, the registry serves as an important tool to inform health care services, project needs for affected populations, and link affected participants to services.

In contrast, the National Exposure Registry was established in 1989 in response to the 1980 Congressional Comprehensive Environmental Response, Compensation, and Liability Act. Although it is regarded as one of the largest data repositories for tracking specific environmental chemical exposures and registrants' health conditions over time, it has several shortcomings. Foremost, the National Exposure Registry lacks individual exposure measures, and therefore, this flawed design does not allow it to effectively assess exposure, as it was intended. Moreover, there is no process for validating self-reported exposures and health outcomes, appropriate control groups are absent, and no biomarkers exist for most hazardous substances; all of these issues reduced the registry's value for research or drawing meaningful conclusions. The numbers and rates of many observed health effects that were reported to be high compared with national norms might be explained by this registry's design and methodological shortcomings (Schultz et al., 2010).

The next section briefly summarizes the two main types of registries and their application by VA and DoD to obtain exposure- and health-related information.

VA AND DOD ENVIRONMENTAL HEALTH REGISTRIES

Two primary types of registries are used to address environmental health issues: disease- or outcome-based, and exposure-based. Generally, both types gather the same sorts of information. The difference between them is the issue that is the focus of the effort—the potential determinants of a health outcome versus the possible consequences of an exposure.

In disease- or outcome-based registries, the eligibility to participate may be dependent on particular signs or symptoms, a diagnosis, or vital status. Depending on the design, inclusion criteria, and extent of information collected, disease registries can be used to estimate the prevalence and incidence of a condition of interest, track or estimate health care resource utilization, study disease progression, or serve as sampling frames for selecting populations for additional studies (Rabeneck, 2001).

The AH&OBP Registry is an exposure-based registry, and this type of registry is the focus of the discussion in the following sections. Exposure-based registries collect information about persons potentially at risk for adverse health outcomes due to specific occupational or environmental exposures encountered. The exposures of interest may be chemical, biological, radiological or nuclear agents, or they may be the result of environmental conditions (for example, airborne dust, extreme heat, or natural or man-made disasters). Exposure registries can be used to help evaluate potential health outcomes or latent conditions resulting from an exposure, especially if the exposure is not common or its potential effects are not well characterized.

VA Environmental Health Registries

VA has established several registries, most of them in response to Congressional directives, along with related programs to monitor the health of veterans of particular conflicts or who were potentially exposed to specific environmental agents during military service. Exposures differ by military conflict, but include herbicides (most prominently, Agent Orange); depleted uranium or embedded fragments; ionizing radiation; and such hazards as smoke from oil-well fires, wind-borne dust, or chemical pollution. Four of the seven active VA exposure registries are targeted at 1990–1991 Gulf War (Operation Desert Shield and Operation Desert Storm) or post-9/11 Gulf region military operations. Table 2-1 lists examples of active VA exposure or event-based environmental health-related registries along with a few of their salient characteristics.

Although each of the VA registries was designed to focus on particular exposures and populations, the eligibility criteria often changed over time, generally in the direction of being more inclusive. For example, the Agent Orange Registry was established in 1978 to monitor veterans' health concerns that may have resulted from their exposure to herbicides in Vietnam. However, the registry has since expanded its scope and now includes veterans of Korea and Thailand who believed they were exposed to these herbicides. Similarly, the Persian Gulf Registry Health Examination Program was established in 1992 and later expanded to cover Operation Iraqi Freedom (OIF) and Operation New Dawn (OND) veterans and renamed the Gulf War Registry.

Because different veterans were subject to different potential exposures, the types of questions asked to ascertain the relevant information differ, as do the types of laboratory tests that may be offered and the areas of emphasis during the medical history and clinical exam, which leads to variations in the type, amount, and detail of data collected by VA registries (VA, 2016b,c). Registry data generally constitute the sum total of information on participants available for analysis, although they are occasionally supplemented by other sources, such as military records.

To be eligible to participate in a VA registry, a veteran must sometimes complete a clinical examination conducted by a VA provider and supply self-reported exposure and other information. The registry examination may

TABLE 2-1 Active (as of June 2016) VA Environmental Health Registries

Registry Name	Period of Military Service	Enrollees	Target Population	Data Source(s)
Ionizing Radiation	1940s–1950s; 1960s	24,550	Approximately 400,000 veterans are eligible to enroll (VA, 2004). Veterans who participated in a test involving atmospheric detonation of a nuclear device; participated in the occupation of Hiroshima or Nagasaki (August 6, 1945, to July 1, 1946); were interned in Japan during World War II; received radium irradiation treatments while on active duty; or were involved in "radiation-risk activities" (VA, 2004, 2010, 2015a).	Ionizing radiation exam; voluntary.
Agent Orange	1960s–1970s	573,000	Veterans with service in Vietnam between 1962 and 1975; or who served in a unit or near the Korean Demilitarized Zone April 1, 1968–August 31, 1971; served on Thailand bases between February 28, 1961, and May 7, 1975; or who were otherwise exposed to herbicides during a military operation or as a result of testing, transporting, or spraying herbicides for military purposes (VA, 2012, 2015b).	Voluntary comprehensive health exam, which includes an exposure history, medical history, physical exam, and other tests if needed.
Depleted Uranium Follow-up	1990s–present	79	Veterans of the Gulf War, Bosnia, Operation Desert Shield, Operation Desert Storm, and Operation Iraqi Freedom (OIF) conflicts "who were on, in or near vehicles hit with 'friendly fire'; rescuers entering burning vehicles, and those near burning vehicles; salvaging damaged vehicles; or near fires involving DU munitions" (McDiarmid, 2011; VA, 2015c).	Screening via an exposure questionnaire and urine test; physical exams and clinical tests for exposed persons.
Gulf War	1990s–present	150,000	Veterans who served in the Gulf during Operation Desert Shield, Operation Desert Storm, OIF, or Operation New Dawn (OND). The registry had enrolled 150,000 1990–1991 Gulf War veterans as of January 2015 and more than 29,000 OIF/OND veterans as of February 2014 (VA, 2015a,e).	Voluntary health evaluation, including exposure and medical history, laboratory tests, and a physical exam.
Toxic Embedded Fragments	2000s–present	9,450	Veterans with active duty service in Operation Enduring Freedom, OIF, or OND. The veteran must have, or likely have, an embedded fragment as the result of injury received while serving in an area of conflict (VA, 2016a).	Responses to screening questions; heath and exposure-related information, such as fragment composition data; and, urine biomonitoring results from electronic medical record systems and follow-up screening.

provide baseline health information, but unless the pathognomonic diagnostic findings and test results have been associated with exposure, an exam cannot determine causation or improve diagnostic accuracy or quality of care. Clinical examinations provide objective information on health status, in contrast to relying solely on self-reported disease information, which is difficult to validate. However, clinical examinations are cross-sectional, point-in-time data and cannot confirm or quantify exposures that may have occurred as a result of military service except in circumstances where unambiguous evidence is present, such as is true for participants in the Depleted Uranium and Toxic Embedded Fragments registries. With few exceptions,[1] follow-up exams are not conducted.

How well VA registries serve to provide new information to add to the scientific knowledge base on a particular topic, contribute to hypothesis generation, or improve programs in the Veterans Health Administration is open to question. The committee failed to identify any publications that used data from the Ionizing Radiation Registry for disease surveillance or in epidemiologic investigations. However, others such as the Agent Orange and the Depleted Uranium Follow-Up registries serve as both databases of health surveillance for their respective populations for VA and sources of data used in epidemiologic studies. For example, the Agent Orange Registry population was used in at least five studies examining the health conditions of these veterans. Bullman et al. (1991) conducted a case-control study using a subset of registry participants to compare demographic and military characteristics of veterans who did and did not have posttraumatic stress disorder. Bullman and Kang (1994) used the registry population to assess the risk of mortality due to traumatic causes for Vietnam veterans who had posttraumatic stress disorder. Other studies used subsamples of the registry population to examine the risk of testicular cancer (Bullman et al., 1994) and nonmelanoma invasive skin cancers (Clemens et al., 2014) with presumed exposure to dioxin and other herbicide contaminants. The Depleted Uranium Follow-Up Registry continues to publish findings on the long-term health consequences in veterans exposed to depleted uranium (Hodge et al., 2001; McDiarmid et al., 2001, 2011; Shvartsbeyn et al., 2011; Squibb et al., 2005). In addition to surveillance, VA also uses some of its registries as an outreach tool, providing registry participants with newsletters and updated information on issues of interest. For example, participants of the Agent Orange Registry receive the *Agent Orange Newsletter* quarterly, which provides selected research findings, a summary of exposure locations, and other information relevant to the registry population, such as how to apply for a disability claim and new studies of Vietnam veterans (VA, 2012, 2016d).

DoD Environmental Health Registries

Similar to VA, DoD has also established registries in response to particular exposures that service members might have encountered; these are briefly summarized in Table 2-2. DoD previously offered an examination program that was similar to VA's Persian Gulf Registry Health Examination Program, called the Comprehensive Clinical Evaluation Program (NIH, 2016; VA and DoD, 2002), for persons serving on active duty. The program began in 1994 and was discontinued on June 1, 2002 (NARA, 2002). Self-reported data were collected on the use of pyridostigmine bromide, having experienced infectious diseases, and exposure to pesticides; chemical and biologic agents; multiple vaccinations; depleted uranium; and airborne hazards from sand, dust, smoke, burn pits, and oil-well fires (IOM, 2010). An Institute of Medicine (IOM) committee had evaluated this registry and the corresponding uniform case assessment protocol with regard to the protocol used, program implementation and administration, outreach efforts to veterans, and provider education, and that committee recommended several improvements to the protocol, referral process, evaluation feedback mechanism, consistency of data reporting, and approach to systematically updating patient information in the registry (IOM, 1998).

DoD's Force Health Protection Program is focused on protecting individuals from hazardous physical, chemical, and biological agents in the air, water, and soil. Within this program, DoD identified the need for environmental health surveillance registries for service members with occupational and environmental health exposures that could cause illness and for any exposure that was not expected to cause illness but that could provide individual-level exposure data (DoD, 2016a). Two registries are currently monitoring occupational and environmental health

[1] For example, participants in the Ionizing Radiation Registry are eligible for follow-up examinations if they report developing new health problems (VA, 2015a).

TABLE 2-2 Active (as of June 2016) DoD Environmental Health Registries

Registry Name	Period of Military Service	Enrollees	Target Population	Data Source(s)
Operation Tomodachi	March 12–May 11, 2011	75,000	All identified individuals from the target population are included in the registry: service members, civil servants, Department of Defense (DoD) contractors, and dependents of service members and DoD civilian employees who were on the four main islands of Japan or on U.S. Navy-affiliated ships near Japan at any time between March 12 and May 11, 2011.	Radiation measurements were taken on military installations and in areas where service members were engaged in humanitarian missions.
Gulf War Oil Well Fire Smoke	August 2, 1990–February 28, 1991*	750,000	Members of the armed forces exposed to fumes of burning oil in connection with Operation Desert Storm.	Daily oil-well fire smoke exposure for an individual was estimated from environmental and other data and included in registry data.
Comprehensive Clinical Evaluation Program	1990–1991	32,876	Veterans of the 1990–1991 Gulf War.	In-person survey and two-stage health evaluation; voluntary participation.

* These dates reflect the period of Operation Desert Storm, although oil-well fires were burned between February 2, 1991, and October 29, 1991.

exposures among select groups of service members: the Operation Tomodachi Registry and the Gulf War Oil Well Fire Smoke Registry.

The Operation Tomodachi Registry was established in response to the March 11, 2011, earthquake and tsunami in Japan and the subsequent release of radiation at the Fukushima Daiichi Nuclear Power Station (DoD, 2016b). Its purpose was to monitor U.S. service members who were on or near the mainland of Japan in the 2 months following the incident by creating a comprehensive database of exposures and health outcomes. DoD also tested water, air, and soil and used the data to calculate potential exposure doses for 13 areas (installations and major cities) on the Japanese mainland where the majority of the DoD population (about 58,000 persons) was stationed. In total, the registry contains data on about 75,000 DoD-affiliated individuals, including nearly 17,000 who were associated with U.S. Navy fleet-based operations (DoD, 2016b). Information includes locations and estimated whole body and thyroid radiation doses (Dunavant et al., 2013).

The Gulf War Oil Well Fire Smoke Registry was established in response to Public Law 102-190, which required DoD to "establish and maintain a record relating to members of the Armed Forces who were exposed to the smoke/fumes from burning oil wells" (DoD, 2016c). More than 750 oil wells were set on fire while Iraqi forces were retreating from Kuwait during the 1991 conflict. These fires caused a marked decrease in air quality and were a health risk for a large part of the country, especially for those in their immediate vicinity. The registry includes more than 750,000 DoD personnel who served in the Gulf War during the time the oil-well fires were burning. Exposures were estimated using information submitted by registry participants, troop location data, the DoD personnel registry, satellite images, and meteorological models. Modeling data were used to estimate smoke exposure, and daily estimates were combined to provide an overall risk estimate.

LIMITATIONS OF REGISTRY DATA

There are significant inherent limitations in the use of registries to draw inferences regarding the presence or strength of an association between an exposure and a health outcome. This section provides an overview of some of the weaknesses of registry data and the various sorts of problems one encounters in using registries to make scientific conclusions. The overview begins with a discussion of the biases introduced by selective participation, which affects all registries that rely on voluntary participation. Other potential biases (misclassification, recall, and reporting) and their implications on data quality are then described, and Table 2-3 provides a summary table of the biases and other limitations of registry data and their implications for data analysis. Subsequent sections address the challenge of identifying an appropriate comparison population to use in evaluations of health outcomes and discuss the cumulative effect these weaknesses exert and how they limit the extent to which registry data may be used to evaluate exposure–health outcome associations. Chapter 3 offers additional observations on how these considerations affect the scientific value of information from the AH&OBP Registry.

Selective Participation and Bias

Several factors influence participation in a registry, including its perceived relevance to the respondent, the importance or scientific credibility of the registry or the sponsor, the survey length and time commitment required to complete it, the degree to which the respondent believes participation will yield a benefit, and the respondent's altruism (Groves et al., 1992). This is especially true for registries that depend on voluntary participation, where those considerations are weighed with the other risks and burdens of participation as well as with any incentives for participation (Raftery et al., 2005). Monetary incentives, for example, have been shown to increase health survey response rates in U.S. veterans (Coughlin et al., 2011).

A registry involving an exposure–health outcome relationship in which people choose to participate may selectively include those who were more highly exposed than the average within the eligible population or those who are more concerned about the potential health effects resulting from such an exposure because these individuals have a greater stake in the issue. For example, Smith et al. (2002) found that 1990–1991 Gulf War veterans who were exposed to the heaviest fighting in theater and had served longer deployments were more likely to participate in a DoD or VA registry than veterans who were deployed for shorter time periods and experienced less intense combat. This is important because a registry population with nearly universal reporting of an exposure or outcome is unlikely to be representative of the full, eligible population. To the extent that these and other factors differ between participants and nonparticipants, such selective participation may seriously undermine the potential utility of the registry to fulfill the objectives and answer the questions that it was intended to address (Hernán et al., 2004).

Nonrandom differences in participation are a specific form of selection or nonresponse bias—a type of systematic error that occurs when the study sample differs from the target population of the study in a way that makes it

TABLE 2-3 Limitations of Registries and Resultant Effects

Limitation	Potential Effect
Biases	
Self-selection	Effects representativeness so that findings may not be generalizable to the broader, target population
Misclassification	May result in exaggerated or underestimated estimates of an effect
Recall	Threatens internal validity and distorts the magnitude of estimates of an effect
Self-report	May result in exaggerated or underestimated estimates of an effect
Lack of active follow-up	May lead to incomplete ascertainment of outcomes
Passive data collection	May lead to missing data
Enrollment of ineligible registrants	May weaken the generalizability of findings
Large numbers of participants	May lead to inflated "statistical significance" but not necessarily clinical relevance

unreflective of the exposures, health outcomes, or exposure–health outcome associations present in the population of interest (Rothman, 2012). Registries that rely on completely voluntary participation and where efforts to contact, recruit, and persuade eligible persons to enroll and participate are not targeted to the full eligible population are especially prone to selection or nonresponse bias. The potential for bias is dependent on the rates of participation (generally lower rates result in a greater potential for bias) and on the extent to which various key variables such as exposures and health outcomes systematically differ between the participant and target populations. For example, persons who perceive themselves as exposed to hazards or those who are experiencing health problems may be more likely to participate than persons who do not consider themselves ill or at risk, thereby resulting in a sample of participants that is not representative of the target population and therefore introducing selection bias.

It is time consuming for participants to provide large amounts of information or similar information for multiple events, and research suggests that a lack of time is a factor in low response rates in surveys of military personnel (Miller and Aharoni, 2015). Another form of selection bias can result if persons who have more events to report exceed the time they have allotted themselves or have been allotted to complete the study, become fatigued, or lose interest. In such circumstances, respondents may not report all eligible events or information related to them, or they may game their answers to avoid having to answer follow-up queries (Egleston et al., 2011). The more onerous an instrument is to complete, the more likely those with greater motivation to participate (such as those who believe that they had high exposure, are ill, or both) will be overrepresented in the study population.

The effects of selective participation bias may be mitigated through an improved representation of participants obtained through changes to messaging or more targeted outreach and communications and by providing incentives to respond—targeting eligible persons who were potentially exposed, but who are not currently experiencing adverse health outcomes, for example. Response fatigue may be minimized by such steps as limiting repetitive questions, using previous responses to eliminate later questions that are no longer salient, and making sure that the survey content is perceived as relevant by respondents (Rolstad et al., 2011).

Misclassification Bias

Misclassification bias results when the information collected about or from respondents is inaccurate and leads to respondent being placed in an incorrect category. It can occur for either an exposure—such as classifying people as exposed when they were not—or a health outcome (Rothman, 2012). Misclassification of participants can be either differential or nondifferential. Differential misclassification occurs when categorization errors for one variable of interest (exposure, for example) are related to their status in another variable of interest (health outcome, for example). This would be the case if respondents who experience shortness of breath were more likely to erroneously report having been more highly exposed to burn pits than persons who do not have shortness of breath. If the pattern of error in reporting exposure is not related to the presence of the outcome, then the misclassification is nondifferential. Differential misclassification can either exaggerate or underestimate an effect; for dichotomous comparisons, nondifferential misclassification generally biases the result toward the null, that is, toward finding no association between an exposure and an outcome. For multiple categories of exposure, nondifferential misclassification among exposure categories can produce the appearance of a more linear or monotonically increasing relationship when in fact the underlying relationship is nonlinear.

Recall Bias

Data collection that is based on self-report rather than objective measures introduces the potential for recall and reporting biases. Recall bias results, for example, if respondents who self-report health problems report their exposure experience differently than those without health problems, thereby threatening the internal validity of the study (Hassan, 2005). When exposed and nonexposed (or greater or lesser exposed) respondents report events or health outcomes in a manner that is different between the two groups, it can lead to differential misclassification that can then distort the magnitude of the measure of association toward or away from the null, depending on the magnitude and direction of the bias (Hassan, 2005).

For many registries, the time between the event of interest, design and implementation of the registry, and recruitment of eligible persons is a factor in recall bias. Given that it has been 25 years since the 1990–1991 Gulf War, veterans of this conflict may not correctly recall all potential exposures or specific details (such as how many hours they were exposed to smoke or fumes) requested in the AH&OBP Registry questionnaire. Recall bias would result in stronger observed associations if persons who were experiencing health problems remembered and reported military exposures to a greater extent than persons who were not experiencing such problems. Similarly, persons who do not perceive that an exposure has affected their current health status may be less likely to recall the exposure or to report related symptoms or diagnoses.

Self-Report Bias

In a registry where participants are asked to assess their own types and levels of exposures, these data are not comparable to, nor as specific as, air monitoring data or similar objective measures of exposure. Respondents report whether they believe they were exposed to various chemical, environmental, or biological agents, but quantifying actual exposure intensity or differentiating among specific chemical components is much more difficult. Furthermore, subjective exposure reports are strongly influenced by recall bias (persons who are ill, or believe themselves highly exposed, may differentially overestimate past exposures, for example).

Similarly, self-report bias influences respondents' reports of health outcomes in registry data. Frequently, individuals cannot accurately recall specific names of medical diagnoses or the dates when such diagnoses were made. Likewise, individuals often evaluate past health in relation to their current health; persons who are currently ill may, for example, differentially overestimate the length of their illness or may mistakenly omit earlier illnesses which they would now consider minor in relation to current illness. Other incentives may also influence self-reports of health data. For example, a belief—accurately or not—that participation in the registry or registry data could influence access to health care or other services or key decisions regarding future exposures or deployment practices might affect how an individual appraises and reports his or her health conditions. To reduce the effect of self-reported bias, researchers often attempt to validate self-reported exposure and outcome data against objective measures, such as air monitoring data or medical records, respectively.

IDENTIFICATION OF AN APPROPRIATE COMPARISON GROUP

To assess the degree to which an exposure may cause a specific health problem, an appropriate comparison group is needed. Ideally, this group should resemble the study group as closely as possible in terms of the characteristics that are related to the risk of the health outcome of interest so that differences in outcomes can be attributed to the factor of interest (such as an exposure) rather than to other factors (confounding factors or biases). The characteristics of interest should be available to be known for all who are eligible to participate in the study or at least for large and well-defined subgroups of eligible persons. Generally, such characteristics include basic demographic information such as sex, age, education level, race, ethnicity, and marital status. For military populations, additional service characteristics such as branch, component, deployment dates and locations, and military occupational specialty are desirable.

A registry that collects self-reported data on both exposures and health outcomes is inherently influenced by same-source bias. An example would be people who believe they were highly exposed overestimating their exposures or those who believe they were not exposed underestimating their exposures. Similarly, people who believe that their health conditions are a result of an exposure of interest at any level are more likely to participate in such a registry. Comparing self-reported exposures and health outcomes provides a quantitative assessment of whether individuals tend to attribute their own conditions to the exposure in question. As such, the exposures and health outcomes are considered in a complementary way rather than based on whether there is an objective or true association between the exposures and outcomes of interest. To determine whether a true association between exists between an exposure and outcome of interest, a well-designed epidemiological study, with objective exposure and outcome metrics, is needed.

The committee was not charged with designing an epidemiologic study, rather it was asked to perform an analysis of "how [AH&OBP] registry participants differ in demographic or exposure status (to the extent available data allows) from non-participant groups, such as all deployers or appropriate U.S. comparison populations." Ultimately, the only comparisons the committee makes in this report are comparisons among registry participants who have different levels of exposure potential because internal comparisons mitigate some of the biases of the sample (although the registry participants constitute a very self-selected group). However, to be responsive to the statement of task, this section identifies some potential comparison groups and the strengths and weaknesses of each. Even if a representative comparison group were available, self-report bias for both exposures and health outcomes would continue to be a concern because both types of bias influence participation.

Potential Comparison Groups

A matter that complicates the identification of an appropriate comparison group for any study of health outcomes of service members who participated in the Southwest Asia theater of military operations during the 1990–1991 Gulf War and thereafter is the inherent differences in demographic make-up, the conditions and characteristics of deployment, and potential exposures experienced between individuals who served in that theater during that time and those who did not. To give just one example, most participants in the 1990–1991 Gulf War operations had a single deployment that lasted less than 1 year. In contrast, as of December 2011, 47% of all Operation Enduring Freedom (OEF), OIF, and OND active-duty service members, 35% of reservists, and 35% of National Guardsmen had deployed more than once, with their cumulative lengths of deployment averaging 15.2–17.6 months depending on branch of service and component (IOM, 2013). Methods exist to adjust for differences in such factors to make the different veteran populations more comparable, but there are limits to the effectiveness of statistical adjustments when the differences are so extensive. Alternatively, each deployment cohort could be considered separately, but this lessens the power of statistical testing.

U.S. Civilian Population

One possible comparison group might be a demographically adjusted population of individuals who completed the National Health Interview Survey (NHIS), from which many of the questions on health behaviors and conditions included in the AH&OBP Registry were drawn. In principle, comparing registry participants with a group of individuals with similar demographic makeup from the general U.S. population might result in some useful inferences. However, such an approach would introduce large problems that would make the results highly questionable. First, military personnel are specifically excluded from participating in the NHIS, which covers only the noninstitutional household population of the United States, raising the question of whether there are systematic differences between the groups that would make comparisons unreliable. While veterans are included in the NHIS, an analysis showed that the 2013 NHIS included only 932 veterans who had served since 1990, 523 of whom served overseas and 409 were nondeployed, thereby further limiting the NHIS as a viable comparison group (May and Haider, 2014). Second, any comparison analyses would be subject to significant selection bias because persons who are able to serve in the military and deploy to combat zones are healthier and fitter than the general population (the "healthy warrior effect") that is sampled by the NHIS or any other national survey (Miller et al., 2012). Third, the NHIS is administered as an in-person household interview survey, and research suggests that in-person interviews yield different responses to the same questions than either telephone- or Web-administered surveys (Dillman et al., 2014).

Millennium Cohort Study

VA and DoD have conducted several surveys of military and veteran populations that have included some components of the eligible population defined for the AH&OBP Registry and some of the content. However, none of those surveys are well suited for comparisons with the registry population, except for possibly the Millennium Cohort Study, a prospective longitudinal survey of post-9/11 service members and veterans explicitly designed to collect data on and assess relationships between potential exposures and health outcomes (Ryan et al., 2007). The IOM recommended that DoD conduct prospective epidemiologic research (IOM, 1996, 2000), such as the

Millennium Cohort Study, in order to assess the impact of deployment and exposures on the long-term health outcomes of military service members. The prospective population design of the Millennium Cohort Study mitigates the inherent deficiencies of collecting retrospective registry data. However, since it is limited to post-9/11 deployed service members and veterans who deployed to post-9/11 only (which make up the majority of registry participants; see Chapter 4), it is not an appropriate comparison group for 1990–1991 Gulf War or stabilization period deployers (January 1992–September 2001). The Millennium Cohort Study does not contain questions on the same exposures or health outcomes as the AH&OBP Registry, which would also affect the types of comparisons that could be made. Responses from the baseline and follow-up surveys are routinely matched with Defense Manpower and Data Center records data, and potential participants in the Millennium Cohort Study who meet the eligibility for the AH&OBP Registry could be identified. If that were possible, demographic and military service characteristics distributions, as well as exposures and health outcomes collected by Millennium Cohort Study, could be compared with those reported by post-9/11 registry participants. There are a number of limitations and barriers to gaining access to and using these data, which prevented them from being analyzed by the committee, but in principle it may be a suitable resource for comparison of post-9/11 registry participants.

Nondeployed Veterans and Service Members

Because deployment itself is not the exposure of interest—as it often is with other studies of 1990–1991 Gulf War and OEF/OIF/OND veterans and service members—nondeployed or deployed-elsewhere groups are other potential comparison groups. Unlike the general U.S. population, this group is more similar to registry participants because they had to meet the same types of standards to be accepted into military service, and deployment records exist for all registry-eligible time periods. Deployment for a given period is generally determined from administrative data defined by a given timeframe—1990 to 1991 for example. A service member may be labeled as "nondeployed" during that time frame but then deployed later, creating misclassification.

There is evidence that persons who deploy are different than those who do not, based on characteristics such as military occupational specialties, readiness, and other factors that define "deployability." The majority of service members in each service branch have deployed in support of OEF/OIF/OND (Baiocchi, 2013). Thus, using nondeployed service members and veterans is an especially problematic comparison group for the specific subset of those deployed and eligible for the registry. Those deployed but otherwise not eligible for the AH&OBP Registry likely have similar problems of eligibility. A more representative group might consist of service members or veterans who were eligible to deploy but did not. However, only 4% (20,000) of active duty soldiers met that requirement as of December 2011 (Baiocchi, 2013).

VA Health Care Users

In principle, veterans who are not participants in the registry and use VA services could serve as a comparison group for registry participants who use VA services. However, information on exposures is not collected or available in these sources. Furthermore, such comparisons would be limited to VA users only and would exclude veterans who are not using VA services. Approximately 46% of deployed and 36% of nondeployed Gulf War veterans and 61% of deployed OEF/OIF/OND veterans use VA services (NASEM, 2015; VA, 2015d). In the past, veterans who used VA health care services were, in general, older, had lower incomes, and had more health problems than nonusers. Therefore, users and nonusers of VA health care might differ in important characteristics that might compromise comparisons between them (NASEM, 2015).

DRAWING INFERENCES FROM REGISTRY DATA

For the reasons noted above, registry data are not likely to provide the information necessary to evaluate the strengths of associations or potential cause-and-effect relationships between an exposure and a health outcome. Whether or not the data suggest an association, their value in assessing the impact of exposure on health is limited. Self-reported data may be useful for recording individual stories of experiences and signs or symptoms that may have developed that are indicative of a particular exposure. However, in most instances it is not possible to translate this information into quantitative data that are suitable for making scientific inferences. Therefore, the committee believes that VA post-deployment health registries are primarily useful as a mechanism to create a roster of concerned individuals and provide outreach and health risk communication to potentially exposed and concerned veterans.

It is understandable that some might view a registry as a means to generate disease incidence or prevalence data among participants and want to use it to determine whether the frequency of such reports is different from what would be expected in a population that is otherwise similar but was not exposed. However—for the reasons previously discussed—it is not possible to confidently draw conclusions regarding this from the information collected. The motivation to participate in voluntary registries often is a result of personal experience, so that those who have suffered health problems—particularly problems potentially attributed to the exposures of interest—are more likely to enroll than those individuals who do not experience such outcomes. The data, therefore, reflect this selective participation, resulting in a rate of adverse outcomes among participants that is uninformative for comparisons to other populations.

In epidemiologic studies, assessing whether there is an association between an exposure and a health outcome requires a comparison of the presence of the outcome in an exposed population versus the presence in a comparable population lacking such exposure. As noted earlier in the chapter, because of incomplete and likely unrepresentative participation, the calculation of disease rates among the enrollees does not reflect the rates in the total population of those exposed. With information on health experience only among those who chose to participate, little benefit can come from comparing the experience of this group to some other population to address the question of whether the disease rate was elevated as a result of exposures.

Given the limitations of registries, the data from them may support an evaluation of the possibility of a relationship but cannot be used to determine whether such a relationship actually exists. Because such relationships are often of great interest to both registry participants and sponsors, it sets the stage for disappointment when enrollment and data analysis are completed.

Registry data may, however, motivate epidemiologic studies that would be better designed as a result of the information they generate. For example, a well-designed questionnaire that captured participants' self-reported information could signal the presence of an unusual or atypical health outcome. Variations in outcomes as a function of specific elements such as locations of deployment, military occupation, and time periods of deployments might also yield targets for rigorous study. Such applications make use of the registry data in a way that takes advantage of the information generated without exceeding the limitations imposed by the quality of the data.

SYNOPSIS AND CONCLUSIONS

Registries are one method for ascertaining information about potential exposures and health outcomes in a defined population. Well-designed registries may be useful for identifying rare conditions in a population of interest, driving hypothesis generation, and informing larger and more rigorous study designs. VA and DoD have established several registries with the intent of collecting and monitoring information on health effects that may be a result of deployment-related exposures. However, registries have several inherent limitations, including selection and misclassification biases that preclude their use in evaluating associations between exposures and health outcomes. The many limitations of registries and the inherent biases associated with the data collected prevent their use in evaluating statistical associations or drawing conclusions regarding whether a particular health outcome results from a specific exposure. Registries may yield information useful in determining which exposure–health outcome issues should be investigated using more rigorous data collection and analysis methods.

The next chapter—Chapter 3—extends and deepens this discussion, focusing on the AH&OBP Registry and addressing how these issues affect the interpretation of its data.

REFERENCES

Antão, V. C., L. L. Pallos, Y. K. Shim, J. H. Sapp II, R. M. Brackbill, J. E. Cone, S. D. Stellman, and M. R. Farfel 2011. Respiratory protective equipment, mask use, and respiratory outcomes among World Trade Center rescue and recovery workers. *American Journal of Industrial Medicine* 54(12):897–905.

Antão, V. C., O. I. Muravov, J. Sapp, 2nd, T. C. Larson, L. L. Pallos, M. E. Sanchez, G. D. Williamson, and D. K. Horton. 2015. Considerations before establishing an environmental health registry. *America Journal of Public Health* 105(8):1543–1551.

Baiocchi, D. *Measuring Army deployments to Iraq and Afghanistan*. 2013. Santa Monica, CA: RAND Corporation. http://www.rand.org/pubs/research_reports/RR145.html (accessed December 2, 2016).

Bowler, R. M., H. Harris, and J. Li. 2012. Longitudinal mental health impact among police responders to the 9/11 terrorist attack. *American Journal of Industrial Medicine* 55(4):297–312.

Brackbill, R. M., S. D. Stellman, S. E. Perlman, D. J. Walker, and M. R. Farfel. 2013. Mental health of those directly exposed to the World Trade Center disaster: Unmet mental health care need, mental health treatment service use, and quality of life. *Social Science and Medicine* 81:110–114.

Bullman, T. A., and H. K. Kang. 1994. Posttraumatic stress disorder and the risk of traumatic deaths among Vietnam veterans. *Journal of Nervous and Mental Disease* 182(11):604–610.

Bullman, T. A., H. K. Kang, and T. L. Thomas. 1991. Posttraumatic stress disorder among Vietnam veterans on the Agent Orange Registry. A case-control analysis. *Annals of Epidemiology* 1(6):505–512.

Bullman, T. A., K. K. Watanabe, and H. K. Kang. 1994. Risk of testicular cancer associated with surrogate measures of Agent Orange exposure among Vietnam veterans on the Agent Orange Registry. *Annals of Epidemiology* 4(1):11–16.

Clemens, M. W., A. L. Kochuba, M. E. Carter, K. Han, J. Liu, and K. Evans. 2014. Association between Agent Orange exposure and nonmelanotic invasive skin cancer: A pilot study. *Plastic and Reconstructive Surgery*-133(2):432–437.

Coughlin, S. S., P. Aliaga, S. Barth, S. Eber, J. Maillard, C. M. Mahan, H. K. Kang, A. Schneiderman, S. DeBakey, P. Vanderwolf, and M. Williams. 2011. The effectiveness of a monetary incentive on response rates in a survey of recent U.S. veterans. *Survey Practice* 4(1). http://www.surveypractice.org/index.php/SurveyPractice/article/view/91/html (accessed December 2, 2016).

Dillman, D. A., J. D. Smyth, and L. Melani Christian. 2014. *Internet, mail, and mixed-mode surveys: The tailored design method, 3rd edition*. Hoboken, NJ: John Wiley & Sons.

DoD (Department of Defense). 2016a. Environmental health surveillance registries. https://registry.csd.disa.mil/registryWeb/DisplayAbout.do (accessed September 26, 2016).

DoD. 2016b. About the Operation Tomodachi Registry (OTR). https://registry.csd.disa.mil/registryWeb/Registry/OperationTomodachi/DisplayAbout.do (accessed September 26, 2016).

DoD. 2016c. About the Gulf War Oil Well Fire Smoke Registry. https://registry.csd.disa.mil/registryWeb/Registry/OWFSR/DisplayAbout.do (accessed September 26, 2016).

Dunavant, J. D., M. Chehata, D. R. Case, M. Mckenzie-Carter, J. Cassata, R. Marro, R. Ranellone, K. Knappmiller, G. Falo, L. Alleman, and P. Blake. 2013. *Operation Tomodachi Registry: Radiation data compendium*. Fort Belvoir, VA: Defense Threat Reducation Agency. https://registry.csd.disa.mil/registryWeb/docs/registry/optom/DTRA-TR-13-044.pdf (accessed September 26, 2016).

Egleston, B. L., S. M. Miller, and N. J. Meropol. 2011. The impact of misclassification due to survey response fatigue on estimation and identifiability of treatment effects. *Statistics in Medicine* 30(30):3560–3572.

Farfel, M., L. DiGrande, R. Brackbill, A. Prann, J. Cone, S. Friedman, D. J. Walker, G. Pezeshki, P. Thomas, S. Galea, D. Williamson, T. R. Frieden, and L. Thorpe. 2008. An overview of 9/11 experiences and respiratory and mental health conditions among World Trade Center Health Registry enrollees. *Journal of Urban Health* 85(6):880–909.

Gliklich, R. E., N. A. Dreyer, and M. B. Leavy (eds.). 2014. *Registries for evaluating patient outcomes: A user's guide, 3rd ed*. Rockville, MD: Agency for Healthcare Research and Quality. https://www.ncbi.nlm.nih.gov/books/NBK208619 (accessed December 2, 2016).

Groves, R. M., R. B. Cialdini, and M. P. Couper. 1992. Understanding the decision to participate in a survey. *Public Opinion Quarterly* 56(4):475–495.

Hassan, E., 2005. Recall bias can be a threat to retrospective and prospective research designs. *Internet Journal of Epidemiology* 3(2):1–7.

Hernán, M. A., S. Hernández-Díaz, and J. M. Robins. 2004. A structural approach to selection bias. *Epidemiology* 15(5):615–625.

Hodge, S. J., J. Ejnik, K. S. Squibb, M. A. McDiarmid, E. R. Morris, M. R. Landauer, and D. E. McClain. 2001. Detection of depleted uranium in biological samples from Gulf War veterans. *Military Medicine* 166(12 Suppl):69–70.

IOM (Institute of Medicine). 1996. *Health consequences of service during the Persian Gulf War: Recommendations for research information systems*. Washington, DC: National Academy Press.

IOM. 1998. *Adequacy of the VA Persian Gulf Registry and uniform case assessment protocol.* Washington, DC: National Academy Press.

IOM. 2000. *Protecting those who serve: Strategies to protect the health of deployed U.S. forces.* Washington, DC: National Academy Press.

IOM. 2010. *Gulf War and health, volume 8: Update of health effects of serving in the Gulf War.* Washington, DC: The National Academies Press.

IOM. 2013. *Returning home from Iraq and Afghanistan: Assessment of readjustment needs of veterans, service members, and their families.* Washington, DC: The National Academies Press.

May, L., and J. Haider. 2014. Preliminary assessment of NHIS data for providing a burn pit registry comparison group. Prepared for the Committee on the Assessment of the Department of Veterans Affairs Airborne Hazards and Open Burn Pit Registry. November 20.

McDiarmid, M. A., K. Squibb, S. Engelhardt, M. Oliver, P. Gucer, P. D. Wilson, R. Kane, M. Kabat, B. Kaup, L. Anderson, D. Hoover, L. Brown, and D. Jacobson-Kram. 2001. Surveillance of depleted uranium exposed Gulf War veterans: Health effects observed in an enlarged "friendly fire" cohort. *Journal of Occupational and Environmental Medicine* 43(12):991–1000.

McDiarmid, M. A., S. M. Engelhardt, C. D. Dorsey, M. Oliver, P. Gucer, J. M. Gaitens, R. Kane, A. Cernich, B. Kaup, D. Hoover, A. A. Gaspari, M. Shvartsbeyn, L. Brown, and K. S. Squibb. 2011. Longitudinal health surveillance in a cohort of Gulf War veterans 18 years after first exposure to depleted uranium. *Journal of Toxicology and Environmental Health Part A* 74(10):678–691.

Miller, L., and E. Aharoni. 2015. *Understanding low survey response rates among young U.S. military personnel.* Santa Monica, CA: RAND Corporation. http://www.rand.org/content/dam/rand/pubs/research_reports/RR800/RR881/RAND_RR881.pdf (accessed December 2, 2016).

Miller, M., C. Barber, M. Young, D. Azrael, K. Mukamal, and E. Lawler. 2012. Veterans and suicide: A reexamination of the National Death Index–Linked National Health Interview Survey. *American Journal of Public Health* 102(Suppl 1):S154–S159.

NARA (National Archives and Records Administration). 2002. Request for records disposition authority: Comprehensive Clinical Evaluation Program (CCEP). https://www.archives.gov/records-mgmt/rcs/schedules/departments/department-of-defense/office-of-the-secretary-of-defense/rg-0330/n1-330-02-001_sf115.pdf (accessed September 26, 2016).

NASEM (National Academies of Sciences, Engineering, and Medicine). 2015. *Considerations for designing an epidemiologic study for multiple sclerosis and other neurologic disorders in pre and post 9/11 Gulf War veterans.* Washington, DC: The National Academies Press.

NIH (National Institutes of Health). 2016. National Cancer Institute Division of Cancer Control & Population Sciences. Department of Defense Comprehensive Clinical Evaluation Program (CCEP). http://epi.grants.cancer.gov/pharm/pharmacoepi_db/ccep.html (accessed September 26, 2016).

Rabeneck, B. 2001. Using the national registry of HIV-infected veterans in research: Lessons for the development of disease registries. *Journal of Clinical Epidemiology* 54(12):1195–1203.

Raftery J., P. Roderick, and A. Stevens. 2005. Potential use of routine databases in health technology assessment. *Health Technology Assessment* 9(20):1–106.

Rolstad, S., J. Adler, and A. Rydén. 2011. Response burden and questionnaire length: Is shorter better? A review and meta-analysis. *Value in Health* 14(8):1101–1108.

Rothman, K. J. 2012. *Epidemiology: An introduction, 2nd ed.* New York: Oxford University Press.

Ryan, M. A., T. C. Smith, B. Smith, P. Amoroso, E. J. Boyko, G. C. Gray, G. D. Gackstetter, J. R. Riddle, T. S. Wells, G. Gumbs, T. E. Corbeil, and T. I. Hooper. 2007. Millennium cohort: Enrollment begins a 21-year contribution to understanding the impact of military service. *Journal of Clinical Epidemiology* 60(2):181–191.

Schultz, M. G, J. H. Sapp, C. D. Cusack, and J. M. Fink. 2010. The National Exposure Registry: History and lessons learned. *Journal of Environmental Health* 72(7):20–25.

Shvartsbeyn, M., P. Tuchinda, J. Gaitens, K. S. Squibb, M. A. McDiarmid, and A. A. Gaspari. 2011. Patch testing with uranyl acetate in veterans exposed to depleted uranium during the 1991 Gulf War and the Iraqi conflict. *Dermatitis* 22(1):33–39.

Squibb, K. S., R. W. Leggett, and M. A. McDiarmid. 2005. Prediction of renal concentrations of depleted uranium and radiation dose in Gulf War veterans with embedded shrapnel. *Health Physics* 89(3):267–273.

Smith, T. C., B. Smith, M. A. K. Ryan, G. C. Gray, T. I. Hooper, J. M. Heller, N. A. Dalager, H. K. Kang, and G. D. Gackstetter. 2002. Ten years and 100,000 participants later: Occupational and other factors influencing participation in U.S. Gulf War health registries. *Journal of Occupational and Environmental Medicine* 44(8):758–768.

VA (Department of Veterans Affairs). 2004. *Ionizing Radiation Review* 2(1) [Entire issue]. http://www.publichealth.va.gov/docs/radiation/irr_newsletter_dec04.pdf (accessed October 12, 2016).

VA. 2010. *Ionizing Radiation Review* 4(1) [Entire issue]. http://www.publichealth.va.gov/exposures/radiation/publications/index.asp (accessed October 12, 2016).

VA. 2012. *Agent Orange Review* 26(1) [Entire issue]. http://www.publichealth.va.gov/docs/agentorange/reviews/newsletter-winter2012.pdf (accessed October 13, 2016).

VA. 2015a. Ionizing Radiation Registry health exam for veterans. http://www.publichealth.va.gov/exposures/radiation/benefits/registry-exam.asp (accessed December 2, 2016).

VA. 2015b. Agent Orange Registry health exam for veterans. http://www.publichealth.va.gov/exposures/agentorange/benefits/registry-exam.asp (accessed December 2, 2016).

VA. 2015c. Depleted Uranium Follow-Up Program. http://www.publichealth.va.gov/exposures/depleted_uranium/followup_program.asp (accessed December 2, 2016).

VA. 2015d. *Analysis of VA health care utilization among Operation Enduring Freedom (OEF), Operation Iraqi Freedom (OIF), and Operation New Dawn (OND) veterans*. Washington, DC: Veterans Health Administration.

VA. 2015e. Gulf War research strategic plan 2013–2017: 2015 update. http://www.research.va.gov/pubs/docs/GWResearch-StrategicPlan.pdf (accessed December 2, 2016).

VA. 2016a. Toxic Embedded Fragment Surveillance Center. http://www.publichealth.va.gov/exposures/toxic_fragments/surv_center.asp (accessed June 23, 2016).

VA. 2016b. Public health: Military exposures. http://www.publichealth.va.gov/exposures (accessed June 21, 2016).

VA. 2016c. Environemntal exposures programs and services for veterans. http://www.publichealth.va.gov/docs/exposures/registry-evaluation-brochure.pdf (accessed December 2, 2016).

VA. 2016d. Agent Orange newsletter: Information for Vietnam-era veterans and their families. http://www.publichealth.va.gov/docs/agentorange/reviews/ao-newsletter-summer-2016.pdf (accessed December 2, 2016).

VA and DoD (Department of Veterans Affairs and Department of Defense). 2002. Department of Defense Comprehensive Clinical Evaluation Program (CCEP). In *Combined analysis of the VA and DoD Gulf War clinical evaluation programs: Study of the clinical findings from systematic medical examinations of 100,339 U.S. Gulf War veterans.* http://www.gulflink.osd.mil/combined_analysis/gulf_war_clinical_evaluation_programs.htm#ccep (accessed August 26, 2016).

3

The Airborne Hazards and Open Burn Pit Questionnaire and Registry

When developing a registry, it is important to clearly define its goals and expected benefits while being mindful about its limitations. To maximize its use and value, developers need to take into account the various logistical constraints (time, expertise, money) to appropriately designing, implementing, and maintaining the registry as well as to conducting outreach to eligible users and making optimal use of the registry. The rationale for and process used to develop the registry both need to be fully explicated to allow registry participants and users of registry information to take these factors into account.

This chapter focuses on the development and implementation of the Airborne Hazards and Open Burn Pit (AH&OBP) Registry and its key element, the self-assessment questionnaire. The chapter begins with a brief review of the salient recommendation from an initial report of the long-term health consequences of exposure to burn pits in Iraq and Afghanistan and the resulting decisions and directives from Congress. Following a discussion on the registry's development, the chapter offers evaluations of the design of the questionnaire and the types of questions used in it, including considerations for how the questions might be improved. Although registries cannot substitute for well-designed epidemiologic studies, improvements to the questionnaire could result in an improved collection instrument. The ability to supplement registry data by linking the registry to other Department of Veterans Afffairs (VA) and Department of Defense (DoD) data sources is next considered. Finally, the recruitment and enrollment of registry participants is discussed.

SCIENTIFIC BASIS AND CONGRESSIONAL DIRECTIVE

Large expenditures of time and money—and substantial levels of experience and technical and scientific expertise—are required to establish and maintain an exposure registry. One justification for making such expenditures is that the exposure or exposures of concern may present a clear health risk. The authoring committee of *Long-Term Health Consequences of Exposure to Burn Pits in Iraq and Afghanistan* recommended that "a prospective study of the long-term health effects of exposure to burn-pit emissions in military personnel deployed at [Joint Base Balad]" be conducted (IOM, 2011, p. 8). Notwithstanding this recommendation for conduct of a prospective epidemiologic study, in January 2013 Congress passed Public Law 112-260 (reproduced in Appendix A) directing VA to establish the AH&OBP Registry within 12 months of the law's enactment. Within this narrow window of time, VA was tasked to develop a process and structure to create a comprehensive and targeted exposure and health outcomes questionnaire and to make it available for veterans' use.

The 12-month schedule required for developing and beginning the implementation of a well-designed and tested registry to assess and track the exposures of interest and their possible health effects was short, considering the complexity of the registry's intent and the tasks involved. Briefly, there are many undertakings required to establish any registry of military personnel or veterans, but the largest involve

1. defining those eligible for the study (based largely on data from DoD records that are not readily structured for research purposes);
2. developing a comprehensive outreach and recruitment strategy for reaching the full population of potentially eligible individuals;
3. determining the most effective mode for enrollment, vetting eligibility, and capturing the information required;
4. determining what information on deployments, exposures, health, and other indicators (including military and demographic characteristics) needs to be measured;
5. determining the best sources of data for capturing this information (for example, self-report or military records);
6. establishing the best mode (such as in-person, online, or computer-assisted interviews) of data collection for self-report assessments;
7. examining previous registries, epidemiologic and health surveys, and other sources to determine the best questions to use or adapt that are consistent with the selected mode;
8. testing of the selected and developed questions to ensure that they are well understood by participants and that they effectively capture the key information required while also minimizing burden;
9. selecting appropriate mechanisms for supporting and maintaining a registry over its expected lifetime that is consistent with best current practices;
10. implementing tests, evaluations, and revisions to ensure that all components of the registry system work together as intended; and
11. developing a strategy to use the collected information for the stated purpose of the registry.

All of these processes need to be conducted within the regulatory constraints and ethical considerations that go into any effort that involves the acquisition and management of personally identifiable information. These tasks are difficult to accomplish successfully even when time is not a factor, and there are limitations to the time savings that can be realized with the application of greater amounts of money, people, or effort. It is thus open to question whether the time period allotted by Congress was realistic.

Public Law 112-260 includes provisions specifying that the registry would be for "eligible individuals who may have been exposed to toxic airborne chemicals and fumes caused by open burn pits" and would include any information that VA determined as "necessary to ascertain and monitor the health effects" of individuals who served in the Armed Forces and reported exposure to toxic airborne chemicals and fumes caused by open burn pits. The law directs that registry participants are to be notified of significant developments in the study and in the treatment of conditions associated with exposure to toxic airborne chemicals and fumes caused by open burn pits.

VA has, in various forums, articulated multiple goals and intents for the AH&OBP Registry. The registry website states that the data collected will be used to help monitor health conditions affecting eligible veterans and service members, to help veterans and service members who report deployment-related exposure concerns, and to improve VA programs. It then states the following benefits of participation: creating a point to identify changes in health over time, using the completed questionnaire to discuss concerns with a health care provider, and learning about follow-up care and VA benefits (VA, 2016a). VA also stated that it intends to use the registry to generate potential hypotheses about exposure response relationships but acknowledges that subsequent studies would be needed to test these hypotheses (VA, 2014a). In a presentation to the committee, VA said that data from the registry will also be used more generally to improve programs in the Veterans Health Administration (VHA) and to provide outreach to veterans who may have experienced adverse health outcomes as a result of their exposures (Ciminera, 2015a). The lack of a consistent message makes it difficult to evaluate the degree to which the registry is meeting its stated intents and suggests a lack of focus that—as the report will later detail—is reflected in information gathering that does not appear to serve a sound research purpose.

To determine whether the registry is fulfilling its intent or purpose—as defined by VA or otherwise—the committee performed a comprehensive assessment, the results of which make up the remainder of this chapter and are further addressed in Chapters 4–6. The assessment begins with a review of the registry's development, including the design and construction of the registry questionnaire and initial testing. The next section provides a detailed description and evaluation of the registry questionnaire, including the appropriateness of topics covered, layout, structural features, and directions. This is followed by a discussion of how data from other sources could be linked with registry data to provide a more complete picture of veteran and service member health. Eligibility criteria for participation, a description of the communications and outreach efforts used by VA to advertise the availability of the registry, and the process for enrolling and completing the questionnaire are detailed in the final section.

DEVELOPING THE REGISTRY

Personnel and Expertise

The AH&OBP Registry consists of responses to an online questionnaire completed by eligible veterans and service members and is housed in the Office of Patient Care Services within VHA. Similar to other Congressionally mandated registries, such as the Agent Orange Registry and the Gulf War Registry, there is no anticipated closing or end date for the AH&OBP Registry (Lezama, 2015). Historically, the number of new participants declines over time but can increase with new media or scientific reports on the health consequences of the target exposures.

Developing a registry of this nature is a major challenge given the large, diverse population of interest, the complexity of the exposures and health outcomes of interest, the requirement to have it be easily accessible to all interested and eligible veterans, and the desirability of having it be completed online (Ciminera, 2015b; VA, 2014a). VA relied on two working groups of VA and DoD subject-matter experts to advise it on the content and design of the questionnaire and registry. The working groups were tasked with developing "clinical guidance, a registry analysis plan (to include DoD exposure data) and supporting information technology requirements" (VA, 2013) and developing additional questions related to environmental exposures unique to military service in contingency operations (VA, 2014a). Staff within the Post-Deployment Health Group of the VHA Office of Public Health (now the Office of Patient Care Services) were consulted for statistical expertise.

The first working group meeting took place in October 2012 and had nine members with expertise in primary care, pulmonology, and public or environmental health. That working group was responsible for developing and implementing clinical evaluation guidance for primary care and other providers to support the interagency Airborne Hazards Action Plan, under which the AH&OBP Registry falls. That working group was also tasked with recommending methods to disseminate educational materials based on the guidance it developed and to develop recommendations to improve collaboration among specialists with the goal of improving the consistency of specialist evaluations and interagency situational awareness for unusual cases or clusters of cases (VA, 2012).

The second working group met in January 2013 and was composed of 10 VA and DoD subject-matter experts in public or environmental health and other specialties who had little overlap with the previous working group participants. The purpose of this working group was to develop an exposure assessment instrument that would be integrated into the registry questionnaire (VA, 2013). Meeting notes, draft or final products, or other materials related to the outcomes of the two working groups are not available, so it was not possible for the committee to further evaluate the work of these groups. VA also conducted usability testing beginning in October 2013, and it included a human factors analysis by its Office of Informatics and Analytics (Ciminera, 2015a,b).

The committee appreciates that the VA staff members who were involved with developing and implementing the registry appear to have been experts in military occupational health and were conscientious in their approach to designing and implementing the questionnaire, especially given the short timeframe. However, the committee believes that the approach of using personnel in VHA's Office of Public Health Post-Deployment Health Group and establishing two working groups of VA and DoD subject-matter experts—but not experts in survey design or survey research methods—to advise on questionnaire content and design was insufficient for developing a registry of this scale. Expanding the expertise to include specialists in fields such as survey design and research methods would have provided input on many issues that do not appear to have been considered in the design, testing, and implementation of the registry. For example, depending on the goal of the registry (surveillance or hypothesis generation), consulting with experts in exposure and disease ascertainment would ensure that, within the constraints of the data collection approach, the most accurate and complete data would be generated. Survey design experts would consider the characteristics of the population of interest (educational and cultural background, incentives and disincentives to participate, and so forth) and, especially, the distinctive features of Web-based survey design and data collection, such as more rapid use of the data, the ability to tailor the survey to individuals through automatic application of fill-ins and skip patterns, increased data reliability, and a reduction in survey costs. Thus, the committee concludes that many of the problems in design and implementation (discussed under Questionnaire Quality) could have been anticipated and ameliorated had experts in survey research been consulted.

In brief, a fairly standard process has evolved across the survey research industry for developing and testing questionnaires, including

- questionnaire formatting that is consistent with and takes advantage of the specific mode(s) of the survey carried out;
- expert independent review of draft questionnaires to identify and resolve problems with question clarity and how questions, potential responses, and instructions are worded;
- standard questionnaire appraisal systems to evaluate and improve survey questions in order to identify potential problems in the wording or structure of questions that may lead to difficulties in questionnaire administration, miscommunication, or other failings (Institute for Social Research, 2016); and
- cognitive interviewing: conducting one-on-one interviews with members of the targeted population that probe respondents' understanding of questions and how they form their responses, the burden of providing a response, and their willingness to provide high-quality responses.

In the case of Web-based surveys, evaluation is typically conducted using cognitive interviewing techniques such as "think aloud" sessions with subjects as they complete a draft survey and usability testing with respondents to determine how they use the Web-survey program to maneuver through the instrument.[1] It is unclear whether or to what extent such work was carried out during the AH&OBP Registry pilot phase, but the committee believes that some of the problems with the questionnaire that it observed might have been avoided if such processes had been rigorously applied. Any registry of this type will benefit from the conduct of a thorough, standardized assessment of the instruments used for data gathering before launch.

VA contracted with two private-sector firms for Web implementation and information technology support of the registry and to design the required database architecture to ensure the information could be accessed, stored, linked to other database systems, and extracted. The Web-based format allows for real-time performance monitoring and quality improvement initiatives to be part of the system architecture. VA stated that this capacity makes it possible to monitor weekly metrics such as the numbers of new registrants and registrant user status, to monitor the registry's status for accessibility, and to log helpdesk calls that provide technical assistance to users (Montopoli, 2016a). VA indicated that the agency has also implemented better integration of the AH&OBP Registry database with the health care and enrollment data available in the VHA Corporate Data Warehouse.

In addition to the Web-based version of the questionnaire and registry, VA has implemented a version that can be accessed with a mobile application accessible on Android, Blackberry, iPad, iPhone, and Windows Phone platforms. The mobile application format was introduced September 1, 2015, and had been used by more than 16,000 individuals through September 15, 2016 (representing ~10% of all individuals who accessed the registry as of that time). An internal analysis of time to complete the questionnaire using the mobile app version found that the average completion time for users who started and completed the questionnaire on the same day was 61 minutes (Personal communication, Michael Montopoli, Director, Post-9/11 Era Environmental Health Program, VA, September 15, 2016).

---

[1] This topic is further discussed in the section titled Open Comment Period and Pilot Testing.

DoD's Defense Health Agency and Army Public Health Command store results from questionnaires from active-duty participants, and VA stores completed questionnaires for all other participants (Ciminera, 2015a). The data are collected and analyzed by the Post-Deployment Health Group of the VHA Office of Public Health (VA, 2014a).

At the end of August 2015, an update to the registry was released and implemented. No changes were made to the content of the questionnaire, but several system updates were implemented. The updates included migrating the database platform from the original MongoDB to SQL software, enhancing the VHA staff portal to make VA health care users' registry data more easily accessible to VA providers and facilities and adding capabilities for ad hoc reporting, creating a "data mart"[2] in VHA's Corporate Data Warehouse for internal analyses of the raw registry data, and integrating with eBenefits[3] to allow access to the registry from the eBenefits website (Lezama, 2016).

Questionnaire Development

The AH&OBP Registry is distinctive from previous registries established by VA in that it was designed to be completed using an online interface only (no paper forms or computer assisted interviews). An online questionnaire was proposed because it could potentially improve access and population monitoring while limiting the burden of participation (Ciminera, 2015a). Veterans and service members of the recent conflicts have high levels of Internet access, and the committee appreciates VA's effort to attempt to include a more representative group of participants by designing the registry to be an online format that can be accessed from nearly anywhere there is an Internet connection. However, although a Web-based survey may confer benefits over more traditional methods of mail surveys, in-person interviews, or computer-assisted telephone interviewing, not all eligible persons have access to a Web-enabled device like a computer or smart phone. The Pew Research Center (2011) notes that "[p]eople with lower incomes, less education, living in rural areas or age 65 and older are underrepresented among internet users and those with high-speed internet access." And researchers have noted that Web-based surveys tend to have lower response rates than other types of surveys (Fan and Yan, 2010).

The questionnaire was designed to take about 30 minutes to complete (*Federal Register*, 2013a) in order to collect relevant information while limiting participation burden. While the questionnaire includes many new questions tailored to the registry and its purpose, it makes extensive use of relevant questions used in another ongoing survey, the National Health Interview Study (NHIS). Recycling questions used in and validated by the NHIS rather than developing new questions may be considered a strength of the registry questionnaire, but it is also one of its key weaknesses. VA chose to use some existing NHIS questions for health conditions and symptoms. These were taken from four different sections of the 2013 NHIS Adult Core module: conditions, health status and limitation of activity, health behaviors, and health care access and utilization (Personal communication, Michael Montopoli, Director, Post-9/11 Era Environmental Health Program, VA, August 18, 2016). However, not all questions for each section were used or presented in the same order as the NHIS, and taking them out of the surrounding context may affect their validity.

A comparison between the registry questions and the 2013 NHIS performed by the committee's subcontractor showed significant discrepancies in wording for questions on respiratory conditions, with only 3 of 13 questions showing an exact match, 8 questions with no match, and 2 questions with changes in reference period. The correspondence is better for questions on cardiovascular conditions and health behaviors: 4 of 6 and 11 of 13 exact matches in wording, respectively. Only 5 of 13 questions on cancer history and other conditions had exact matches. Changing the wording of the questions and the order in which they are presented weakens any assumption that the registry questions have been validated.

Furthermore, the NHIS was designed to be a cross-sectional household interview survey of the civilian[4]

---

[2] A data mart is a subset of a data warehouse focused on a specific set of information.

[3] "eBenefits is a joint VA/DoD web portal that provides resources and self-service capabilities to Veterans, Service members, and their families to research, access and manage their VA and military benefits and personal information" (VA, 2014b).

[4] Active-duty military personnel (but not veterans) are explicitly excluded from the NHIS, unless another family member is a civilian and eligible. However, when data are collected on military personnel, they are limited to familial factors and given a final weight of zero so that their individual characteristics are not counted in national estimates (CDC, 2015).

noninstitutionalized U.S. population and is conducted using face-to-face interviews. As a result, most of the questions, formats, and responses used in the AH&OBP Registry questionnaire were developed for use in interview-based rather than Web-based surveys. In survey interviews, questions and response choices are typically read by interviewers to respondents and answers recorded by the interviewer. While these interviews are generally now computer-assisted, they are not self-administered. Using a different means to administer the questions also calls into doubt whether they can be considered to be validated for information-gathering purposes.

The committee appreciates that under the time constraints given by Congress, using existing validated questions from previous surveys made sense. However, the NHIS was not designed for or validated using active-duty or veteran populations and the designers should have considered surveys that have been used or validated in studies of veterans and service members and those that were explicitly designed for Web-based administration.

For example, the Millennium Cohort Study began in 2001 and was designed to be a 21-year longitudinal study to survey U.S. military personnel from all service branches to evaluate the health risks of military exposures, including deployment on short- and long-term health outcomes. Through 2011, the study had enrolled more than 200,000 participants (Crum-Cianflone, 2013). It was designed to link survey responses with several data sources, including VA and DoD data, to complement the self-reported questionnaire responses. Before implementing the survey, the developers conducted several focus groups that discussed the instrument, and they tested it in a pilot study of 1,000 participants, which resulted in systematic validation of the instrument. Quality control processes, including methods to encourage nonbiased responses and retention, have been established and are used to make improvements and correct errors as the study progresses. The Millennium Cohort Study has been found to be a reasonably representative sample of the U.S. military, and its survey data have been confirmed as having excellent reliability in several publications (Crum-Cianflone, 2013).

The most recent iteration of the Millennium Cohort Study instrument is designed to be completed either online or on paper by mail, consists of about 450 questions, takes approximately 30–45 minutes to complete, and includes validated questions on militarily relevant issues, including a specific question on exposure to smoke from burning trash and/or feces (Crum-Cianflone, 2013). The 9 questions on environmental exposures experienced during deployment were derived from a subset of the 23 environmental exposure questions included in VA's National Health Survey of Persian Gulf War Era Veterans, first conducted in 1995 (Kang et al., 2000). The Millennium Cohort Study questionnaire collects information on mental, physical, behavioral, and functional health and incorporates several standardized instruments, such as the 36-Item Short Form Health Survey, Patient Health Questionnaire, Posttraumatic Stress Disorder Checklist–Civilian Version, and the so-called CAGE questions for alcohol problems (NIAAA, 2016; RAND, 2016; VA, 2016b). In sum, the Millennium Cohort Study illustrates that there are other approaches to eliciting information of the type sought by the AH&OBP Registry that better align with validated health survey instruments and that yield results that are more useful in research.

Optional In-Person Exam

VA sought to improve access and participation in the registry while limiting the burden on participants in part by offering an optional in-person exam instead of requiring that the exam be a criterion for participation, as was done for other VA registries. The reasoning was that prospective subjects might be more likely to participate if they were not required to go to a specific place to register or be inconvenienced by scheduling and completing an exam that they might not feel they need. As mentioned elsewhere, one value of VA registries is that they generate a roster of concerned individuals; other uses include outreach, surveillance, and health-risk communication to potentially exposed veterans. Therefore, not requiring an in-person physical evaluation at a VA medical center as a requisite for participation increases the potential for greater representation among service members and veterans, especially those who are not enrolled in VA, who have already seen a non-VA provider for a condition or concern potentially related to the exposure, who would have to travel great distances for an exam, who are unable to miss work, or who do not have or wish to take the time to make an appointment and receive an exam. On the other hand, removing the requirement to undergo a physical exam results in a loss of valuable objective health and functional status information and a potential mechanism to verify self-reported information for at least a subsample of the registry population.

After completing and submitting the questionnaire, participants are given the option to print and save a copy of their responses and may schedule the optional in-person clinical evaluation. VA health care providers conduct the exam for veterans and members of the reserves and National Guard who are not currently activated. Participants who are enrolled in the VA health care system can make an appointment with their primary care provider or patient-aligned care team. Veterans and non-activated service members not enrolled in the VA health care system need to first contact a VA environmental health coordinator (Sharkey et al., 2014). Active-duty service members and members of the reserve and National Guard who are on active duty orders for more than 30 days may request a medical evaluation through their designated medical treatment facility or DoD primary care manager. When requesting an appointment, service members are instructed to indicate that the appointment is for "health concerns related to the Airborne Hazards and Open Burn Pit Registry exposures" (Sharkey et al., 2014).

Both VA and DoD have prepared fact sheets for participants as well as clinical guidance for health care providers conducting follow-up exams. The in-person exam is not standardized, which makes it difficult to assess various aspects of its process or quality. General broad guidance is available to VA clinicians via a training webinar, and the committee was provided with a copy of the National Note Airborne Hazards and Burn Pit Initial Evaluation Clinical Template. The guidance advises physicians to use "their own evidence based knowledge, expertise, and skills to guide a patient-centered evaluation and management" and adds that additional diagnostic tests or specialty consultations may be appropriate (VA, 2016c). The committee notes that although there are no known conditions that are directly attributable to burn pit exposure, the clinical exam is useful in that it allows participants an opportunity to be connected with a provider, articulate any health concerns they may have, and, if warranted, undergo appropriate diagnostic testing or referral and begin treatment to improve symptoms.

DoD does not require a specific form for the clinical assessment, but providers are encouraged to review the service member's questionnaire; to take a medical history that focuses on occupational and environmental exposures, airborne hazards, and smoking history; and to determine the person's primary concern or complaint. The provider may perform a physical examination if indicated based on symptoms or concerns, order additional diagnostic tests, or refer the service member to a specialist for further evaluation. The examination, diagnoses, and any referrals are to be fully documented in the service member's medical record (Defense Health Board, 2015). The U.S. Army Public Health Command has developed provider education and training—including a downloadable provider registry exam toolbox—concerning the registry and the associated clinical exam for physicians, physician assistants, and nurse practitioners who may be seeing these DoD participants (Montopoli, 2016a). The training is intended to familiarize military providers with the registry and its purpose, provide coding and recording guidance, and provide additional optional clinical tools for assessment that have been developed by VA and DoD, such as an algorithm for primary care providers that details a clinical approach to the participant requesting the clinical exam (Ciminera, 2015b). VA was given the opportunity to review the DoD materials during their development. The U.S. Army Public Health Center has hosted multiple, internal video teleconferences with provider groups and has produced other marketing efforts in person, in print, and online (Montopoli, 2016a).

The computerized patient record system note template within the VA electronic health record was developed by VA's New Jersey War Related Illness and Injury Study Center with input from VA's Office of Public Health, patient-aligned care team primary care providers, and environmental health clinicians. The note template was created to standardize the clinical evaluations conducted by VA, collect information on health outcomes, and capture administrative data for registry monitoring and improvement. It allows clinicians to view an individual's questionnaire responses online through a secure portal and provides links to additional information about airborne hazards, health conditions possibly related to those exposures, and guidance for completing an appropriate evaluation. The note also allows clinicians to document the person's chief complaint; medical, social, family, and substance-use history; a physical exam, documenting positive and negative findings by system; diagnostic tests and evaluations performed to date with applicable results; and an overall assessment, recommendations, and follow-up orders. The computerized patient record system patch was installed in late November 2014 (Ciminera, 2015b; Montopoli, 2016a), but the software to indicate that a clinical evaluation related to registry participation took place was not finalized until November 2015 (Lezama, 2015). The template is intended as a clinical progress note and not as a standardized data collection tool.

Registry participation may be entered by the provider in a patient's medical record, but there is no standard "flag" indicating that a patient is in part of the registry. When a patient is seen for an in-person exam related to participation in the AH&OBP Registry, the provider may choose to use the recommended standard AH&OBP Registry note template to capture complaints, abnormalities, and other data from that exam, but that use is not required. The AH&OBP Registry questionnaire data and VHA clinical data can be aggregated within the VHA Corporate Data Warehouse for use by registry staff or VA researchers (Montopoli, 2016a).

As of fall 2016, VA was in the process of developing new patient appointment scheduling software that would allow the registry to interface with that system, allowing registry participants who are enrolled with VA to request the clinical examination directly through the registry (Montopoli, 2016a). By November 30, 2015, about 28,800 individuals had completed the questionnaire and indicated that they were interested in having a health exam, but only 750 participants (2.5% of those interested) had received the health exam. The reasons for the low proportion of interested respondents completing an in-person clinical exam are unknown. An analysis of the chief complaints (participants may have indicated more than one) for the 543 participants who underwent a clinical evaluation showed that the three most common complaints were shortness of breath (57.5%), decreased exercise ability (47.8%), and chronic sinus infections (47.3%) (Lezama, 2015). However, the chief complaints cited during the clinical evaluations have not been matched with the most prevalent conditions noted on the questionnaire to validate responses or to determine whether the persons undergoing evaluation are reporting different outcomes or conditions, reporting a greater severity or differences in other parameters, or reporting conditions at different frequencies than are observed among all registry participants.

One strength of the AH&OBP Registry is that the completed questionnaire generates a record of potential exposures and health concerns that is recorded in the participant's VA electronic health record that can be accessed by military and veteran health care system providers, and that can be downloaded and printed by the participant for his or her reference and the use of other health care providers. Extending this functionality would provide an even greater benefit. The lack of time available for health care providers to do thorough clinical work-ups is well documented (Chen et al., 2009; IOM, 2013; NASEM, 2015b). Something as simple as a one-page document that would extract relevant exposures and reported health outcomes, including potential functional impacts, would provide participants with a means of quickly and efficiently educating their providers on their concerns and would give clinicians information allowing them to better tailor their care to the individual's health care needs.

Open Comment Period and Pilot Testing

Following an initial draft of the questionnaire, VA announced a 60-day public comment period in a June 5, 2013, *Federal Register* posting (*Federal Register*, 2013a) to address the following issues with regard to the registry's information collection:

> (1) Whether the proposed collection of information is necessary for the proper performance of VHA's functions, including whether the information will have practical utility; (2) the accuracy of VHA's estimate of the burden of the proposed collection of information; (3) ways to enhance the quality, utility, and clarity of the information to be collected; and (4) ways to minimize the burden of the collection of information on respondents, including through the use of automated collection techniques or the use of other forms of information technology.

The comment period was later extended an additional 15 days (*Federal Register*, 2013b). VA also submitted the questionnaire to the Office of Management and Budget for review and comment (on September 6, 2013), and a 30-day public comment period followed that activity (*Federal Register*, 2013c). In total, VA received approximately 300 comments from individuals and veterans' advocacy groups (Ciminera, 2015a).

Several veterans service organizations responded to VA's request for comments on the questionnaire and registry during the open comment period; some of these groups addressed the committee with their concerns during the May 2015 workshop. For example, the Sergeant Thomas Joseph Sullivan Center (SSC) submitted multiple-page letters to both VA and the committee detailing its concerns about the questionnaire and registry (SSC, 2013, 2015). Several of their and other advocacy groups' concerns focused on changes that they believed would enhance the quality, utility, and clarity of the information to be collected. SSC's letter to VA stated that the questionnaire should permit respondents to disclose objectively reportable symptoms, diagnoses, and functional limitations covering all organs and bodily systems potentially affected by airborne exposures, rather than being limited to primarily the respiratory and cardiovascular systems. The letter also stated that adding a comprehensive checklist to the questionnaire and an open field to capture additional participant comments and concerns would best address these concerns. The SSC also sought improvements to the questionnaire to capture information on exposures during deployment and on what diseases veterans have or developed post deployment. According to the veterans service organizations that attended the committee's workshop,[5] their recommendations to VA made in the open-comment periods were not implemented in the final version of the questionnaire.

Moreover, VA stated in its justification to the Office of Management and Budget (OMB; the federal agency that reviews and approves government collections of information) that "[n]o testing of the instrument employed to collect self-reported data will be done in more than 10 individuals" (VA, 2014a). In the same application, VA stated that it "performed extensive one-on-one software usability testing with eight Veterans to improve the web application user interface" and that additional technical testing of the system would be conducted in a production environment to ensure system functionality under varying system loads once OMB approval was granted. OMB approved the questionnaire in March 2014. However, despite the statements about restricting the number of individuals who participated in pilot testing the questionnaire and registry, the pilot testing effort was more extensive.

Pilot testing for the registry questionnaire took place for a little less than 2 months, from April 25 to June 18, 2014, at three VA sites: Detroit, Indianapolis, and New Jersey (Ciminera, 2015b). During this time there were 321 participants: 72 persons consented to participate but did not complete the questionnaire, 194 completed the questionnaire, and the remaining 55 pilot phase users consented and completed the questionnaire after the pilot period had ended (Ciminera, 2015c). There is no information available on how these pilot testers were selected; on what follow-up, if any, was conducted to determine why 72 consenters did not complete the questionnaire; or on the details of the experiences of and the lessons learned by those who did complete the questionnaire. VA told the committee that several changes were made to the questionnaire following the pilot phase (Ciminera, 2015a), but no specifics were provided. No changes have been made to the questionnaire since the registry opened nationally on June 19, 2014 (Ciminera, 2015b; *Federal Register*, 2014; Lezama, 2016; VA, 2015a).

In an effort of this scale, a thorough piloting of survey components and enrollment processes would be expected and needed. No matter how carefully planned the approach might be, initial efforts at implementation always reveal new challenges that call for refinements. This includes a qualitative assessment of the participants' experience, often through focus groups, and an examination of the data initially collected to ensure that the questionnaire is working as desired.

RECRUITMENT AND ENROLLMENT

Eligibility

Public Law 112-260 (see Appendix A) specified that individuals who participate in the registry must have deployed on or after September 11, 2001, in support of a contingency operation while serving in the Armed Forces (whether active duty, reserve, or National Guard) and during their deployment must have been based or stationed at a location where an open burn pit was used. *Open burn pit* was defined in the law as an area of land located in Afghanistan or Iraq that was designated by the Secretary of Defense to be used for disposing solid waste by burning in the outdoor air and that does not contain a commercially manufactured incinerator or other equipment specifically designed and manufactured for the burning of solid waste. VA later modified this definition to allow participation by a much larger pool of veterans and service members. First, the location of deployment was expanded beyond Iraq and Afghanistan to include the entire Southwest Asia theater of operations: Kuwait, Saudi Arabia, Bahrain, Djibouti, Oman, Qatar, and United Arab Emirates; the Gulf of Aden, Gulf of Oman, Persian Gulf, Arabian Sea, and Red Sea; and the airspace above all of the listed countries and bodies of water. Second, it extended the timing of eligible deployments to begin on August 2, 1990, for the Southwest Asia theater of opera-

[5] The individuals and organizations that presented during the committee's workshop are listed in Appendix B.

tions (except Afghanistan and Djibouti, in which an eligible deployment began on or after September 11, 2001). The decision to expand the eligible population to include 1990–1991 Gulf War veterans was made because these service members experienced some environmental exposures that were similar to those experienced during the post-9/11 conflicts. VA indicates that a small number of persons who were not eligible under these criteria have also been permitted to submit questionnaire responses (Ciminera, 2015a).

With the expanded eligibility pool, VA estimates that approximately 3.5 million service members and veterans are eligible to participate in the registry. Based on previous experiences with other congressionally mandated VA environmental health registries, VA estimates that 10% (350,000) of the eligible target population may participate over a 10-year period. Of the projected 350,000 participants, VA estimates that 200,000 of them will request an in-person clinical evaluation (Lezama, 2016).

To participate in the registry, an eligible service member or veteran must first have a Premium DoD Self-Service Logon Level 2 account. To obtain the account, an individual must meet one of several requirements: have a DoD Common Access Card with an accessible reader; have a Defense Finance and Accounting Service myPay account; or be a veteran, dependent of a veteran, survivor of a veteran, or registered in the Defense Enrollment Eligibility Reporting System (DEERS). The DEERS account provides secure, self-service identification that is used to access several account and password-protected websites as well as access to VA eBenefits. A registry help desk is available to service members and veterans who are experiencing difficulties registering for an account or accessing the online questionnaire (VA, 2014b). No information was available to the committee on whether or how these access requirements affect participation in the registry.

Eligibility to participate is confirmed using the VA Defense Information Repository database, where the information is derived from DoD sources. This process to confirm eligibility is an advantage of the AH&OBP Registry over other VA registries in that the eligible population is well-defined by reference to records on periods of deployment and deployment locations compiled by the Defense Manpower Data Center (DMDC), which also maintains a broad array of demographic and military characteristics information for all of these eligible veterans and service members (further discussed in Chapter 4).

Process for Participation

After eligibility is confirmed, an individual may access the questionnaire. The first section of the questionnaire lists the individual's eligible deployment segment records and gives the participant the opportunity to either confirm the system data or correct or enter additional deployment history information. Individuals are able to modify the dates of deployment, add missing deployments, and select or enter the bases they served at while deployed. However, the committee heard from veterans who had participated in the registry that the process of updating deployment information and entering the names of bases and dates was difficult and frustrating. While Section 1.1 of the registry is clearly crucial, it would benefit from design modifications that make it easier and more user-friendly for registry participants.

On the database side of the registry, "userEntered" and "userVerified" fields indicate, respectively, whether the record was entered by the registrant and whether the user indicated the information is accurate (Lezama, 2016), which allows researchers to examine how well this data entry system works. An analysis of deployment segments of registry participants found that 20% of all deployment segments provided by DoD were not verified by respondents as correct. When deployment segments were stratified by date (before September 11, 2001, versus September 11, 2001, and after), pre-9/11 deployment segments accounted for 2.7% of total deployments and of these, 61% were not verified, compared with 19% of post-9/11 deployment segments that were not verified. Of all verified deployment segments, 14% were entered by participants. Again stratifying by the era of service, 70% of pre-9/11 deployment segments were entered by participants compared with 12% of deployment segments for post-9/11 (Ciminera, 2015c).

For persons who do not have at least one eligible deployment segment noted in DoD records, VA can issue waivers to allow persons to participate and enter deployment segment information that is not in DoD records. In a few cases, VA reported that participants were identified who had no eligible or validated deployment segments. These participants were found to be eligible based on system data, but they had indicated that the records were incorrect without entering corrected deployment information, and, as a result of the skip patterns, the section on deployment exposures was not completed. Therefore, in these cases potentially eligible persons would have no validated deployment segments, and their data would not be used in comparisons with the eligible population (see Chapter 4).

VA states that the questionnaire takes about 40 minutes to complete (VA, 2016d). However, the veterans who had participated in the registry and attended the committee's workshop stated that in practice the questionnaire took closer to an hour to complete. Along with longer times needed, several reported that the website would freeze and they would have to start again, sometimes requiring multiple attempts before the questionnaire could be completed and submitted. Participants are able to save sections of the questionnaire as they complete them, and they are able to come back to a section to continue or submit it (Montopoli, 2016b).

An analysis of the time required to complete the questionnaire, which did not include the first section of deployment segment verification, found that nearly 37,000 participants had completed the questionnaire and that about 75% of participants completed it in 45 minutes or less (Ciminera, 2015d). The median time of completion was 31 minutes. Further analysis revealed that the time required to complete the questionnaire was directly related to the number of deployment segments for an individual. For example, 51% of participants who had 1 to 3 deployment segments completed the questionnaire in 30 minutes or less, whereas 41% of participants who had 10 or more deployment segments completed the questionnaire in 30 minutes or less. Since the deployment verification section was not included in the time to completion analyses, the times are underestimated and likely proportionate with the number of deployment segments that an individual needs to verify or manually input.

COMMUNICATIONS AND OUTREACH

In the June 25, 2014, *Federal Register* notice, VA stated that it, in coordination with DoD, would conduct extensive outreach to veterans and service members to raise awareness about the registry and to inform eligible individuals of the advantages of participation (*Federal Register*, 2014). The various communication and outreach efforts used by VA and DoD to promote participation in the registry were broad and not specifically targeted to a single subset of the eligible population, presumably with the intention of maximizing the reach to all eligible participants. The communications strategy included using intermediaries within and associated with VA, a social media campaign, and other electronic notifications to publicize and encourage participation in the registry.

The intermediaries used included points of contact in veterans service organizations, public affairs officers, VA environmental health clinicians and coordinators, and OEF/OIF/OND program managers (VA, 2014a). VA medical centers and facilities were sent fact sheets and postcard-sized flyers for distribution. VA environmental health coordinators, OEF/OIF/OND program managers, and clinicians who work directly with veterans were encouraged to inform the veterans about the registry. VA stated that it will continue to coordinate with veterans service organizations to encourage greater communication with a broader set of veterans who may not use VA services (Ciminera, 2015a; VA, 2014a).

The social media campaign was chosen in part because eligible veterans, being younger than veterans from other eras, tend to be more aware of and are more likely to use social media and the Internet. "VA expects that by using social media sites, websites, and postcard/fact sheets to inform veterans about the value of participating in the registry, participation in the registry will be maximized" (VA, 2014a). Because the questionnaire was to be completed online, social media and other website forums to invite participation would be appropriate.

While both of these assertions or suppositions by VA sound reasonable, the committee is not aware of any strong evidence that they are in fact true. It is reasonable to assume that veterans who are eligible for the registry are more likely than older veterans of earlier eras to use social media and the Internet, but this does not necessarily mean that structuring outreach around these types of sources will ensure maximum participation in the registry. First, although many of these veterans may be reached through such channels, it is plausible that others (and perhaps even most) will not be, either because they do not use social media or because they do not access the media channels (such as VA's Twitter feed) that promote the registry. As part of the development and rollout of the registry, an empirical test of these assumptions and a systematic evaluation of the communication plan on which they are based could have been carried out, with a specific focus on veterans' awareness of these channels and messages and their perceptions of their clarity and effectiveness. Because so few of those eligible even attempted to access the registry, and fewer still successfully enrolled and completed the questionnaire (a topic addressed in Chapter 4), a reason might have been that these channels were less effective than expected in stimulating participation.

Second, research indicates that the most effective methods or modes used to approach a target population, communicate with them, and encourage their participation are not necessarily the same as—and are often quite different from—the methods used to interview or otherwise capture the information desired from them (Dillman and Messer, 2010; Dillman et al., 2014). For example, many Web-based surveys initially approach potential subjects by mail and telephone as well as emails and text messaging and very few have been successful in using only indirect or passive communications and recruitment by general social media, message boards, and the like, other than for pretesting during the development and testing of questionnaires and methods (Couper, 2000; Scherpenzeel and Toepoel, 2012; Tourangeau et al., 2013).

Beginning in July and continuing through November 2014, information about and links to the registry were publicized using several avenues in VA and DoD. Such approaches included posts on VA and VHA Facebook pages and Twitter accounts, emails to subscribers to the VHA GovDelivery listserv, inclusion in the VAntage Point Blog, and posts to VA websites (for example, www.va.gov/health/insidevha.asp and www.publichealth.va.gov). VA reported that the Facebook posts reached 458,208 veterans; there were 26,636 "click-throughs;" and there were 10,281 likes, comments, and shares (Ciminera, 2015a). The VHA GovDelivery listserv had 77,927 recipients; 16% opened the message, and 3% clicked on a link in the email. VAntage Point Blog had 18,307 unique page views, and the Inside Veterans Health webpage had more than 10,000 unique page views. In September 2015, Web banners and announcements were posted to several VA websites. An update was included in the digital-only Post-9/11 Vet Newsletter, and this newsletter was shared with veteran service organization liaisons. Announcements regarding the newsletter were shared on VHA Facebook and Twitter accounts with the potential to reach about 214,000 and 71,500 followers, respectively. A message was sent through the GovDelivery listserv to the Public Health's Military Exposures subscriber list of 45,658 recipients, and 17.9% opened the message. VA reported that the webpage for the Post-9/11 Vet Newsletter had received 6,230 views and that the burn pits registry article had received more than 2,500 page views (Lezama, 2016). In addition to publicizing the registry on websites and through social media and dedicated listservs, announcements were included in VA newsletters, such as the Post-9/11 Vet Newsletter and WRIISC Advantage Newsletter (Ciminera, 2015a; VA, 2014a).

DoD publicized the registry to eligible active-duty service members and activated members of the reserve and National Guard. Members of the reserve forces and National Guard who were not activated were under the purview of VA. DoD also used social media (Facebook; Twitter) to advertise the registry, but metrics were not reported. Announcements were posted to DoD and military websites, printed in military-oriented newspapers, and made through communications from each service branch to their active-duty service members (Defense Health Board, 2015).

The registry went live in June 2014, and ongoing communication and outreach efforts appear to have diminished within a few months of the launch. This drop-off took place at a time when there were several instances[6] when the registry was unavailable—sometimes for several days at a time—due to software problems or other issues (Montopoli, 2016a). Few messages about those disruptions appeared on the registry homepage or in any other venue. For example, the committee observed in September 2015 that the questionnaire was unavailable for more than a week before a message was posted indicating that the registry was experiencing difficulties and advising interested persons to check back at a later time. During the committee's open session in December 2015, VA acknowledged that ongoing communication with potential registrants was an issue (Lezama, 2015). VA provided the committee with a copy of its communications strategy for the registry for 2016. The plan consisted of several avenues including continued social media messaging, GovDelivery listserv emails, a Twitter chat, posts to the VAntage Point blog, and announcements at various meetings and calls (Lezama, 2016). However, for most of these activities the time frame was not stated.

As opposed to generic posts and shares on VHA social media sites and broad communications and outreach initiatives that may or may not reach the intended population of service members and veterans, a more focused

[6] VA reports that more than 30 outages occurred between the time the registry went public and May 30, 2016.

approach would help ensure that—to the maximum extent possible—persons eligible to participate are receiving the communications. For example, in the same way that VA uses the registry as an outreach mechanism to mail post-participation fact sheets to participants who have submitted questionnaires (Ciminera, 2015b), one area that VA could explore would be to target communications at participants who start but do not complete or submit the questionnaire (discussed in Chapter 4). These follow-up communications could offer reminders or direct individuals to sources that may help answer any questions they may have.

A second approach might be to more intensively and systematically target specific groups of eligible persons in an attempt to produce a more representative cohort of those eligible. For example, several surveys of veterans and service members, including the Millennium Cohort Study and VA's National Health Surveys of 1990–1991 Gulf War era veterans and OEF/OIF veterans, oversample women and reserve/National Guard personnel (Crum-Cianflone, 2011; Smith et al., 2007). The Millennium Cohort Study also specifically targeted persons with recent past deployments in its first baseline assessment (Crum-Cianflone, 2011). Making such populations a larger part of the registry would increase the confidence with which one could draw generally applicable conclusions from the data it contains. A formal analysis of the demographic and other characteristics of non-respondents could provide clues that would help VA target specific groups through more appropriate channels or contact methods.

Encouraging enrollment by eligible persons who were not exposed to burn pits or who were exposed to burn pits but are not experiencing any adverse health outcomes will also likely lead to a broader representation and therefore possibly lead to improving the estimates and generalizations that can be made using registry data. Another method to improve representation might be to offer incentives to participants who are selected and targeted in a manner to enhance representativeness.

The main webpage for accessing the registry states that its purpose is for persons "to report exposures to airborne hazards (such as smoke from burn pits, oil-well fires, or pollution during deployment), as well as other exposures and health concerns" (VA, 2016e). However, based on the registry name, the emphasis in communications and messaging regarding it, and the number of questions in it relating to burn pit exposure compared with other exposures, VA has highlighted exposure to burn pit emissions as its primary interest. It is therefore not surprising that 96% of participants in the database available for committee review reported being exposed to a burn pit on at least one of their deployments (Chapter 5). If VA wishes to gain a greater understanding of exposures and health concerns in the entire population of Southwest Asia theater of operations veterans, then outreach efforts should target all eligible persons, regardless of whether they were exposed to a burn pit, and the messages should encourage all eligible persons to participate, emphasizing participation for persons not experiencing symptoms of poor health and those who were not exposed specifically to burn pits.

If a purpose of the registry is hypothesis generation related to exposures to airborne hazards and health concerns, there would be benefit in a targeted outreach to those persons who are likely to have been among the most highly exposed. These persons may be identified through additional linkages with DoD records of deployment locations, number of deployments, length of deployments, and, potentially, military occupation specialty. However, such targeted efforts are limited by the registry's architecture and compatibility with additional sources and databases (a topic discussed in the section titled "Linking Other Data to Registry Data"). It would thus be appropriate to pilot test any such effort in order to determine whether it is achieving the intended goals.

Veterans service organizations and military service organizations have an interest in military occupational exposures and are a valuable resource for getting information out to their membership. VA should consider how it can better work with these organizations on an ongoing basis to increase awareness of the registry and encourage participation by concerned individuals.

QUESTIONNAIRE QUALITY

The quality of a registry is dependent on "the confidence that the design, conduct, and analysis . . . can be shown to protect against bias (systematic error) and errors in inference" (AHRQ, 2010, p. 307). The value of the collected information also relies on the quality of that data as well as its use and purpose for decision making. This section first considers the design of the questionnaire separately from the actual questions it contains. In particular, the discussion begins with an examination of various aspects of the questionnaire's design, including its layout, directions, and flow of questions. The second part of the section provides a broad description and evaluation of the types of questions included and of the areas where improvements (such as the relevance and appropriateness of the questions) could be made that would enhance the overall value of the registry.

The intent of this section and the examples it provides is to be illustrative of general categorical problems with the structure of the questionnaire and formats of questions. It is not an exhaustive or an item-by-item assessment, but rather a general overview of the types of problems that were observed. The committee was not tasked with redesigning the questionnaire and is not in a position to offer alternative wordings or approaches, which would require testing to validate. Rather, its charge was limited to suggesting changes that could improve the instrument.

Registry Questionnaire Basic Characteristics

The complete AH&OBP Registry questionnaire, version 15 (December 2014) is reproduced in Appendix C. Box 3-1 contains an outline of the questionnaire's sections and topic areas. The questionnaire contains approximately 140 questions. A participant may answer more or less than this number depending on the applicability of skip patterns and the number of eligible deployment segments he or she has indicated. For example, the Tobacco Exposures section of the questionnaire consists of 10 questions, but persons who report that they have not smoked more than 100 cigarettes in the first question skip the next four questions, because those questions are not applicable to them. On the other hand, respondents are instructed to answer the same nine questions on location-specific deployment exposures for each eligible deployment segment, with some respondents having multiple eligible deployment segments. Respondents were required to answer every question ("don't know" and "refused" options were provided) in order to submit the questionnaire and therefore be included in the registry.

A limitation of this evaluation of the questionnaire is that it is based on the paper version of the questionnaire and on the committee's understanding of the online version, but the committee was unable to have an interactive review of the questionnaire as it is coded to appear and is implemented online because there was no mechanism to allow the full committee access to the online version. VA indicates that there are only minor editorial difference between the paper version and the text of the online form, but to the extent that the paper and online versions are discordant or otherwise different, some of these comments may not reflect user experience.

Questionnaire Design

A Web-based questionnaire can offer advantages, and the AH&OBP Registry questionnaire exploits these to some extent. For example, it uses previous responses to fill in content for questions that follow, although not consistently. However, there are other advantages of the online format that the registry fails to make use of. These and other problems identified by the committee are detailed below.

Layout

The questionnaire is divided into several sections that are inconsistently labeled and numbered. Several of the sections, such as between Deployment History and Symptoms and Medical History, cover multiple topics in varying degrees of detail. The order of the questions primarily follows the information objectives of the registry rather than being presented in a flow that might better optimize the interest and engagement of participants. While transitions between major sections are generally used, such as between Deployment History and Symptoms and Medical History, transitions for groups of questions within a section (for example, Location Specific Deployment Exposures and General Military Occupational Exposures) are rarely used.

Directions and Clarity

Directions and clarifying instructions are rarely provided throughout the questionnaire. In the interview-based NHIS—from which several of the AH&OBP Registry questions were adopted—all possible responses are read to respondents (with some exceptions, such as "refused"). Reading the responses may help respondents who do

BOX 3-1
Airborne Hazards and Open Burn Pit Registry Questionnaire Sections and Topics

1 Deployment History
1.1. Deployment data from the VA Defense Information Repository (VADIR) and DMDC
1.2. Location-specific deployment exposures
1.3. General military occupational exposures
1.4. Environmental exposures, regional air pollution

2 Symptoms and Medical History
2.1. Functional limitations and reported cause
2.2. Health conditions
2.2.1. Respiratory conditions
2.2.2. Cardiovascular conditions
2.2.3. Other conditions
2.3. Height and weight
2.4. Cancer history
2.5. Tobacco exposure
2.6. Deployment smoking history
2.7. 12-month alcohol use

3 Health Concerns

4 Places You've Lived

5 Work History
5.1. Current occupational status
5.2. Main occupation
5.3. Dust exposures
5.4. Gas, smoke, vapors or fumes exposures
5.5. Asbestos exposure

6 Home Environment and Hobbies

7 Health Care Utilization

8 Contact Preferences

not recognize a word or condition in written form but recognize it when spoken, allowing them to provide a valid response to a question rather than skipping it or answering "don't know." This may be particularly relevant for some of the medical conditions that respondents are asked about.

If the NHIS questions are to be used in a self-administrated Web-based format, they should have been changed to account for these differences. Additionally, because specific instructions and clarifications by question are available to NHIS interviewers, these could be made available to the respondents either in the text or available by a keystroke to help them better understand the intention of specific questions.

A variety of different time reference periods are used throughout the questionnaire, sometimes within the same group of questions, which can easily lead to misunderstanding or confusion. Although sometimes the key reference period words are bolded (for example, **never** versus **past 12 months**), the questions are not otherwise differentiated from each other and do not use transition language to provide increased clarity.

Certain formatting strategies, such as skip patterns, are typically used in survey design to reduce respondent burden and confusion and, in turn, increase the quality of responses. The AH&OBP Registry questionnaire uses complex skip patterns (for example, in the Tobacco Exposure section [2.5]) that have more potential to confuse respondents than help them navigate the series of questions, sometimes with very little apparent gain. Since the committee could not access the online version of the questionnaire, it is not possible to comment on the format as seen in real time.

Evaluation of Questions

General Observations

The questionnaire appears to be at an appropriate reading level. The committee did not have access to the software used by survey organizations to assess the questionnaire, but used the functions available in Microsoft (MS) Word to conduct a cursory assessment. The Flesche reading ease score (an indicator of readability based on an algorithm that uses the number of words per sentence and the number of syllables per word) is 70.0 on a scale of 100.0, suggesting that the questions are easy to read. The Flesche-Kincaid grade level is computed to be 5.7, which indicates that a U.S. fifth-grader should be able to read the questionnaire. Of note, MS Word caps reading level at grade 12 (high school senior), and any reading level above that is reported as grade 12. All persons who serve in the military must have completed high school or have an equivalent GED, so a questionnaire written at a fifth-grade level is acceptable. However, the committee notes that neither of these measures is an indicator of comprehension.

Each question included in the survey should serve a purpose, whether to elicit details on potential exposures, symptoms and health conditions, or to investigate factors that may influence associations and therefore need to be adjusted for in analyses. Just as important as what is asked is how it is asked, because poorly worded or confusing questions will not elicit useful information. The following discussion provides some examples of the types of questions used throughout the questionnaire by topic area (for example, exposures, symptoms and conditions, and other behavior and effect modifiers) and how they could have been improved.

The committee noted some general issues related to how the questions are phrased. Examples of these types of questions are compound questions, bundled questions, and "check-all" formats. The use of these formats is confusing and problematic, but each is used several times in the questionnaire.

A compound question addresses more than one issue in the same question, but allows for only one answer. Alternatively, a question is also compound if it presents more than one issue in the available responses. An example of a compound question is 1.4.A:

> Did you do anything differently during your deployment(s), when you thought or were informed air quality was bad (for example during dust storms or heavy pollution days)? 1. Yes, 2. No, 3. Never thought of this, 4. I was not informed or aware of bad air quality, 5. I do not wish to answer, 6. Don't know.

This question is compound because the exposure and outcome information are mixed in the question. Respondents are not given the opportunity to indicate whether they had encountered any circumstances where they thought the air quality was bad or to note when they were first informed of that fact before they are asked whether they took a different action. For persons who endorsed yes, the possible responses in the follow-up question (1.4.B) also appear to be problematic in that some (such as response 6 to this question, *Spent less time in convoy*) would be out of one's control and at least one other potential response (response 11, *I did not [or could not] do anything differently*) would seem to be an invalid choice because it contradicts the "yes" response to 1.4A.

A second example of a compound question is 1.4.F, which attempts to determine severity of common symptoms likely related to poor air quality:

During your deployment(s), did you seek medical care for wheezing, difficulty breathing, itchy or irritated nose, eyes or throat that you thought was the result of poor air quality? 1. Yes, 2. No, 3. I do not wish to answer, 4. Don't know.

This question is not useful because it groups several symptoms of differing degrees of importance (that is, wheezing and difficulty breathing are much more likely to have medically important implications than itchy or irritated nose or eyes) while attempting to elicit severity (for example, bothersome enough to require medical care).

"Bundled" questions seek answers to a series of questions that a respondent may endorse or not, followed by a summary question(s) referring to all of those previously endorsed, such that the response to it cannot be associated directly with any one of those endorsed. This type of question format may have a similar effect as, but is not the same as, a compound question.

Instead of presenting separate questions for each activity, disease, or symptom of interest, the questionnaire frequently includes long lists of questions and responses formatted as "check all that apply." This design decision was presumably intended to limit respondent burden, but the use of a forced-choice format (yes or no to each item) rather than these so called "check-all" formats is widely regarded as a superior method for collecting these data, especially for self-administered Web-based surveys (Smyth et al., 2006). Check-all formats are problematic because it is unclear whether an unchecked response means "no," is overlooked, or is otherwise missing, and this issue is magnified when the lists are presented across multiple pages, as often happens on the AH&OBP questionnaire.

Question 1.4.B, referenced in the compound question discussion, is one example of a question presented in the check-all format. Survey methodologists have shown that respondents endorse more options under a fixed-choice format, but that this format does not result in fatigue or acquiescence bias (responding the same to all questions in that section) or high levels of nonresponse (Callegaro et al., 2015; Mooney and Carlson, 1996; Rasinski et al., 1994; Smyth et al., 2006, 2008). The increased response using a forced-choice format is thought to require more cognitive processing, and, therefore, respondents take more time to carefully consider each item individually and the appropriate response on this type of question. More responses do not necessarily indicate better quality of responses, but studies have shown that a forced-choice format likely produces more accurate responses than the check-all-that-apply format (Ericson and Nelson, 2007; Feindt et al., 1997).

Exposure Questions

A questionnaire should be designed and laid out in a manner that lessens the chance that respondents will be fatigued or unengaged before they get to the questions of greatest importance. The time required to complete the AH&OBP Registry questionnaire is critically dependent on the number of deployment segments an individual has and when those deployments occurred—which is the first item that respondents are asked about. As shown in Table 4-3, about 37% of registry respondents made available to the committee had five or more deployment segments, and approximately 11% had 10 or more. Therefore, given the time required for verifying and correcting deployment information, especially for pre-9/11 deployment segments (the majority of which are user entered), participation in the registry can be a very time-consuming process which may contribute to not all of the eligible segments being completed, biased responses (reporting the same high levels of exposure for all deployments, for example), or to dropout at that or later stages, introducing additional forms of selection or nonresponse bias. The questionnaire presents eligible deployment segments in order from oldest to most recent. Reversing the order so that the most recent segments are presented first might improve data collection and reliability, especially for participants with multiple segments to verify, add, and answer the same questions about. Furthermore, individuals are likely to recall recent deployments more accurately than older ones, and if survey fatigue occurs and respondents decide to skip inputting some segments, more accurate information would have been obtained for the most recent subset.

The second section of the questionnaire (1.2: Location Specific Deployment Exposures) consists of nine questions about possible exposures encountered for each deployment segment that was verified or added. The first question asks whether persons who served during 1990–1992 were exposed to soot, ash, smoke, or fumes from the Gulf War oil-well fires. This question is not displayed if the person did not serve in Operation Desert Shield or Operation Desert Storm in 1990–1991. The second and third questions ask where the participant spent most of his or her time, using a list of base names or text entry. Exposure to a burn pit is not introduced until the fourth question (near a burn pit during the deployment dates, defined as on the base or close enough to the base to see the smoke). The next question collects information on who was in charge of the burn pit (U.S. forces or contractor, coalition forces, or host nation). Neither the purpose of including this question nor how it can be used in analyses of registry data is clear. Next is a yes-or-no question on whether the participant's duties included the burn pit. The seventh question in this module asks for the number of hours (0–24) the respondent had exposure to smoke or fumes on a typical day between worksite and housing. Given that the primary purpose of the registry, as stated by its title, is to collect information on health outcomes potentially related to burn pit exposures, these questions should have been asked before other airborne-hazard exposure questions, such as soot, ash, smoke, or fumes from the Gulf War oil-well fires. Additionally, if burn pit exposure is the crux of the registry, more than four questions are needed on that topic.

Some of the questions, presumably those of greatest importance, are poorly formatted and written. For example, Question 1.2.F requires a yes-or-no response as to whether a person's duties during deployment included the burn pit, with some examples of what these duties might entail. Inquiry on this topic could be strengthened by adding more specific questions such as "Did you personally throw anything into the burn pit?" and an open ended question for the respondent to list what she or he saw in the burn pit (plastic water bottles, uniforms, munitions, medical supplies, and the like). No information is collected on the size of the burn pit, whether incinerators were in use, or other proxies for the level and types of individual exposure, such as intensity, smoldering versus flame exposure, the proximity of work or housing locations to the burn pit, and the like. Collecting this type of information would likely strengthen the inferences that could be made using the registry data. Furthermore, the questionnaire does not distinguish between the large military burn pits that are the object of investigation and civilian trash-burning activities, possibly resulting in exposure misclassification as some respondents recall their experiences near civilian trash burning instead.

Other questions in this section ask respondents to specify the number of hours in a typical day (0–24) that they may have been subject to a particular exposure. Using a broad range of numeric responses instead of grouping possible responses (for example, 0, 1–5, 6–10, . . .) requests a level of precision that likely exceeds the capacity of many participants to recall. Moreover, people tend to recall events that were unusual over the mundane. For example, if there was a period within a deployment of particularly intense smoke, this will likely have a greater influence on recall and lead to a reporting bias toward reporting an average exposure across the entire deployment that is closer to the short-term intense exposure than to the more mild exposure that was experienced the rest of the time. Questions eliciting information about both typical and peak exposures would have improved the questionnaire.

Section 1.2 begins with participants being presented with a list of base names from which to select and indicate where the majority of their time during deployment was spent or they may use text entry if they do not see the base listed. The committee heard from its data analysis contractor that the list is not exhaustive, and although a provision is made for "other" write-ins, it is likely that participants may know or recall certain locations by different names. This section is also problematic for data cleaning and analysis because the locations of some bases were classified—specifically, smaller forward operating bases—and service members may not precisely know or be able to spell the names of the places to which they were deployed (Szema, 2015). Thus, editing and reconciling these responses requires a great deal of time and effort.

The final question of the locations-specific deployment exposures section asks the number of hours on a typical day that the person was near sewage ponds. No definition or additional description of "sewage ponds" is given. It is unclear to the committee why sewage pond exposure (which had an item response of "don't know" that was more than 38%) was thought to be important enough to be asked for every eligible deployment, but exposures to dust and airborne hazards (the other half of the registry's title) were grouped and later assessed as exposure that occurred on any deployment. Questions of general military occupational exposures, environmental exposures, and regional air pollution also do not differentiate between experiences for individual deployment segments and include exposures (for example, number of days a month a person performed pesticide duties) that appear to be outside the purview of the registry. Grouping military occupational exposures, environmental exposures, and regional air pollution is a shortcoming because the experiences of one deployment do not necessarily translate to others, and grouping them together makes it more difficult to estimate exposure level or duration.

Furthermore, if a purpose of the registry is to elucidate information that may potentially be used for hypothesis generation and health care improvement, additional, well-designed questions are needed to collect that level of detail. Instead, the questionnaire attempts to elucidate some of the exposure information through compound questions. While the questions may have been intentionally phrased in an attempt to reduce participant burden, it introduces the possibility of confusion, misinterpretation, or logical inconsistency (if the respondent believes that the response to one part contradicts another). Instead, such questions should more appropriately be broken into multiple parts, using skip patterns where appropriate. For example, in Question 1.3.D, instead of asking "In a typical month, how many days did you perform refueling operations?" the first question could be "Did you ever perform refueling operations?" with yes-or-no response options. If the person answers yes, then the next question could be "In a typical month, how many days did you perform refueling operations?" with possible responses grouped into ranges of days instead of allowing any whole number between 0 and 31.

Finally, given that the AH&OBP Registry is intended to be a military exposure-based registry, several important potential exposures are missing. For example, the questionnaire does not ask about other sources of combustion products, such as exposure to burning trash and other materials in the absence of large burn pits. Nor are there questions about high-risk jobs other than combat. Other important sources of military exposures are also missing, such as exposure to diesel exhaust, welding, paint, or other chemical fumes or to organic dusts such as those associated with wet or water-damaged indoor environments during deployment.[7] Gathering information on these types of potential exposures would likely enrich the registry.

Health Condition Questions

Section 2: Symptoms and Medical History has two primary weaknesses: the wording or phrasing of many of the individual questions, and the fact that the questionnaire does not capture all diagnoses of concern and that for those diagnoses that are included, the questionnaire does not capture information with the specificity that would be necessary to draw inferences about the presence or absence of specific diagnoses in the registry population. The section begins with five questions on functional limitations. These questions were taken from the NHIS, although they are only a subset of all the questions on functional limitations asked in that survey and are presented in a different order. In the NHIS the functional-limitations questions are intended to be used to assess severe dysfunction and not to capture a range of functional limitations of varying severity. Because no information is collected about predeployment functional status, the utility of these questions is further limited.[8] Moreover, Question 2.1.F—which asks that the respondent indicate, for any question in the series that was endorsed as "difficult," the condition that causes the difficulty with those activities—is formatted as check-all-that-apply as opposed to forced choice. Many of the conditions listed are neither necessary nor appropriate as explanations for the functional limitations elicited in the questionnaire and could be eliminated, such as birth defect, diabetes, fibromyalgia/lupus, hearing problem, hernia, migraine headaches, multiple sclerosis, other developmental problem, polio, senility, thyroid problems, ulcer, varicose veins, and hemorrhoids. Moreover, the reasons for each functional limitation may vary, and for persons who indicate one or more difficulties with functional activities, the questionnaire is not structured to allow the respondent to supply different reasons for difficulties with different activities. For example, the primary reason that a person might have difficulty jogging a mile may be different from the reason the person has difficulty climbing a flight of stairs.

None of the questions on functional limitations, symptoms, or health conditions ask about onset or severity following individual deployment segments. The discussion that follows illustrates examples of the many questions that are overly general, are unable to be answered accurately, or appear to have limited relevance. The health conditions questions range from non-specific symptoms (such as fatigue, hay fever, and allergies) to very specific diagnoses or technical terms (idiopathic pulmonary fibrosis, constrictive bronchiolitis, angina pectoris). The questionnaire does not collect objective measures or validated diagnoses. As a proxy for diagnosis, all the questions eliciting specific diagnoses begin with "Have you ever been told by a doctor or other health care professional that you had . . . ?"

---

[7] Some of these are asked with regard to exposures outside the military (Section 5.4).

[8] Arguably, persons with functional limitations before deployment would not have been deployed, but this is an assumption without information to back it.

There are many examples of awkwardly and poorly worded questions or responses that likely result from changes made to a number of the standardized questions used in the NHIS (May and Haider, 2014). Using Section 2.2.1 (Respiratory Conditions) as an example, it is not clear why filter Question 2.2.1.F

> Have you ever been told by a doctor or other health care professional that you had some lung disease or condition other than asthma, emphysema, chronic bronchitis or COPD? 1. Yes, 2. No, 3. I do not wish to answer, 4. Don't know.

is needed other than to skip two questions on specific lung diseases—constrictive bronchiolitis and pulmonary fibrosis. Furthermore, it appears that responses to Questions 2.2.1.A–H could be used to tailor the wording in 2.2.1.I

> [If B–F = yes] When you were told you had asthma, emphysema, chronic bronchitis, COPD or some other lung disease by a doctor or other health care professional, were you told before, during, or after deployment? (check all that apply.) 1. Before deployment, 2. During deployment, 3. After deployment, 4. I do not wish to answer, 5. Don't know.

to include only those conditions endorsed by the respondent rather including the whole list because, for persons who endorse multiple conditions, it is unclear for which diagnosis the timing refers. Or persons who indicated that the onset of the respiratory condition or conditions was before deployment, a follow-up question (2.2.1.J) asks whether the lung disease got better, worse, or about the same during deployment. This is poorly worded—it assumes a respondent has a single lung disease, and does not indicate whether any change was determined by a doctor or by the respondent's subjective assessment. Additionally, because many of the respondents have had multiple deployment segments, specifying before, during, or after deployment does not clarify the temporality of condition onset (whether it might have occurred after one deployment but before the next, for example). The same type of problematic questions are repeated in both the Cardiovascular Conditions (2.2.2) and Other Conditions (2.2.3) sections.

Moreover, the respiratory and cardiovascular outcomes addressed in the questionnaire are limited to the following doctor diagnosed respiratory conditions: hay fever or allergies (to pollen, dust, or animals only); asthma; emphysema; chronic bronchitis; chronic obstructive pulmonary disease; lung disease other than asthma, emphysema, chronic bronchitis, or COPD; constrictive bronchiolitis; and pulmonary fibrosis or idiopathic pulmonary fibrosis. If a respondent endorses "other lung condition," no additional details or information is collected. Other important respiratory conditions that, if included, might have strengthened the ability of the registry data to be used for hypothesis generation include reduced lung function, eosinophilic pneumonia, other lung infections (such as tuberculosis, fungal pneumonia, community-acquired pneumonia), lung scarring or fibrosis (a more inclusive diagnosis than idiopathic pulmonary fibrosis), bronchiolitis other than constrictive bronchiolitis (respiratory or obliterative), sarcoidosis/hypersensitivity pneumonitis, rhinosinusitis, and vocal cord dysfunction.

Other wording problems with specific questions occur throughout the questionnaire. For example, Question 2.2.1.K does not define how "currently" should be interpreted. Question 2.2.1.L appears to be a follow-up to 2.2.1.K and asks about the same nine symptoms or conditions experienced in the past 12 months (as opposed to currently). Both questions are formatted as check-all instead of forced choice, which generates better results for questions of this type. These questions could potentially be used to check for internal validity of the questionnaire because persons who endorsed any current symptoms should also endorse the same symptoms for the past 12 months. The questions specifically ask respondents whether they have any of a list of symptoms, but not all of the possible responses are symptoms (chronic sinus infection/sinusitis, for example, is a diagnosis), and others are too vague (a "decreased ability to exercise" may be due to musculoskeletal problems or deconditioning) or compound ("chest pain, chest discomfort or chest tightness" may be due to cardiac, respiratory, or musculoskeletal conditions) to attribute directly to a respiratory condition. For persons who endorsed shortness of breath or breathlessness in the past 12 months (2.2.1.L), 2.2.1.M is a follow-up question which seeks to elicit additional details on the severity of this symptom, and participants are directed to choose one response that best corresponds to their level of severity. However, the possible responses are not mutually exclusive.

Other examples of poorly phrased questions occur in Section 3: Health Concerns. Questions 3.E and 3.I ask respondents to rate their *level of* concern that something they breathed during deployment has already affected their health or will affect their future health. This is a leading question; the 3-point response scale—"not at all," "a little," or "very concerned"—is not optimal, and the order presented will likely skew responses toward the middle (Choi and Pak, 2005). Reversing the order and adding an additional option for an even number of possible responses, such as "very concerned," "somewhat concerned," "a little concerned," and "not at all concerned," would improve the questions and potential responses. In this same section, questions mix past and present tense. The high rates of "don't know" responses for 3.B (During your deployment(s), do you believe that you were sick because of something you breathed?) and 3.C (Do you **currently** have a sickness or condition you think began or got worse because of something you breathed during deployments(s)?)—29% and 31%, respectively—likely indicate confusion or uncertainty. Question 3.K asks which exposure the individual thinks had the biggest overall effect on his or her health. This is an important and difficult question to ask, and it requires a judgment call that is difficult to make. The list does not appear to be complete or exhaustive, and no "other specify" option is provided. Additionally, the fourth response in Question 3.K ("military jobs while I'm not deployed") is inconsistent with the focus of two previous questions in this section (3.C and 3.H) which specify "during deployment."

There remains much scientific uncertainty about the conditions and diseases that may result from deployed service members' exposure to airborne hazards in Iraq and Afghanistan and some studies have shown that other organs and organ systems are also affected (IOM, 2011). A questionnaire focused primarily on respiratory and cardiovascular outcomes is not sufficient for hypothesis generation and surveillance. Because the committee surmises that the developers were intending to include a broader set of health outcomes (as reflected in questions in Section 2.2.3: Other Conditions), the questionnaire could be improved by adding more questions eliciting details for outcomes related to these other conditions of interest as well; single questions of whether a respondent had "problems" or "conditions" related to broad categories of health outcomes are insufficient and not useful for evaluating issues regarding potential exposures and these outcomes. For example, although no other questions about the digestive system are included, Question 2.2.3.D (inexplicably placed between a question on immune system problems and a question on doctor-diagnosed chronic multisymptom illness) asks about a doctor-diagnosed liver condition within the past 12 months. In a related omission, no questions are included on comorbidities.

The health concerns section appears primarily composed of questions with little or no testing or validation. In the case of certain questions and responses, the wording or phrasing is unclear, awkward, or nonspecific. For example, in Question 3.A, as well as in Sections 2.2 and 2.3, "predeployment" (or "before deployment" for Sections 2.2 and 2.3) is not defined. This term could be interpreted as meaning immediately before deployment or at any time prior to deployment. For other questions, such as 3.F and 3.J, the list of possible health concern responses is not all inclusive. While a response of "other problem" is provided, people tend to avoid using it (Bradburn et al., 2004, p. 58). and usually this option is phrased as "other specify"; however, there is not an option to write in an area of concern that is not listed. Also included in this set of responses is the concern of "effect on children or ability to have children," which is different from how the other response options are worded because it is phrased as a personal issue as opposed to a more general reproductive problem. Its placement is also odd, and given its current wording, it should either follow the cancer section or be examined separately as it covers two different problems (that is, a compound response).

The questionnaire contains seven questions related to cancer (Section 2.4). No transitional language or instructions are provided to make respondents aware of the change in topic from the previous section that requests height and weight (Section 2.3). Respondents are asked about up to three types of cancer diagnoses, but they are not informed of this at the beginning of the section. Respondents who answer that they have ever been told by a doctor or other health professional that they have cancer or a malignancy of any kind are then presented with a check-one format for the type of cancer. This type of format is inconsistent with the question intent. If the registry designers were interested in only three types of cancer, the question could be reworded to identify the three types by, for example, order of onset, then obtaining the age at diagnosis for each. Questions 2.4.C, 2.4.E, and 2.4.G (age at primary, secondary, and tertiary cancer diagnosis) in this section are open-ended and accept whole numbers between 0 and 99, leading to unnecessary errors. By using date of birth and current date, the acceptable range could be reduced considerably.

Questions Regarding Other Factors That Might Influence Health Outcomes

Tobacco use has been shown to affect respiratory and cardiovascular conditions and may also affect associations with exposure to burn pits and other airborne hazards. Thus, it is important to collect information on it and other potential confounders and effect modifiers for inclusion in analyses. Other factors may also influence the relationship between exposure and health outcome, but the degree of influence is unknown, so questions about these factors, such as height and weight, are presumably included "to be safe" (and can be used to calculate body mass index). Other questions have been included that appear to have no scientific relevance, such as places of residence.

The tobacco and smoking questions included are relatively standard, but they could be improved by instituting tailored range checks, which can easily be done in an online survey format. For example, the response to Question 2.5.D (*How long has it been since you quit smoking cigarettes?*) should be between the age reported in Question 2.5.B (*How old were you when you first started to smoke fairly regularly?*) and the individual's current age. Instead of asking for the number of cigarettes a person smokes, packs of cigarettes may be a better measure and easier for a person to answer. Whereas the first five questions in this section ask specifically about cigarette smoking, Questions 2.5.F through 2.5.I ask about the use of other tobacco products. To highlight this change in focus, certain words could be bolded, and definitions and lists could be made clearer to avoid confusion and improve the questions. Two questions specific to smoking while deployed are separated from the more general smoking questions and are not well coordinated with the general tobacco exposure questions.

Only one question (2.7.A) is included to assess alcohol use

> In the PAST YEAR, how often did you ever drink any type of alcoholic beverage [Included are liquor such as whiskey or gin, beer, wine, wine coolers, and any other type of alcoholic beverage]? "On average, how many days per week did you drink?" 1. Never, 2. Less than one, 3.1–7 days per week, 4. I do not wish to answer, 5. Don't know.

This question is badly worded, confusing, and compound. The response options are also poor in that persons should be allowed to enter the number of days between 0 and 7, as is the case for other questions in the questionnaire on number of hours or days of exposures. Instead, the responses group the possible answers into 0, less than 1, and 1–7; essentially assigning the same severity score for consuming alcohol one day per week and every day. Additional questions asking the average number of drinks per day when an individual consumes alcohol would improve this section.

Other questions cannot be answered accurately by a respondent, or they request information that does not appear to have a relationship to evaluating the effects of wartime airborne exposures. For example, Questions 2.2.3.H (*How often do you snore?*) and 2.2.3.I (*How often do you have times when you stop breathing during your sleep?*) appear to have little salience, as evidenced by the high rate of item nonresponse (39% responded "don't know" to Question 2.2.3.I). It is not clear that many people can answer these questions accurately or what the purpose of collecting this information would be as these are not health conditions that would be related to burn pit or other airborne hazards exposures.

Following the sections on health concerns, the questions in the last one-third of the instrument (Sections 4 through 6) ask about current residence, place of longest residence, main occupation outside of the military, dust and other exposures in civilian jobs, and home environment and hobbies. These questions are presumably included to gather additional information on other exposures that may affect deployment-specific exposures and reported symptom and condition outcomes. However, many of these questions are problematic, and it is unclear how any of this information would be analyzed.

Section 4 begins by stating, "Poor air quality in places where you've lived may impact how deployment exposures affect you." This is a broad statement and implies that the person's prior residence is somehow to blame for exposures experienced on deployment and for any health conditions they may be experiencing. It is unclear how such information could be used for analysis and what assumptions, if any, about exposure risk can be made from broad domains of city, state, zip code, duration of residence, and "address where you lived the longest before age 13."

The intent of Question 5.2 on main occupation outside of the military appears to be to identify those individuals in specific occupations relevant to the focus of Sections 5.3 (Dust Exposures), 5.4 (Gas, Smoke, Vapors, or

Fumes), and 5.5 (Asbestos Exposure) which follow, rather than creating a classification of civilian occupations per se. About 12% of respondents endorsed working "in any dusty job outside the military," and most of these respondents appeared to try to fit or force their occupations into these restricted categories, as evidenced by less than 5% selecting "other." This makes analysis problematic and indicates that the answers are likely subject to considerable response error.

Ultimately, the questions attempting to elicit exposures from main nonmilitary occupations are vague and of little use. For example, in Question 5.3.A, "dusty job" is a vague, undefined term and likely to be interpreted differently by participants. The occupational exposure questions do not consistently use the same language in the questions. For instance, Questions 5.3.B.1 and 5.4.B.1 mix the terms "biggest exposure" and "longest exposure" within the question, and these terms are not necessarily defined the same way in people's minds. These questions would be improved if the terms were consistent and well-defined. Similar to the questions asked in the health conditions, Questions 5.3.B.2 and 5.4.B.2 should be formatted as forced-choice rather than check-all-that-apply responses. Finally, Questions 5.3.B.3 and 5.4.B.3 suffer from limitations similar to those of the deployment exposure questions in Section 1.2; the ranges of years provided are too broad (0–99) and subject to error; and it is not clear that this level of precision is actually required. If so, a range check edit is needed to flag and verify or correct apparently invalid or unlikely values, or one could create a categorical response scale with ranges of years combined.

The questions on asbestos exposure (Section 5.5), inquire about combined civilian and military occupations and exposures—something that is not done elsewhere in the questionnaire. There is no explanation or preamble to the section or its first question. The fact that 34% of the respondents indicated "don't know" to this question (5.5A: *Have you* ever *worked in a job with asbestos exposure, including military service?*) suggests that it has serious problems in comprehension or reflects a genuine lack of knowledge about potential past exposures. It also is not clear why asbestos exposure was not included in military-specific occupations and separately for civilian occupations. Additionally, 5.5.B is awkwardly worded (including asking respondents to *circle* their answers, which is not compatible with an online instrument[9]) and allows for multiple responses instead of the question having been designed to be clearer through forced-choice.

The section on home environment and hobbies (Section 6) begins with a single sentence of introductory text, "*Exposures in your home environment or hobbies may impact how deployment exposures affect you.*" Three questions ask whether or not respondents live with or visit traditional farm animals (not otherwise defined), have had mold in their home, or lived in a home with elevated levels of radon (6.A, 6.B, 6.C). Respondents are then asked to select from a detailed but incomplete list any hobbies that they participate in and to indicate how many hours per week, on average, they spend on those hobbies (6.D, 6.E). Aside from these questions being non-specific, the section is particularly problematic in that questions mix reference periods between "ever" (6.B and 6.C) and the present tense and in its use of wording such as "on a regular basis" (6.A) without making this clear. Moreover, Sections 5 and 6 ask questions that are not central or relevant to the focus of the registry and add substantially to the amount of time needed to complete the questionnaire without adding value.

The questionnaire contains a substantial number of questions on non-military, nondeployment variables, which are of little relevance to the stated purpose of registry. While critical disease influences such as smoking need to be considered, there is little basis for trying to address potential or subtle influences associated with other jobs, environments, and lifestyle factors. VA states that the registry's primary purpose is to record self-reported exposures and health outcomes and to explore possible associations or determine potential health effects from exposure to airborne environmental hazards in service members and veterans. This purpose is not served by the poorly worded, nonspecific questions aimed at nondeployment-related factors. Such information might have utility in the context of an epidemiological analysis but, as discussed elsewhere, the registry data are inappropriate to use for that purpose.

[9] Question 1.2.E also asks respondents to circle their responses.

LINKING OTHER DATA TO REGISTRY DATA

VA designed the AH&OBP Registry with the intent of integrating multiple VA and DoD data sources to supplement questionnaire information and to provide a more complete picture of long-term health associated with exposure to burn pits. The registry was designed to link to and incorporate data from several other sources, including DoD's DMDC, the Veterans Benefits Administration, and VA health records, including Medicare and mortality data (Ciminera, 2015).

Currently, the VA information technology system is able to link questionnaire responses to VA clinical data and other VA administrative data (Montopoli, 2016a). VA has added a template note to the electronic health record which allows a provider to indicate that a clinical evaluation related to registry participation has occurred. However, the registry design and architecture do not allow for information, once submitted, to be updated, either in the form of changes to individual items in the questionnaire itself—for example, additional information or new diagnoses gleaned through the clinical evaluation—or by linking with sources other than basic DMDC deployment information. The capability to add supplemental information, such as from follow-up questionnaires, has reportedly been added to the system, which would allow any future follow-up questions to supplement or update information in a participant's record, but information on plans for conducting follow-up on the population is unavailable. VA medical records and other administrative data, such as vital status, can be linked to the registry data to provide additional information that might be used to identify health status or mortality or support registry operations (Montopoli, 2016a). The ability to link and update participant responses and health records for VA users would allow VA to validate responses, initiate longitudinal follow-up of VA users, and conduct sub-analyses of respondents to determine whether those with clinical evaluation data are reporting different outcomes, more severe health outcomes, or other factors that may differ from other registry participants.

VA data are easier to link and incorporate with questionnaire information than data from other sources, specifically DoD. Although VA has stated that it plans to gather longitudinal data on registry participants and has added that capability to the system, the committee found no details or information on methods of how this might be operationalized. For veterans, some additional information such as enrollment in VA, use of any VA benefits, use of VA health and mental health services, service-connected disabilities, and the like could also be examined by extracting those data and appending them to the registry information. The committee was told that DoD data are much harder to link and incorporate and are currently limited to DMDC data that provides deployment, demographic, and military characteristics information; VA has no plans to append DoD medical records to the registry (Montopoli, 2016a).

To the extent possible, linking individual participants with information on them that is available from VA or DoD databases could both increase the accuracy of the registry data and reduce respondents' burden. Linking registry data with VA and DoD medical records, including hospitalization data, would allow for the evaluation of both subjective and objective health outcomes and validate self-reported conditions. Additionally, if DoD personnel records could be linked to the registry, they might provide additional information on pre-deployment conditions and exposures. For example, DoD has begun piloting the use of personal monitoring devices to link monitored exposures with individual health outcomes. These devices measure and record vitals, location, and external exposures (such as radiation and organic vapors) (Hartman et al., 2016). In the future, such devices could collect information on exposures that could potentially be linked to reported health outcomes in medical records or to health outcomes reported by registry participants. It is possible that other data sources may eventually be linked to the registry data, but because the registry was not designed specifically to link to ancillary data sources, the structure or validation of those data may shape or limit their use compared with the primary VA and DoD data linkage sources for which the registry was specifically designed.

Before linking or appending information or databases to the registry, several compatibility issues need to be considered. Primary among these would be what is being gained by the additional information. For example, linking to additional VA sources will only provide additional information for VA users. Less than half (46%) of deployed 1990–1991 Gulf War veterans and 64% of OEF, OIF, and OND veterans use VA for their health needs (NASEM, 2015a; VA, 2015b). Even among VA health care users, not all use VA for all of their health needs. Should it be determined that the additional information is advantageous to include, the second issue will be to determine whether the registry database architecture and data structure are compatible with the form, structure, and availability of the intended sources to be linked. Third, validation of data quality of the additional sources should be conducted prior to a proposed linkage. Such quality criteria might include the type of data (self-report, clinical exam, laboratory tests), temporal information (longitudinal, cross-sectional), and whether measures of validity and reliability were embedded in a data source during data gathering. Fourth, the type of statistical techniques that will be used for linking data records (for example, deterministic matching, probabilistic matching, or another method) needs to be considered, since it will depend, in part, on the type of data being linked. Finally, the timing of and consequences related to updating information need to be considered. Timing includes how often the linking source information will be updated, for example, a single instance versus periodic or as-needed occurrence. For linked source information that differs from the registry, which source will be considered primary? Who or what algorithm will be used to validate the differing information? For sources that will be updated more than once, one consideration will be what will happen to the existing information when there is an update—whether it will be overwritten or a new field will be created. Answering these questions and implementing linkages across multiple data sources is critical for ensuring maximum utility of the registry.

SYNOPSIS AND CONCLUSIONS

Based on the information presented in this chapter, the committee has reached the following findings, conclusions, and recommendations related to the actions taken by VA to a design and implement a registry for the purpose of collecting health outcomes related to potential exposure to burn pits and other airborne hazards. Subsequent chapters revisit some of these issues in greater detail and offer additional observations based on analyses of registry data.

The 12-month timeframe directed by Congress for VA to establish a comprehensive and targeted exposure and health outcomes questionnaire and registry and to make it available for veterans' use was quite short, considering the complexity of its intent and the tasks involved, including development and testing. Among other issues, the short timeframe did not allow for designers to consult needed expertise to implement a well designed instrument that could handle the complexities required of it in terms of information to be collected and how the information could best be managed and used. It is open to question whether the time period allotted by Congress was realistic.

The AH&OBP Registry questionnaire is problematic in many respects. This chapter identifies flaws in its layout, directions, and in the flow of the questions, and it cites several examples of poorly worded questions and questions that are not relevant to the stated goal of the registry or reflective of the limitations of the instrument. Questions that were designed for other types of surveys (notably, the NHIS) were seemingly used without regard to whether taking them out of that context affected their validity. And the process of verifying DoD-supplied information on locations where a respondent had served at (deployment segments)—something that was no doubt intended to be a time-saving element of the questionnaire—was instead burdensome to those with large numbers of deployments and may have led to response fatigue.

These shortcomings likely stem in part from the developers of the registry having not consulted with external experts in questionnaire design while developing the instrument and having not used specialized Web-based survey software. Thus, **the committee recommends that VA involve external survey experts experienced in Web-based instruments in any restructuring of the registry questionnaire.**

The question of how the registry questionnaire should be changed at this point in time depends critically on what VA intends for the registry to accomplish going forward. VA states that the data collected by the registry will be used for several purposes: to help monitor health conditions affecting eligible veterans and service members, to improve VA programs aimed at helping veterans and service members with deployment exposure concerns, to generate potential hypotheses about exposure response relationships, to improve programs in the VHA, and to provide outreach to veterans who may have experienced adverse health outcomes as a result of their exposures. However, except for the outreach activities and perhaps hypothesis generation, it does not appear that the registry—in general, or the data collected by it—is useful for addressing the stated purposes. For example, if it were to be used to help monitor health conditions affecting eligible veterans and service members, there should be a mechanism to ensure the periodic follow-up of participants. Or if burn pit exposure is to be the focus, more than four appropriately worded questions would be required to elicit information on the duration and intensity of exposure. Instead, the registry's most productive use appears to be as a forum to allow eligible veterans and service members to register their concerns about potential exposures and current health effects. If this more modest goal were to be adopted, the process of enrollment could be markedly simplified and data collection streamlined with little loss of information. Therefore, **the committee recommends that VA eliminate the questionnaire sections addressing locations of previous residences (Section 4), non-military work history (5), and home environment, community, or hobbies (6), all of which collect data that might only be useful in epidemiologic studies of the population.** Removing these sections would result in a less burdensome instrument with little if any loss of usable information for any of the stated purposes of the registry. Having a clear, consistent message about the purpose of the registry will allow it be tailored to focus on the issues of most importance and minimize the burden of completing the questionnaire.

VA stated that, as directed in the public law, it will use the recommendations from an independent scientific organization review (this study) to improve the registry program, including improving the questionnaire as necessary (VA, 2014a). However, the committee notes that addressing the problems identified in this chapter will not be sufficient to overcome the fundamental design flaws of the registry and that the registry will continue to have little value as a scientific tool for research or monitoring veterans' health or improving the delivery of VA health care services.

As previously noted, the registry's design and architecture do not allow for information, once submitted, to be updated. However, such a Web-based system does have the capacity to add supplemental information from participants—for example, from longitudinal or follow-up questionnaires, if conducted—and to link to VA and DoD data sources such as medical records, other administrative data (including benefits and vital status), and selected DoD information. Linkages with these other sources have the potential to reduce future participant burden, increase data quality (by avoiding recall and other biases), and increase the utility of the registry database. The AH&OBP Registry has the potential to be an advance in design over other VA registries if better use is made of this capacity. **The committee recommends that once VA clarifies the intent and purpose of the registry, it develop a specific plan for more seamlessly integrating relevant VA and DoD data sources with the registry's data, with the goals of reducing future participant burden, increasing data quality by restructuring questions to minimize recall and other biases, and improving the usefulness of the registry database as an information source for health care professionals and researchers.**

Although a Web-based questionnaire may confer several benefits over more traditional methods, not all eligible persons have access to a computer or the Internet. Offering additional formats is not a trivial matter, but it would potentially improve access to the registry. Steps should therefore be taken to ensure that this subset of eligible persons has the opportunity to participate in the registry. **The committee recommends that alternative means of completing the questionnaire such as a mail-in form or via a computer-assisted phone interview be offered in order to ensure that the subset of eligible persons who do not use or are not facile with the Internet have the opportunity to participate in the registry.** Eligibility could be checked when a potential participant contacts VA. It would be necessary to put work into developing a system that would minimize the burdens on both the respondent and VA but the challenges are surmountable.

Other additional outreach efforts are also desirable. The relatively small number of respondents to date suggests that generic posts and shares on VA social media sites and broad communications and outreach initiatives are not reaching the full intended population of service members and veterans. One means to expand the participant pool would be to use data on current respondents to determine the characteristics of the people who are not signing up and then tailor messages and media on the basis of this information. Clarifying the purpose of the registry would help inform the question of how best to ensure that—to the greatest extent possible—persons eligible to participate are aware of it.

A strong point of the AH&OBP Registry—as a record of the respondent's potential exposures and health concerns—should be taken further advantage of. This information can already be accessed by military and veteran health care system providers tied into VA's electronic medical records and can be downloaded and printed by respondents. **The committee recommends that VA enhance the utility of the registry by developing a concise version of questionnaire responses focused on information that would be most useful in a routine clinical**

encounter and make it available for download. A one-page synopsis, for example, could facilitate conversations between patients and providers about medical concerns, leading to more productive visits and better focus on health care needs.

Finally, VA should investigate why so few of the respondents who say they would be interested in the in-person medical exam offered in conjunction with the registry have actually arranged for one. **The committee recommends that VA continue its efforts to make it easier for participants to schedule and get the optional health examination offered as part of the registry—such as through targeted follow-up of respondents who indicate interest—and that it investigate the reasons that such a small percentage of respondents who indicate interest in an exam (~2.5%, to date) request one.** Adding a means of scheduling an exam as part of the questionnaire—a capability that the committee understands is being implemented—will be a useful first step.

REFERENCES

AHRQ (Agency for Healthcare Research and Quality). 2010. Registries for Evaluating Patient Outcomes: A User's Guide, 2nd ed. Edited by R. E. Gliklich and N. A. Dreyer. Rockville, MD. P. 307.

Bradburn, N. M., S. Sudman, and B. Wansink. 2004. *Asking questions: The definitive guide to questionnaire design—for market research, political polls, and social and health questionnaires, 2nd ed.* San Francisco, CA: John Wiley & Sons.

Callegaro, M., M. H. Murakami, Z. Tepman, and V. Henderson. 2015. Yes–no answers versus check-all in self-administered modes: A systematic review and analyses. *International Journal of Market Research* 57(2):203–223.

CDC (Centers for Disease Control and Prevention). 2015. *2014 National Health Interview Survey (NHIS) public use data release.* Hyattsville, MD: Department of Health and Human Services, Centers for Disease Control and Prevention, National Center for Health Statistics. http://ftp.cdc.gov/pub/Health_Statistics/NCHs/Dataset_Documentation/NHIS/2014/srvydesc.pdf (accessed December 3, 2016).

Chen, L. M., W. R. Farwell, and A. K. Jha. 2009. Primary care visit duration and quality: Does good care take longer? *Archives of Internal Medicine* 169(20):1866–1872.

Choi, B. C. K. and A. W. P. Pak. 2005. A Catalogue of biases in questionnaires. Preventing Chronic Disease 2(1). http://www.cdc.gov/pcd/issues/2005/jan/04_0050.htm (accessed July 12, 2016).

Ciminera, P. 2015a. Charge to the Committee on the Assessment of the Department of Veterans Affairs Airborne Hazards and Open Burn Pit Registry. PowerPoint Presentation Presented at the Meeting, Washington, DC. March 13.

Ciminera, P. 2015b. VA Airborne Hazards and Open Burn Pit Registry. VA mobile discussion series, August 28. PowerPoint presentation. https://mobile.va.gov/sites/default/files/VMH-093AugustDiscussionSeries.pdf (accessed September 29, 2015).

Ciminera, P. 2015c. *Deployment segment analysis from the Airborne Hazards and Open Burn Pit Registry memo.* Prepared for the Department of Veterans Affairs Office of Public Health, Post-Deployment Health Group by Westat. April 2015.

Ciminera, P. 2015d. *Time to complete self-assessment questionnaire.* Prepared for the Department of Veterans Affairs Office of Public Health, Post-Deployment Health Group by Westat. May.

Couper, M. P. 2000. Review: Web surveys: A review of issues and approaches. *Public Opinion Quarterly* 64(4):464–494.

Crum-Cianflone, N. F. 2013. The Millennium Cohort Study: Answering long-term health concerns of U.S. military service members by integrating longitudinal survey data with military health system records. In J. Amara and A. Hendricks (eds.), *Military health care: From pre-deployment to post-separation.* New York: Routledge.

Defense Health Board. 2015. Deployment Pulmonary Health. Office of the Assistant Secretary of Defense. Falls Church, Virginia. http://www.health.mil/Reference-Center/Reports/2015/02/11/Deployment-Pulmonary-Health (accessed February 9, 2017).

Dillman, D. A., and B. L. Messer. 2010. Mixed-mode surveys. In J. D. Wright and P. V. Marsden (eds.), *Handbook of survey research.* San Diego, CA: Elsevier. Pp. 551–574.

Dillman, D. A., J. D. Smyth, and L. M. Christian. 2014. *Internet, phone, mail, and mixed-mode surveys: The tailored design method, 3rd ed.* Hoboken, NJ: John Wiley & Sons.

Ericson, L. and C. Nelson. 2007. A comparison of forced-choice and mark-all-that-apply formats for gathering information on health insurance in the 2006 American Community Survey Content Test. In Federal Committee on Statistical Methodology Research Conference, Arlington, VA. Available online at: https://fcsm.sites.usa. gov/files/2014/05/2007FCSM_Ericson-VI-A. pdf (accessed September 29, 2016).

Fan, W., and Z. Yan. 2010. Factors affecting response rates of the Web survey: A systematic review. *Computers in Human Behavior* 26(2):132–139.

Federal Register. 2013a. Department of Veterans Affairs [OMB Control No. 2900–NEW] Proposed information collection (Open Burn Pit Registry Airborne Hazard Self-Assessment Questionnaire) activity: Extension of comment period. Notice: 44625 Vol. 78, No. 108. Published June 5.

Federal Register. 2013b. Department of Veterans Affairs [OMB Control No. 2900–NEW] Proposed information collection (Open Burn Pit Registry Airborne Hazard Self-Assessment Questionnaire) activity: Comment Request. Notice: 33895 Vol. 78, No. 142. Published July 24.

Federal Register. 2013c. Department of Veterans Affairs agency information collection (Open Burn Pit Registry Airborne Hazard Self-Assessment Questionnaire) activities under OMB review. Notice: 54956 Vol. 78, No. 173. Published September 6.

Federal Register. 2014. Department of Veterans Affairs establishment of the Airborne Hazards and Open Burn Pit Registry. Notice: 36142 Vol. 79, No. 122. Published June 25.

Feindt, P., I. Schreiner, and J. Bushery. 1997 Reinterview: a tool for survey quality improvement. *Proceedings of the Joint Statistical Meeting, Survey Research Methods Section*. Washington, DC: American Statistical Association. Pp. 105–110.

Hartman, R., S. Jones, and K. Phillips. 2016. DoD's development of the total exposure health (TEH) initiative. Presentation to the National Academy of Sciences, Engineering, and Medicine Board on the Health of Select Populations. June 14.

Institute for Social Research. 2016. Guidelines for best practice in cross-cultural surveys, fourth edition. University of Michigan, Ann Arbor. http://ccsg.isr.umich.edu/images/PDFs/CCSG_Full_Guidelines_2016_Version.pdf (accessed February 9, 2017).

IOM (Institute of Medicine). 2011. *Long-term health consequences of exposure to burn pits in Iraq and Afghanistan*. Washington, DC: The National Academies Press.

IOM. 2013. *Best care at lower cost: The path to continuously learning health care in America*. Washington, DC: The National Academies Press.

Kang, H. K., C. M. Mahan, K. Y. Lee, C. A. Magee, and F. M. Murphy. 2000. Illnesses among United States veterans of the Gulf War: A population-based survey of 30,000 veterans. *Journal of Occupational and Environmental Medicine* 42(5):491–501.

Lezama, N. G. 2015. Presentation to the Committee on the Assessment of the Department of Veterans Affairs Airborne Hazards and Open Burn Pit Registry. December 15.

Lezama, N. G. 2016. Responses from the Department of Veterans Affairs to questions from the Committee on the Assessment of the Department of Veterans Affairs Airborne Hazards and Open Burn Pit Registry. Received February 18.

May, L., and J. Haider. 2014. Preliminary assessment of NHIS data for providing a burn pit registry comparison group. Prepared for the Committee on the Assessment of the Department of Veterans Affairs Airborne Hazards and Open Burn Pit Registry. November 20.

Mooney, G. M. and B. Lepidus Carlson. 1996. Reducing mode effects in "mark all that apply" questions. *Proceedings of the Survey Research Methods Section, American Statistical Association* 614–619.

Montopoli, M. 2016a. VA responses to requests and questions from the committee. Received June 20.

Montopoli, M. 2016b. VA responses to committee's questions following August 24–25, 2015, meeting. Received September 15.

NASEM (National Academies of Sciences, Engineering, and Medicine). 2015a. *Considerations for designing an epidemiologic study for multiple sclerosis and other neurologic disorders in pre and post 9/11 Gulf War veterans*. Washington, DC: The National Academies Press.

NASEM. 2015b. *Improving diagnosis in health care*. Washington, DC: The National Academies Press.

NIAAA (National Institute on Alcohol Abuse and Alcoholism). 2016. CAGE questionnaire. http://pubs.niaaa.nih.gov/publications/inscage.htm (accessed September 25, 2016).

Pew Research Center. 2011. Internet surveys. http://www.people-press.org/methodology/collecting-survey-data/internet-surveys (accessed September 29, 2016).

RAND. 2016. RAND Medical Outcomes Study (MOS): 36-Item Short Form Health Survey (SF-36). http://www.rand.org/health/surveys_tools/mos/36-item-short-form.html (accessed September 25, 2016).

Rasinski, K. A., D. Mingay, and N. M. Bradburn. 1994. Do respondents really "mark all that apply" on self-administered questions? *Public Opinion Quarterly* 58(3):400–408.

Scherpenzeel, A., and V. Toepoel. 2012. Recruiting a probability sample for an online panel. Effects of contact mode, incentives, and information. *Public Opinion Quarterly* 76(3):470–490.

Sharkey, J. M., D. K. Harkins, T. L. Schickedanz, and C. P. Baird. 2014. Department of Defense participation in the Department of Veterans Affairs Airborne Hazards and Open Burn Pit Registry: Process, guidance to providers, and communication. *US Army Medical Department Journal* July–September:44–50.

Smith, T. C., I. G. Jacobson, B. Smith, T. I. Hooper, M. A. Ryan, and the Millennium Cohort Study Team. 2007. The occupational role of women in military service: Validation of occupation and prevalence of exposures in the Millennium Cohort Study. *International Journal of Environmental Health Res*earch 17(4):271–284.

Smyth, J. D., D. A. Dillman, L. M. Christian, and M. J. Stern., 2006. Comparing check-all and forced-choice question formats in Web surveys. *Public Opinion Quarterly* 70(1):66–77.

Smyth, J. D., L. M. Christian, and D. A. Dillman. 2008. Does "yes or no" on the telephone mean the same as check-all-that-apply on the Web? *Public Opinion Quarterly* 72(1):103–111.

SSC (Sergeant Thomas Joseph Sullivan Center). 2013. Letter to Ms. Cynthia Harvey-Pryor at the Veterans Health Administration regarding preliminary comments to the OMB application for the AH&OBP self-assessment questionnaire. Provided to the Committee on the Assessment of the Department of Veterans Affairs Airborne Hazards and Open Burn Pit Registry. Washington, DC. May 1.

SSC. 2015. Statement of Peter M. Sullivan to the Committee on the Assessment of the Department of Veterans Affairs Airborne Hazards and Open Burn Pit Registry. Washington, DC. May 1.

Szema, A. 2015. Analysis of VA burn pits registry: Testimony to workshop. Presentation to the Committee on the Assessment of the Department of Veterans Affairs Airborne Hazards and Open Burn Pit Registry. May 1.

Tourangeau, R., F. G. Conrad, and M. P. Couper. 2013. *The Science of Web Surveys*. New York: Oxford University Press.

VA (Department of Veterans Affairs). 2012. Airborne Hazards Clinical Evaluation Working Group charter. Email from Nicolas Lezama, deputy chief consultant, Post-Deployment Health Services Patient Care Services, Veterans Health Administration on February 18.

VA. 2013. Open Burn Pit Registry/Airborne Hazards Exposure Assessment Working Group charter. Email from Nicolas Lezama, deputy chief consultant, Post-Deployment Health Services Patient Care Services, Veterans Health Administration on February 18.

VA. 2014a. Justification template: Airborne hazards and Open Burn Pit Registry self-assessment questionnaire. www.reginfo.gov/public/do/DownloadDocument?objectID=44258503 (accessed September 29, 2016).

VA. 2014b. eBenefits fact sheet. https://www.ebenefits.va.gov/ecms-proxy/document/ebenefits-liferay/dynamic-content/ebenefits/assets/downloads/eBenefits_factsheet.pdf (accessed September 25, 2016).

VA. 2015a. *Report on data from the Airborne Hazards and Open Burn Pit (AH&OBP) Registry.* http://www.publichealth.va.gov/docs/exposures/va-ahobp-registry-data-report-april2015.pdf (accessed September 10, 2016).

VA. 2015b. *Analysis of VA health care utilization among Operation Enduring Freedom (OEF), Operation Iraqi Freedom (OIF), and Operation New Dawn (OND) veterans.* Washington, DC: Veterans Health Administration.

VA. 2016a. Airborne Hazards and Open Burn Pit Registry: About the registry. https://veteran.mobilehealth.va.gov/AHBurnPitRegistry/index.html#page/about (accessed January 28, 2016).

VA. 2016b. PTSD checklist for DSM-5 (PCL-5). http://www.ptsd.va.gov/professional/assessment/adult-sr/ptsd-checklist.asp. (accessed September 25, 2016).

VA. 2016c. U.S. Department of Veterans Affairs clinicians guide to airborne hazards. http://www.publichealth.va.gov/docs/exposures/clinician-guide-airborne-hazards.pdf (accessed September 29, 2016).

VA. 2016d. Airborne Hazards and Open Burn Pit Registry fact sheet. http://www.publichealth.va.gov/docs/exposures/burn-pit-registry-fact-sheet.pdf# (accessed April 4, 2016)

VA. 2016e. Airborne Hazards and Open Burn Pit Registry. https://veteran.mobilehealth.va.gov/AHBurnPitRegistry/#page/home (accessed February 23, 2016).

4

Analysis Methods and Descriptive Statistics

This chapter is the first of three chapters that describe the methods and results of the committee's analysis of the first 13 months of data from the Airborne Hazards and Open Burn Pit (AH&OBP) Registry. It begins with an overview of the data requested and received. This is followed by a discussion of preliminary analyses undertaken to examine the quality of the data and potential biases by quantifying missing information and characterizing the respondents in terms of participation rates and representativeness to the eligible population. Descriptive statistics including demographic and military characteristics of registry respondents are provided, and comparisons with all eligible veterans and service members are made. Detailed methods, analyses, and interpretation of exposure data collected by the questionnaire are presented in Chapter 5, and Chapter 6 provides the methods, analyses, and interpretation of the health outcomes that the committee believes offer the most value. Where applicable, the committee describes challenges encountered throughout its process and the resulting impacts.

DATA REQUESTED AND RECEIVED

The analysis of the AH&OBP Registry data was conducted with the purpose of describing the exposures and health conditions reported by veterans and service members who chose to participate in the registry; it cannot, by nature (see Chapter 2), support any conclusions about health conditions caused by exposure to burn pits. Given the inherent limitations of registries in providing representative and reliable data, the committee took a cautious approach to its analysis of registry data. This section describes the committee's general approach to requesting, collecting, and analyzing the data collected by the AH&OBP Registry.

The data request process began in October 2014 and took more than 13 months to complete. The committee, under the National Academies of Sciences, Engineering, and Medicine (the National Academies), and the Department of Veterans Affairs (VA) signed a Data Use Agreement on December 18, 2015, to allow the committee access to de-identified data for analyses to complete their task. Appendix D contains a list of the variables requested from each data source and the status of each (received or not released).

DATA ON AH&OBP REGISTRY PARTICIPANTS

The committee's dataset included all completed questionnaires (N = 46,404) submitted between June 19, 2014, and July 31, 2015. The committee requested all data contained in the questionnaires. However, no personally identifiable information was included and some variables were modified, as described in the section titled "Unavailable Data."

The extent of the available data was limited in two important ways. First, although analyses would likely have been more informative if the committee had additional data, it was constrained to using data limited to approximately the first 13 months of registry operation. Second, the dataset only included individuals who completed the full questionnaire (referred to as respondents). "Complete" was defined by VA as those questionnaires for which respondents answered all applicable questions ("don't know" and "refused" options were provided) and clicked the "submit" button. The choice of using this definition of "completed" rather than allowing for questionnaires that were only partially completed to be used was made by VA (Nicolas Lezama, VA, personal communication, January 26, 2016). After submission, a user's status changes to "participant" in the database, and the time stamp is recorded in the context of tracking participation. No data on partial completions—that is, data from questionnaires that had been started but the individual never clicked "submit," regardless of the number of sections or questions completed—were made available. It was not clear to the committee why such a strict criterion was used to define participation in the registry, and it likely resulted in a substantial loss of data (see discussion of participation rates) and possibly a less representative sample consisting only of final registry respondents. Such a rigid definition of completion is inconsistent with best practices in survey and epidemiological research (AAPOR, 2016), which call for participants to be as those who have completed some minimal number of sections or questions that are designated as key to information needs and objectives.

Questionnaire data were provided in four files that could be linked using a unique identification code: a main dataset that contained responses to all of the exposure and health condition questions (SAQMain); a data file that contained information on all deployment segments, both verified and added deployments (SAQ Deployment Segment Data); participation statistics, such as date and time started and completed, and year of birth; and other user data, such as dates that the user account was created, consent given, and eligibility status. All four files were extracted from VA's database on July 31, 2015.

In all, the committee's dataset included data from three sources—the registry, the Gulf War Oil Well Fire Smoke Registry, and the Contingency Tracking System (CTS)—which are summarized in Box 4-1. The CTS and Gulf War Oil Well Fire Smoke Registry were current to December 31, 2015. Data from all three sources were able to be linked using a unique, randomly assigned identifier. The Gulf War Oil Well Fire Smoke Registry contains records of all service members who served in Operation Desert Storm and Operation Desert Shield while the oil-well fires were burning (VA, 2015a). The CTS is a subset of the Department of Defense (DoD) Defense Manpower Data Center (DMDC) database of service personnel who were physically located in the Operation Enduring Freedom (OEF) and Operation Iraqi Freedom (OIF) areas of operations or who were specifically identified as directly supporting those missions outside of the designated combat zone, such as aircrew or support personnel located outside the combat zone (Bonds et al., 2010). Because the registry questionnaire does not collect all key demographic variables necessary for the committee's analysis, demographic information for respondents was drawn from the CTS extract for all OIF/OEF service members and from the Gulf War Oil Well Fire Smoke Registry file for all Gulf War service members.

Characterizing Eligibility

The eligible population was defined as service members and veterans who deployed to contingency operations in the Southwest Asia theater of operations (Iraq, Kuwait, Saudi Arabia, Bahrain, Gulf of Aden, Gulf of Oman, Oman, Qatar, United Arab Emirates, Persian Gulf, Arabian Sea, and Red Sea) at any time on or after August 2, 1990, or in Afghanistan or Djibouti on or after September 11, 2001. A key advantage of the AH&OBP Registry is that eligibility is well-defined by period of deployment and deployment locations and documented by DMDC, which also maintains a broad array of demographic and military characteristics information for all service members.

To examine how well the experience of the registry respondents reflects the experience of the larger eligible deployed population, the committee used the data from the CTS and Gulf War Oil Well Fire Smoke Registry to define and characterize the eligible population. Neither the CTS extract nor the Gulf War Oil Well Fire Smoke Registry data included information on service members deployed to the Gulf region during the stabilization period

BOX 4-1
Summary of Available Datasets

AH&OBP Registry Files containing data collected June 19, 2014–July 31, 2015:

- responses to all of the exposure and health condition questions (SAQMain);
- all deployment segments, verified and added deployments (SAQ Deployment Segment Data);
- participation statistics, such as date and time started and completed and year of birth;
- other user data, such as dates that the user account was created, consent given, and eligibility status.

DoD DMDC deployment information to define the eligible population:

- Contingency Tracking System OEF/OIF/OND Roster File provides demographic information on all persons deployed in support of OEF/OIF/OND (post-September 11, 2001);
- Gulf War Oil Well Fire Smoke Registry file provides demographic information on all persons deployed in support of the Gulf War (from 1990 to 1991).

between April 1991 and August 2001. Therefore, these respondents were excluded from the analysis if they did not also deploy during the 1990–1991 Gulf War or post-9/11 conflicts. Because less than 1% of respondents were deployed during the stabilization period, the committee did not feel that their exclusion would significantly affect the results of the analysis. For analyses of exposure and disease stratified by era, respondents were counted in each era for which they had a deployment. Similarly, for analyses that were deployment-segment based, respondents were included in multiple country and year categories if they had deployments in more than one category. The resulting eligible population included 545,383 Gulf War and 2,483,392 post-9/11 service members.

The CTS file was used as the "gold standard" for determining the eligibility of post-9/11 service members and veterans because its data were imported into the registry database to determine an individual's eligibility. However, when the CTS file data were linked to the questionnaire data, 333 respondents either had no deployment records in the CTS or were ineligible according to the CTS and were excluded to reconcile the inconsistency. The number of cases excluded is a small proportion of the respondents. Because the CTS file is supposed to be used to screen for eligibility for the registry, in theory there should be no differences, and the same screening criteria that is used by the registry was applied using the code VA provided. One possible explanation for the discrepancy is that VA allowed some individuals who were determined ineligible to request an eligibility review and complete the questionnaire, but it is unknown how often such waivers are given. Another possibility is that because the CTS is updated monthly, the CTS file that the committee had was more up-to-date than the CTS file that was used to screen for the registry.

Service era is defined on the basis of deployment year only. Therefore, for the stratification analyses by service era, Gulf War era was defined as deployments during 1990–1991, and post-9/11 era was defined as 2001 or later. Because the committee only had data on the deployment year, not the month and day, anyone with a deployment date in 2001 or later was included and considered part of the post-9/11 cohort. Therefore, respondents deployed between January 1, 2001, and September 10, 2001, would have been included with post-9/11 rather than the stabilization period. Table 4-1 shows the number of respondents with deployments in each era based on the questionnaire. Because VA only provided the year of deployment, there may be some inaccuracies in these numbers. Similar discrepancies were observed for the Gulf War Oil Well Fire Smoke Registry file and treated in a similar manner.

TABLE 4-1 Number of Respondents by Era

Era	Number	Percent
Not Mutually Exclusive		
Gulf War (1990–1991)	5,595	12.1
Stabilization Period (1992–2000)	1,356	2.9
Post-9/11 (2001 or later)	42,673	92.0
Mutually Exclusive		
Gulf War only (1990–1991)	3,498	7.5
Stabilization Period only (1992–2000)	120	0.3
Post-9/11 only (2001 or later)	39,804	85.8
Multiple eras	2,982	6.4

Unavailable Data

Several of the committee's data requests were not fulfilled because VA determined that such data could potentially be personally identifiable. Notably, many of these data have been made available to contractors for other analyses. The text below briefly describes the committee's purpose for requesting those data and the effect that not having those data had on the committee's ability to conduct its analyses.

Only information on completed questionnaires was made available to the committee. Information about persons who did not complete the questionnaire would have provided valuable information about the differences between those who attempted and those who completed the questionnaire. Differential completion rates by demographic and military characteristics, reported exposures, and reported health outcomes could reveal other challenges and biases.

The dates and locations of deployment were limited to the year(s) and country or countries. Limiting the deployment dates and within-country location information weakened the committee's efforts to construct reliable measures of exposure potential (see Chapter 5). However, VA provided a variable of deployment length in days for each segment reported in the questionnaire that could then be used to calculate exposure potential and other outcomes of interest.

VA further affected the committee's ability to estimate exposure by not releasing the names or locations of bases where respondents reported spending the majority of their time on each deployment. All results provided to the committee were required to be aggregated, and having such information as the number of respondents who served at a particular base would in no way provide identifying information. DoD has created and selectively made available a list of bases with burn pits in the gulf region (Joseph Gasper, Westat, personal communication, August 11, 2016). The base names reported in the registry combined with that information would have allowed the committee to create a more reliable measure of burn pit exposure by cross-checking the information reported by respondents. For example, if a respondent reported exposure to burn pits, but also reported spending the majority of his or her time at a base without burn pits, this could be adjusted for in the analysis (as done in a separate analysis for VA (Gasper and Katawa, 2015b).

To conduct more complete analyses of potential exposure and health outcomes would require more detailed information on the dates and specific locations of deployment, patterns of participation by periods of service, and objective measures of exposures. Sub-analyses could be conducted to compare respondents to eligible persons known to have documented burn pit exposure—a more appropriate comparison group than potential exposure based on country and year of deployment. Such an exercise would provide additional insight into how characteristics may differ between registry respondents and the eligible population. To accomplish this would require DoD data that documented burn pit sites overall, the dates of use, and the dates when incinerators were implemented (perhaps more accurately on a subset of bases, such as Joint Base Balad, Camp Taji and Camp Speicher); it would also be necessary to link this information to registry deployment and location data. That analysis would result in a subsample of the total eligible population who were potentially exposed to burn pits, but it would also likely produce a more accurate representation of the proportions of the eligible population who served at such locations and who chose to participate in the registry as well as differences in their characteristics. If these data are made available for the full eligible population, the volume of information for a systematic analysis of differences between registry respondents and the eligible population or nonrespondents would be far more extensive than that which is often available in research.

The committee also was unable to examine differences based on separation status—active service members compared to veterans—because the CTS extract file provided by VA did not contain a variable for separation code. Without this information, it was not possible to determine whether differences in respondents versus the eligible population (and potential participation biases) were greater among veterans than among those still on active duty.

No data on health care use or conditions as recorded in medical records through VA or DoD's Military Health System were provided for either registry respondents or the eligible population. The committee sought that data to help answer VA's request for recommendations regarding the means of addressing the medical needs of veterans with health conditions related to burn pit exposure and "associations of self-reported exposures with Veterans Health Administration (VHA) health care experience." Specifically, the VHA Assistant Deputy Under Secretary for Health Enrollment Files contain enrollment, eligibility, demographic, cost, and location information for VHA enrollees and non-enrollees who have received VA care. VHA clinical and administrative data systems contain in-person registry evaluations that are identified based on the presence of a clinical note title or health factors, such as *International Classification of Diseases* codes that may have been useful to the committee in considering data linkages and analyses to enable a response to this request (VA, 2015a).

Mortality data were not available, so service members who died before the registry was implemented could not be removed from the eligible population. Having the number of eligible persons who have died along with their cause of death would have allowed the committee to analyze whether certain primary causes of death were more frequent and are perhaps an area of interest for further study that may or may not have implications on deployments.

ANALYSIS METHODS

Through the National Academies, the committee contracted for all analyses to be performed by Westat, Inc., but the analyses were done under the direction of the committee. Several iterations of analysis were necessary as the committee absorbed the results of an analysis and identified new questions and important areas to examine in the next iteration. Westat was selected to be the committee's subcontractor for several reasons. Its analysts already had intimate knowledge of the registry, the Gulf War Oil Well Fire Smoke Registry, and the CTS data; Westat had previously been approved as a subcontractor for VA-related studies through other contacts with the National Academies, given their authority to operate in VA spaces and with VA data; and Westat had completed similar analyses and issued reports with similar goals to the committee's (Gasper and Katawa, 2015b; Liu et al., 2016; VA, 2015a,b).

The comparisons made in this chapter are of two types. The first comparisons are made between respondents and the eligible population minus the respondents (effectively, the nonrespondents). The second type of comparisons presented are between respondents and the eligible population (which includes the respondents). When results are described, it is clearly noted where one method or the other was used.

Most of the results are presented as frequencies and percentages. Any statistical testing conducted to describe differences between groups used t-tests or chi-square statistics. All significance levels (alpha) were set at 0.05. The methods and the results of the univariate and multivariate analyses used to examine self-reported exposures and health conditions are described in detail in Chapters 5 and 6. All analyses were carried out using SAS version 9.4, a software package commonly used for this type of application and well known for high quality.

No corrections were made for multiple comparisons. Testing multiple associations within a single dataset—for example, testing several exposures against multiple health outcomes—increases the probability that at least one comparison will result in a statistically significant association even if the result is not real. The probability of finding statistical associations increases as the number of comparisons goes up and using analytical methods to attempt to account for this creates problems of its own. Thus, instead of attempting to apply a correction method to its analyses, the committee presents all comparisons and results, and its interpretations are not based on p-values alone, but the overall consistent pattern of results and the plausibility of individual associations.

Assuring Quality Control of Analyses

As the committee's subcontractor, Westat used a variety of systems and procedures to ensure that the committee's analyses were performed accurately and that the results were of the highest possible quality. Such processes included a review of the analysis plan, checking raw data files, and a multistep review of results, the details of which were provided by Westat and are described in the following paragraphs (Westat, 2016).

Before it was implemented, the analysis plan drafted by the committee was first reviewed by senior Westat staff, including a biostatistician, an epidemiologist, and a social scientist with expertise in military health research. Based on clarification questions and other expertise offered by Westat staff, the committee then refined various aspects of the analysis plan. In the months following the receipt of the data, a similar process was followed for all subsequent requests for analysis, ensuring that Westat staff understood the committee's intent and applied the most appropriate statistical methods, especially when there were several possible methods to use, to answer the research questions.

The first step to ensuring the integrity of the data files received from VA—and therefore the accuracy of the committee's data analysis—was to generate frequencies of each item in the data files provided by VA. Westat inspected all administrative and survey data for range errors, logic errors, and other types of errors. All corrections, updates, or additions to the data were recorded and documented in the SAS analysis programs. Full documentation was maintained on all programs used in the process of building the analysis files. This procedure allowed for a rapid rebuilding of the analysis file when additional variables were needed from the raw data received from VA or when variables were recoded.

To ensure that the results of the analysis were of the highest quality, Westat used a multistep process. For each analysis requested by the committee, Westat staff translated the request into detailed specifications for the senior programmer. Specifications included instructions for constructing new variables, dividing the data to specific subgroups (such as deployment by era), and the statistical procedures to be used for running multivariate analyses and performing statistical tests. The specifications were reviewed by an epidemiologist or biostatistician prior to sending them to the programmer. All programs used to run the analysis were fully documented by the programmer.

The committee's analyses required the construction of new variables, including exposure and disease measures. For each constructed variable, cross-tabulations of the raw variables and the constructed variables were made and reviewed to ensure that all of the cases were coded correctly in the constructed variable. As an additional check, a small sample of cases were selected and reviewed a second time to ensure that the constructed variables were properly coded based on the raw data.

Meetings were held among Westat staff at least once a week to review output from the analyses for accuracy and to determine any potential revisions that the analysis required. National Academies staff and committee members were consulted on an as-needed basis. Changes were communicated to the programmer in revised specifications as needed. Before the final written results of an analysis were delivered to National Academies staff and the committee, they were reviewed internally by Westat, typically a senior staff member, to check for accuracy, completeness, and quality and to ensure that the findings were fully supported by the data.

PARTICIPATION RATES

Based on DoD data, as of December 31, 2015, there were approximately 3.5 million individuals who were eligible to participate in the AH&OBP Registry. As of June 30, 2015, 46,498 people had completed the registry questionnaire. Some of those respondents—94—did not verify or add an eligible deployment segment and were removed from the final number of respondents used in the committee's exposure and health outcome analyses. Therefore, the final number of respondents was 46,404.

DoD deployment information (both the CTS and Gulf War Oil Well Fire Smoke Registry files) was used as the gold standard for determining deployment dates and countries and thus determining eligibility and response rates as well. The CTS data included 2,483,392 eligible post-9/11 service members, and the Gulf War Oil Well Fire Smoke Registry file included 545,383 eligible service members. These numbers cannot simply be added together to determine the eligible population denominator because they are not mutually exclusive, as described above and shown in Table 4-1, and the respondents from each era are not necessarily a subset of the populations in the CTS and Gulf War Oil Well Fire Smoke Registry files because a small number of the respondents (those who were deployed during the Stabilization period only) are not in one of those files.

There were 42,665 respondents with an eligible post-9/11 deployment in the registry after excluding 8 respondents who had a post-9/11 deployment in the registry but for whom the country was not eligible according to the questionnaire. It is not clear why these respondents were included. An additional 333 respondents were excluded, 108 because they were not in the CTS and another 225 who did not have an eligible deployment in the CTS. Moreover, 88 respondents who had no eligible post-9/11 deployment in the registry were included because they were determined to be eligible according to the CTS. After these exclusions and inclusions, the final number of post-9/11 questionnaire completers was 42,420, as shown in Figure 4-1.

A total of 5,621 Gulf War veterans completed the AH&OBP Registry questionnaire; of those, 496 were found to be ineligible and were excluded based on data from the Gulf War Oil Well Fire Smoke Registry file. The 801 respondents who were eligible according to the Gulf War Oil Well Fire Smoke Registry file but were ineligible based on questionnaire deployment data were included in the analysis because the DoD data were considered to be the gold standard for determining eligibility. A total of 5,621 Gulf War questionnaire completers were included in the final analysis as shown in Figure 4-2.

The response rates were 1.0% and 1.7% for Gulf War and post-9/11 questionnaire completers, respectively. Because the committee only had access to the year of deployment (not the month or day), one reason for the discrepancies could be the lack of granularity. The reason that the number of eligible respondents is 46,404 and not 48,041—the sum of 42,420 and 5,621—is because the post-9/11 and Gulf War respondents are not mutually exclusive groups; some respondents are included in both groups (see Table 4-1).

During the pilot testing period, from April to June 2014, 194 respondents completed the questionnaire; 181

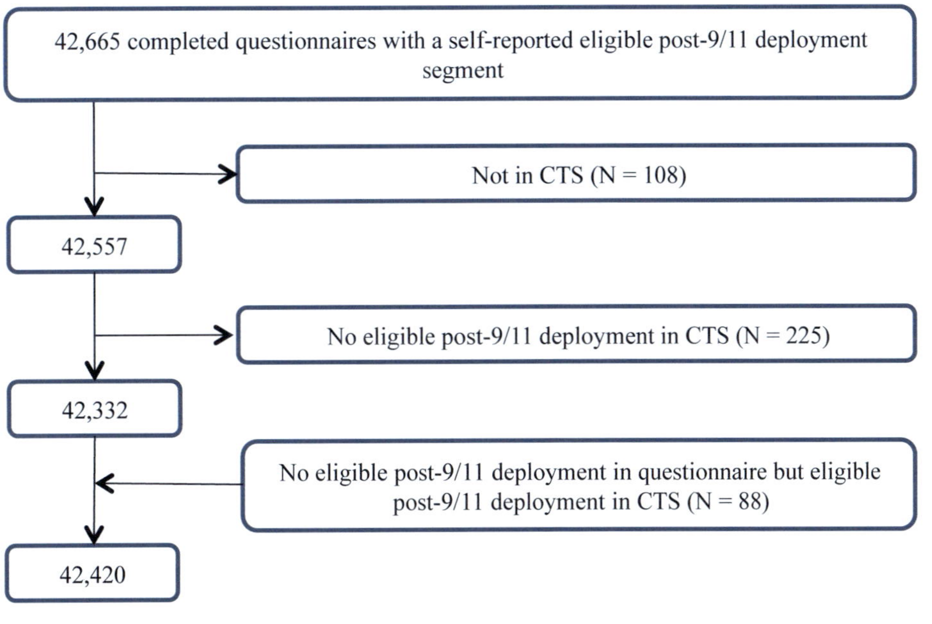

FIGURE 4-1 Eligible post-9/11 questionnaire completers.
NOTE: CTS = Contingency Tracking System.

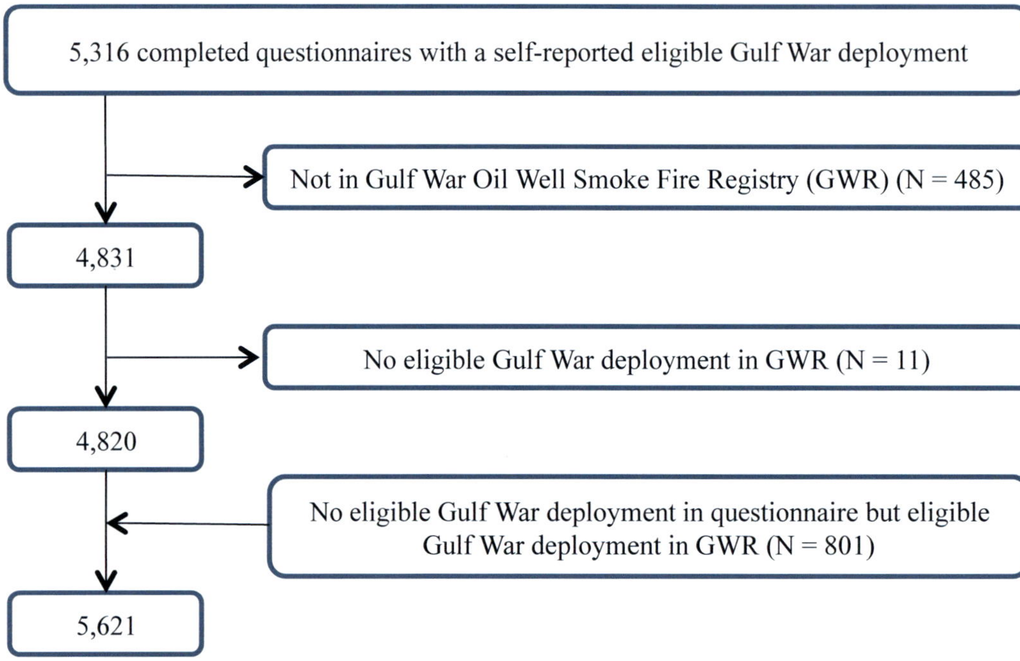

FIGURE 4-2 Eligible Gulf War questionnaire completers.

had eligible deployments and were included in the committee's analyses. These questionnaires are included in the June 2014 numbers. The monthly number of completed questionnaires has remained relatively stable since the registry was opened (see Figure 4-3).

Participation is dependent on having created a user account, completed a Web-based consent form, completed the online self-assessment questionnaire, and had at least one eligible deployment segment, as discussed in Chapter 3. Thus, the committee identified a need to examine potential selection bias created by the process of completing the questionnaire, according to the four stages of questionnaire completion: people who *accessed* the registry and established a user account; people who *consented* to participate; people who *partially completed* the registry's questionnaire; and people who *completed* and submitted the full questionnaire.

Comparisons between the eligible population and those who participated in each stage would give information about the facilitators and barriers to questionnaire completion and shed light on potential selection biases. However, because data on respondents who accessed, consented, and partially completed the questionnaire were not made available to the committee, this exercise could not be conducted.

Nevertheless, some data covering a similar time period (April 25, 2014, to December 31, 2014; N = 28,426) have been reported (VA, 2015b). Table 4-2 shows the number of people who accessed the registry, consented, and completed the full questionnaire (the number of persons who partially completed the questionnaire was not available). Based on that data, 38% of individuals who accessed the registry never completed the questionnaire (N = 17,498). The inclusion of partially completed questionnaires could result in a more representative and larger sample. A dropout rate of nearly 40% raises some significant concerns that merit additional investigation, although the limitations set on the data provided did not allow the committee to carry out such an investigation. In particular, because no information is available on the high numbers of partial completers, it is not possible to determine what effect, if any, successful efforts to retain or recapture these cases in the registry would have on the influence of selective participation bias or what steps might be taken for them to complete the questionnaire. This information would also have provided valuable insights into influences of initial and full participation.

VA internally examined the characteristics of drop-off points for persons who did not complete the question-

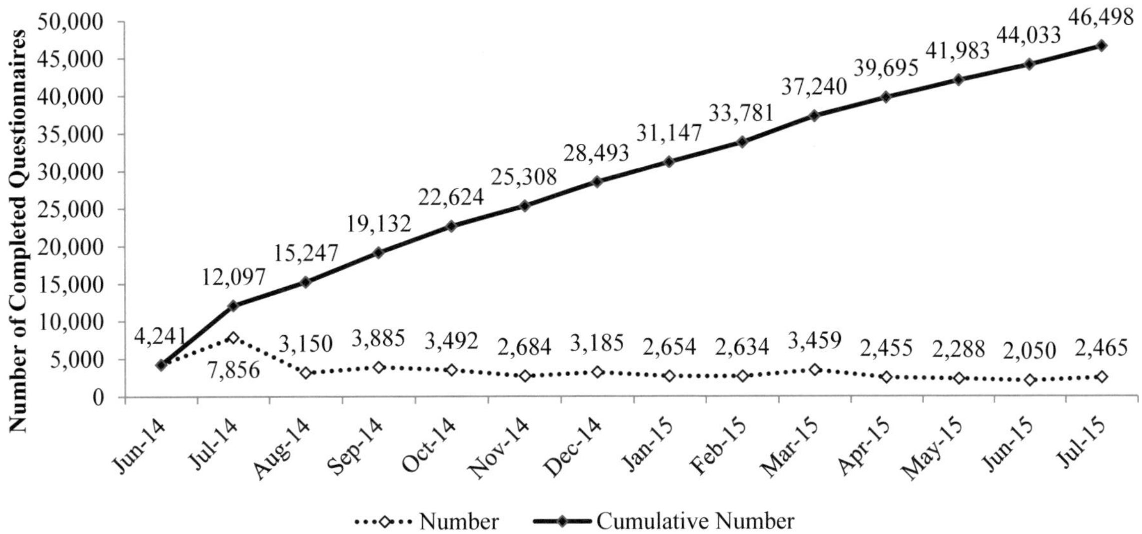

FIGURE 4-3 Number of completed questionnaires by month, June 2014 to July 2015.

TABLE 4-2 Number of Respondents at Each Stage of Participation

Status of Respondents	Total Through December 2014
Total number of user accounts	45,924
Users providing Web consent	43,308
Completed questionnaire	28,426

SOURCE: VA, 2015b.

naire. More than 50% of people who were determined to be eligible and able to access the questionnaire had not started it (defined as no response to Section 1.2, which required respondents to confirm or add deployment segment information). Of those who started the questionnaire (defined as answering at least one question), 38% did not proceed past the first question (Lezama, 2016). Without additional data on the characteristics of those who reached each stage of participation, especially those who started but did not submit a fully completed questionnaire, the committee's ability to carefully examine factors that might have affected participation was limited.

REPRESENTATIVENESS

The degree to which the characteristics and experiences of a group or sample of people reflect those of the larger population is known as representativeness. The more representative a group is of the population, the greater the ability to generalize findings from analyses based on the group to the population. For this reason, it is important to examine how representative the AH&OBP Registry respondents (1.0% and 1.7% for Gulf War and post-9/11 respondents, respectively) are of the eligible deployed veteran population. The representativeness of the respondents is primarily affected by a type of selection bias created by the voluntary nature of participation, which leads to the respondents being self-selected (see Chapter 2). Therefore, to assess the representativeness of registry respondents, the committee compared the demographic and military characteristics of the respondents and the eligible population.

Demographic and Military Characteristics Comparisons

Comparisons of demographic and military service characteristics between respondents and the full eligible population can provide important insights into selective participation among those defined as eligible and into what might have motivated participation, so that the value of the registry for simply describing the exposure and health experience of the total population of those eligible can be evaluated.

The only demographic characteristics collected by the questionnaire were birth date and current age. For the nonrespondents who were included only in the Gulf War Oil Well Fire Smoke Registry file, 25 years (the number of years between 1991—the end of the Gulf War—and the launch of the registry) was added to their listed age in the Gulf War file to estimate the ages they would have been if they had completed the registry questionnaire. For the post-9/11 nonrespondents included in the CTS file, their most current age in the CTS file was used. Other demographic information was obtained by linking data from the CTS and Gulf War Oil Well Fire Smoke Registry to the registry data.

Table 4-3 shows the distribution of select demographic and military characteristics of all respondents included in the committee's dataset. As shown in Table 4-1, respondents consist primarily (86%) of individuals who served only in the post-9/11 era; service members of the Gulf War constitute 8% of respondents; 7% served in multiple periods; and less than 1% served only during the stabilization period. The categories of military characteristics, including pay grade, branch of service, and unit component were created to be mutually exclusive based on the most current information from the last deployment, according to the CTS and Gulf War Oil Well Fire Smoke Registry files. (Therefore, if a person was deployed first as an active-duty service member for one deployment then served in the reserves for the next, the person would be categorized as a reservist because this is the latest information according to the DoD administrative deployment records.) Individuals were counted once in the demographic-characteristic categories because those categories were mutually exclusive. Education level and marital status were based on information from the most recent deployment.

Overall, registry respondents were primarily male (89.1%), white (65.1%), married (58.8%), and enlisted (81.5%); they had primarily served in the Army (65.2%) and on active-duty (64.3%); a majority had completed high school or less (60.3%); and they were an average of 38.7 years of age when they completed the questionnaire. They spent an average of 14.5 months deployed, and 75% reported more than one eligible deployment, with an average of 4.4 deployments.

Comparisons of Differences Between Respondents and Total Eligible Population

Because of the large sample sizes available, the majority of comparisons will show statistically significant differences between groups, even when the differences are small, so the committee examined relative percent differences between the respondents and the eligible population. The relative percent differences are used in order to describe the magnitude of the differences and are calculated for each characteristic by subtracting the percent of the eligible population with that characteristic from the corresponding percent among the respondents, and then dividing by the percent of the eligible population, with the answer expressed as a percent (that is, the percent difference of two percentages). Differences that are positive indicate overrepresentation among registry respondents, while negative differences indicate characteristics that are underrepresented among registry participants. The committee used a cut-off of 10% to identify differences with practical (versus statistical) significance. One sample chi-square tests of independence (and t-tests for average age) were used to determine statistically significant differences in the distributions of registry respondents and the full eligible population for key demographic and military characteristics. Table 4-4 presents these data comparing respondents and all eligible service members (both respondents and nonrespondents combined), stratified by era of service.

Most of the characteristics show relative differences greater than 10%, with many being more than 40%. The greatest relative difference was between post-9/11 respondents and eligible personnel who served in both Iraq and

Afghanistan (80.6%). Differences on some characteristics (e.g., marital status) have little practical significance or meaning, but many others are important to consider as potential confounders of health outcomes; these include sex, age, race, branch of service, cumulative deployment measures, and countries of deployment. People who reported service in multiple eras were included in all categories for which they had deployments (for example, Gulf War and post-9/11). Individuals were counted once in most categories, with two exceptions: country of deployment and era of service, where individuals were counted in multiple categories because all records were considered.

Differences between Gulf War era respondents and the eligible population were not as pronounced as was the case with the post-9/11 group, and the relatively smaller samples sizes for certain categories of respondents make some of the estimates less reliable. Army service was overrepresented among Gulf War respondents compared with the eligible population (76.4% versus 56.3%, a relative difference of 35.7%). Conversely, the proportion of respondents who served in the Navy/Coast Guard was lower than in the eligible population (4.0% versus 19.0%). Likewise, respondents who served in the Air Force and Marine Corps were also underrepresented (relative differences of 25.2% and 19.0%, respectively). The overrepresentation of Army service may be because these veterans were more likely than those serving in other branches to be deployed at land-based locations with or near burn pits.

About half (50.4%) of all eligible Gulf War service personnel had not deployed to either Iraq or Kuwait, compared with only 35.5% of respondents (a relative difference of –29.6%). Greater positive differences, and thus overrepresentation, were observed between the two groups for deployment to Iraq only (relative difference 40.0%) and Iraq and Kuwait (43.4%). The overrepresentation of Iraq/Kuwait locations of deployment is likely because the forward operating bases that used burn pits at that time were located in Iraq and Kuwait. Therefore, veterans and service members deployed to those locations may be more inclined to participate in the registry.

The few differences observed in demographic as opposed to military characteristics included an overrepresentation of women (16.2%) and those with some college or a bachelor's degree (relative difference of 33.3%) and a substantial underrepresentation of those aged 60 and older (relative difference of –56.6%).

While the post-9/11 respondents who chose to participate differed from the eligible population in ways similar to the Gulf War respondents (branch of service and country of deployment), they also differed from the eligible population on nearly every other characteristic examined. Among post-9/11 respondents, service in the Army was overrepresented by 26.9%, while those in the Navy/Coast Guard and Marines were underrepresented (relative differences of –73.0% and –11.0%, respectively). Previous reports of post-9/11 veterans and service members have corroborated this finding and have specifically restricted study populations to Army and Air Force personnel because of the low number of Navy and Marine Corps personnel with deployment locations within a 5-mile radius of the documented burn pits in the sample (AFHSC et al., 2010; Smith et al., 2012).

While the majority of both respondents and the eligible population had deployed to Iraq, Afghanistan, or both countries (86.3% versus 67.1%), registry respondents were substantially overrepresented among those who had deployed to Iraq only (relative difference, 31.8%) and to Iraq and Afghanistan (relative difference, 80.6%). In contrast, respondents were substantially underrepresented among those who did not serve in either Iraq or Afghanistan (a relative difference of –58.0%).

Regarding other military characteristics, post-9/11 respondents who were members of the reserve or National Guard were overrepresented (relative difference, 34.3%), and those on active-duty service were underrepresented (relative difference, –12.8%) in the registry when compared with the eligible population. The distribution of the number of deployments was also markedly different, with respondents reporting more eligible deployment segments than the eligible population. For example, 15.1% of respondents had one eligible deployment compared with 26.0% of the eligible population (relative difference, –41.9%), but 14.0% of respondents had 10 or more eligible deployment segments compared with 8.8% of eligible service members (relative difference 59.1%).

In addition to the differences in military characteristics, post-9/11 respondents and eligible persons from that era differed in all of the demographic characteristics available for assessment. For example, women (–14.3%), those less than 30 years old (–41.0%), and minority race/ethnicities (with several relative differences of –20% or more) were substantially underrepresented among respondents, while those who were married and those with some college or with bachelor's degrees were overrepresented among registry participants (relative differences of 20.9% and 33.3%) relative to all eligibles from this era.

Other comparisons of post-9/11 respondents with all eligible persons conducted for VA (Gasper and Katawa,

TABLE 4-3 Distribution of Select Demographic and Military Characteristics for All Registry Respondents

	N	%
Sex		
Male	41,345	89.1
Female	4,678	10.1
Missing	381	0.9
Age at questionnaire completion[a]	46,404	38.7
<30 years old	7,906	17.0
30–39 years old	17,983	38.8
40–49 years old	14,367	31.0
50–59 years old	5,346	11.5
≥60 years old	802	1.7
Race		
White	30,214	65.1
Black	3,773	8.1
Hispanic	3,372	7.3
Other	1,480	3.2
Missing	7,565	16.3
Marital status		
Married	27,279	58.8
Not married	15,136	32.6
Missing	3,989	8.6
Rank		
Enlisted	37,520	81.5
Officer/Warrant Officer	8,507	18.4
Branch		
Army	30,253	65.2
Air Force	8,514	18.3
Marine Corps	5,087	11.0
Navy/Coast Guard	2,534	5.5
Missing	16	0.0
Component		
Active duty	29,860	64.3
Reserve/National Guard	16,167	34.8
Missing	377	0.8
Education		
High school or less	27,973	60.3
Some college/bachelor's degree	14,356	30.9
Postgraduate	3,105	6.7
Missing	970	2.1
Occupation		
Infantry, gun crews, and seamanship	11,974	25.8
Electronic equipment repair	3,144	6.8
Communications and intelligence	4,013	8.6
Health care	3,074	6.6
Other technical and allied specialists	1,374	3.0
Functional support and administration	6,520	14.1

continued

TABLE 4-3 Continued

	N	%
Electrical/mechanical equipment repair	6,368	13.7
Craft work	1,641	3.5
Service and supply	6,212	13.4
Nonoccupational or missing	2,084	4.5
Months of deployment (from Survey)[a]	46,404	14.5
3 or less	2,082	4.5
4–6	6,744	14.5
7–12	16,298	35.1
13–15	4,989	10.8
16–18	3,487	7.5
19–21	3,422	7.4
22–24	2,847	6.1
More than 24	6,535	14.1
Eligible segments (from Survey)	46,404	4.4
1	11,613	25.0
2	7,519	16.2
3	5,685	12.3
4	4,279	9.2
5	3,727	8.0
6	3,037	6.5
7	2,329	5.0
8	1,765	3.8
9	1,395	3.0
10	1,119	2.4
More than 10	3,936	8.5

[a] Age at questionnaire completion (in years) and months of deployment are presented as mean and standard deviation.

2015b) that were based on a different sample of registry data (with fewer participants) were largely consistent with these results. However, because that report included data pertaining to VHA health care utilization (not available to the committee), some important additional differences were noted. In particular, respondents were substantially more likely to have used VA services than all eligibles from this era (69.3% versus 46.0%, a relative difference of 50.6%), and among VHA enrollees, the registry underrepresented veterans in lower VA priority groups for treatment by 30%.

Gulf War Respondents Versus Post-9/11 Respondents

The demographic makeup of the registry population differs in significant ways between the Gulf War and the post-9/11 respondent populations for most demographic and military characteristics. These differences are most apparent in age, level of education, rank, branch of service, component, and number of deployments. Post-9/11 respondents were younger on average (37.8 versus 48.6 years) and more educated (33.8% versus 4.3% had a bachelor's degree or higher) than Gulf War respondents. As for military characteristics, more post-9/11 respondents were officers (19.3% versus 10.0%), served in the Air Force (19.1% versus 7.7%), and were part of the reserves or National Guard (36.4% versus 17.1%) than Gulf War respondents. Although the majority of both post-9/11 and Gulf War respondents served in the Army, the proportion was smaller for post-9/11 (64.7% versus 76.4%).

TABLE 4-4 Relative Differences of Select Demographic and Military Characteristics of Respondents and Nonrespondents, Stratified by Era of Service

	Gulf War						Post-9/11					
	Respondents (N = 5,621)		All Eligibles (N = 545,383)				Respondents (N = 42,420)		All Eligibles (N = 2,483,392)			
	N	%	N	%	P-value	Relative Difference	N	%	N	%	P-value	Relative Difference
Sex												
Male	5,173	92.0	507,167	93.0	0.0015	−1.1	38,108	89.8	2,188,896	88.1	<0.0001	1.9
Female	443	7.9	37,175	6.8		16.2	4,311	10.2	294,427	11.9		−14.3
Missing	5	0.1	1,041	0.2		−50.0	1	0.0	69	0.0		
Age (years)												
<30					<0.0001		7,575	17.9	743,305	29.9	<0.0001	−40.1
30–39							17,895	42.2	1,049,801	42.3		−0.2
40–49	2,988	53.2	265,696	48.7		9.2	12,361	29.1	469,207	18.9		54.0
50–59	1,461	26.0	148,541	27.2		−4.4	4,121	9.7	188,736	7.6		27.6
≥60	201	3.6	45,140	8.3		−56.6	465	1.1	32,208	1.3		−15.4
Missing	971	17.3	86,006	15.8		9.5	3	0.0	135	0.0		
Mean age (years)[a]	48.6		49.9		<0.0001		37.8		35.2		<0.0001	
Race												
White	3,852	68.5	362,828	66.5	0.0091	3.0	29,219	68.9	1,549,807	62.4	<0.0001	10.4
Black	1,280	22.8	130,311	23.9		−4.6	3,254	7.7	344,065	13.9		−44.6
Hispanic	271	4.8	27,774	5.1		−5.9	3,361	7.9	252,262	10.2		−22.5
Other	208	3.7	22,755	4.2		−11.9	1,406	3.3	128,926	5.2		−36.5
Missing	10	0.2	1,715	0.3		−33.3	5,180	12.2	208,332	8.4		45.2
Marital status												
Married							27,279	64.3	1,320,818	53.2	<0.0001	20.9
Not married							15,136	35.7	1,161,949	46.8		−23.7
Missing							5	0.0	625	0.0		
Rank												
Enlisted	5,058	90.0	490,666	90.0	0.9834	0.0	34,241	80.7	2,137,892	86.1	<0.0001	−6.3
Officer/Warrant Officer	562	10.0	54,569	10.0		0.0	8,179	19.3	345,489	13.9		38.8
Missing			148	0.0					11	0.0		
Branch												
Air Force	435	7.7	56,089	10.3	<0.0001	−25.2	8,117	19.1	446,536	18.0	<0.0001	6.1
Army	4,292	76.4	306,834	56.3		35.7	27,447	64.7	1,265,848	51.0		26.9
Marine Corps	669	11.9	80,219	14.7		−19.0	4,778	11.3	316,317	12.7		−11.0
Navy/Coast Guard	225	4.0	102,241	19.0		−79.0	2,078	5.0	454,691	18.0		−73

Unit Component												
Active Duty	4,657	82.9	460,153	84.4	0.0017	−1.8	26,965	63.6	1,810,150	72.9	<0.0001	−12.8
Reserve/National Guard	964	17.1	85,229	15.6		9.6	15,455	36.4	673,242	27.1		34.3
Missing			1	0.0								
Education												
High school or less	5,320	94.6	520,324	95.4	<0.0001	−0.8	27,446	64.7	1,844,601	74.3	<0.0001	−12.9
Some college/bachelor's degree	224	4.0	16,272	3.0		33.3	12,224	28.8	505,986	20.4		41.2
Postgraduate	15	0.3	2,299	0.4		−25.0	2,136	5.0	92,578	3.7		35.1
Missing	62	1.1	6,488	1.2		−8.3	614	1.4	40,227	1.6		−12.5
Country of Deployment												
Iraq only	1,573	28.0	108,978	20.0	<0.0001	40.0	18,627	43.9	826,062	33.3	<0.0001	31.8
Kuwait only	716	12.7	70,658	13.0		−2.3						
Iraq and Kuwait	1,339	23.8	90,734	16.6		43.4						
Neither Iraq or Kuwait	1,993	35.5	275,013	50.4		−29.6						
Afghanistan only							9,692	22.8	578,646	23.3		−2.1
Iraq and Afghanistan							8,292	19.5	269,378	10.8		80.6
Neither Iraq or Afghanistan							5,809	13.7	809,306	32.6		−58.0
Duty Occupation												
Infantry, gun crews, and seamanship	1,674	29.8	150,788	27.6	<0.0001	8.0	10,978	25.9	635,871	25.6	<0.0001	1.2
Electronic equipment repair	289	5.1	36,942	6.8		−25.0	2,935	6.9	191,457	7.7		−10.4
Communications and intelligence	521	9.3	48,351	8.9		4.5	3,647	8.6	211,558	8.5		1.2
Health care	327	5.8	27,998	5.1		13.7	2,842	6.7	141,102	5.7		17.5
Other technical and allied specialists	134	2.4	11,191	2.1		14.3	1,285	3.0	59,700	2.4		25.0
Functional support and administration	669	11.9	63,188	11.6		2.6	6,059	14.3	323,987	13.0		10.0
Electrical/mechanical equipment repair	905	16.1	101,695	18.6		−13.4	5,826	13.7	419,648	16.9		−18.9
Craft work	162	2.9	18,572	3.4		−14.7	1,552	3.7	105,227	4.2		−11.9
Service and supply	759	13.5	71,123	13.0		3.8	5,710	13.5	312,888	12.6		7.1
Nonoccupational	117	2.1	9,460	1.7		23.5	902	2.1	44,625	1.8		16.7
Missing	64	1.1	6,075	1.1		0.0	684	1.6	37,329	1.5		6.7

continued

TABLE 4-4 Continued

	Gulf War						Post-9/11					
	Respondents (N = 5,621)		All Eligibles (N = 545,383)				Respondents (N = 42,420)		All Eligibles (N = 2,483,392)			
	N	%	N	%	P-value	Relative Difference	N	%	N	%	P-value	Relative Difference
Eligible deployment segments												
1	1,747	31.1	209,544	38.4	<0.0001	−19.0	6,403	15.1	646,351	26.0	<0.0001	−41.9
2	2,148	38.2	192,375	35.3		8.2	6,393	15.1	472,130	19.0		−20.5
3	1,395	24.8	110,727	20.3		22.2	4,830	11.4	284,519	11.5		−0.9
4	285	5.1	25,393	4.7		8.5	3,782	8.9	195,025	7.9		12.7
5	46	0.8	7,344	1.3		−38.5	3,752	8.8	182,661	7.4		18.9
6							3,288	7.8	157,664	6.3		23.8
7							2,821	6.7	122,092	4.9		36.7
8							2,117	5.0	87,434	3.5		42.9
9							1,687	4.0	65,763	2.6		53.8
10							1,419	3.3	51,697	2.1		57.1
More than 10							5,928	14.0	218,056	8.8		59.1

[a] Age at questionnaire completion (in years) is presented as mean and standard deviation.

Likewise, a smaller proportion of post-9/11 respondents were active duty compared with Gulf War respondents (63.6% versus 82.9%).

The distribution of the number of eligible deployment segments also varies greatly between respondents of the two eras. Almost all Gulf War respondents were deployed one, two, or three times (31.1%, 38.2%, and 24.8%, respectively), whereas only 41.6% of post-9/11 respondents reported three or fewer deployments. The maximum number of eligible deployment segments among Gulf War respondents was five, and less than 1% of respondents reported that many. However, 48.6% of post-9/11 respondents (and 35.4% of the eligible population) had five or more eligible deployment segments, and 14% of respondents were deployed 10 times or more as of December 31, 2015.

Because of the short nature of the Gulf War—6 months of lead-up operations and 100 hours of ground combat (Torreon, 2015)—few deployments would have been possible. On the other hand, there has been a well documented trend of multiple deployments among post-9/11 service members (Baiocchi, 2013; IOM, 2013).

In summary, there are several important differences in demographic and military characteristics between respondents and the eligible populations from which they are derived. Among Gulf War respondents, there were few demographic differences, but respondents were more likely to have served in the Army and have been deployed to Iraq or Iraq and Kuwait than the eligible population. Post-9/11 respondents were less racially diverse and older that the eligible comparison population and were overrepresented by service in the Army, service in the reserves/National Guard, deployment to Iraq and Iraq and Afghanistan, and having multiple deployment segments. Comparing the respondents by era shows that the Gulf War era respondents were older, more racially diverse, had less education, were more likely to be enlisted, had a greater proportion that served in the Army, were more likely to have been active duty, and deployed fewer times than post-9/11 respondents.

Overall, the registry (based on the data through July 2015) contains a low proportion of eligible respondents (less than 2% of the full population). However, the results of the comparisons are not unexpected. For example, given the mission and function of the Army, it is likely that soldiers of the post-9/11 era were most likely to have been deployed to locations with burn pits, and therefore, it is not surprising that they have a higher participation rate than respondents who served in other branches. As noted, even with an increased representation, only 2.2% of eligible post-9/11 Army members participated in the AH&OBP Registry. When conducting analyses of registry data, it is important to recognize these differences by era and to present results separately by era of service.

Effect of Nonresponse

To further study the representativeness of the registry data, the committee examined a separate analysis generated for VA that was limited to post-9/11 respondents and eligibles and linked registry data with VA health care records which included information of service use and priority group assignment (Gasper and Katawa, 2015b). The registry respondents were compared with their eligible population to determine the effect of nonresponse from eligible nonrespondents on estimates of disease prevalence. Seven self-reported provider-diagnosed respiratory and cardiovascular conditions and a measure of functional limitations due to a lung or breathing problem were used to determine whether a bias exists.

Nonresponse-adjusted weights were developed and applied based on calculated propensity scores that used all available demographic, military, and health care utilization characteristics that were found to be correlated with nonresponse. Propensity stratification uses a logistic regression model to relate covariates to an individual's probability of response (that is, participation in the registry). The propensity score is the conditional probability of response given the covariates. The propensity scores were used to stratify the participants into propensity classes (cells), which served as nonresponse adjustment cells. The nonresponse weight was calculated as the inverse of the response rate (fraction of respondents in a cell). The eligible population was then divided into cells of the propensity score, and nonresponse weights were calculated for each cell. The weights were then applied to estimates of disease and exposure and compared with unadjusted estimates. It should be noted that this approach does not eliminate bias due to unobserved outcomes or variables that are available in the registry data but not available in VA or DoD administrative datasets (Gasper and Katawa, 2015b).

TABLE 4-5 Comparison of Unadjusted and Nonresponse Adjusted Estimates of Self-Reported Health Conditions from the AH&OBP Registry

Health Condition	Unadjusted (%)	Adjusted (%)	Relative Difference (%)
Allergies	38.5	38.1	1
Asthma	14.8	13.9	6.1
Emphysema, chronic bronchitis, or COPD	14.7	12.9	12.2
Constrictive bronchiolitis	1.2	0.9	25
Idiopathic pulmonary fibrosis	0.2	0.2	0
Functional limitation/breathing problem	27.5	25.0	9.1
High blood pressure	34.7	32.1	7.5
Coronary artery disease	2.9	2.3	20.7

NOTE: AH&OBP = Airborne Hazards and Open Burn Pit; COPD = chronic obstructive pulmonary disease.
SOURCE: Gasper and Kawata, 2015b.

All adjusted estimates of health conditions were consistently lower than (or, in the case of idiopathic pulmonary fibrosis, the same as) than unadjusted estimates (see Table 4-5), with the greatest absolute difference being 2.6% for hypertension and relative differences of 10% or more for emphysema, chronic bronchitis, or chronic obstructive pulmonary disease (COPD); constrictive bronchiolitis; and coronary artery disease. These results suggested that the registry may modestly overestimate disease prevalence. The largest differences between respondents and the eligible population were for VA health care utilization and VA priority group (which is based on a number of factors, including disability). In fiscal year 2013, respondents were more likely to use VA health care (69% versus 46%, a relative difference of 50.6%) and to report a disease than members of the eligible population. An examination of VA health care user status and self-reported health outcomes among registry respondents found higher prevalence rates of most health conditions among VA users compared with nonusers. For example, 37.7% of registry respondents who use VA health care reported hypertension, compared with 28.0% of respondents who do not use VA health care. This suggests that if more non-VA users participated in the registry, the prevalence of many health conditions would be lower (Gasper and Katawa, 2015b).

Committee's Propensity Analysis

The committee conducted a similar analysis of nonresponse bias by comparing demographic and military characteristics of post-9/11 era respondents with those of the eligible population. Information on the demographic and military service characteristics of respondents and eligible non-respondents was obtained by linking registry data with the CTS deployment file.

The committee did not have access to VA medical records, and therefore the models on disease estimates accounting for nonresponse bias were based only on the demographic and military characteristics that were available to the committee and correlated with nonresponse. Nonresponse-adjusted weights were developed using propensity stratification and a method similar to the one used for the VA analyses described above. Whereas 5 cells were used for the VA analysis, 10 cells were used for the committee's analysis because this resulted in the least amount of bias for the post-9/11 respondents.

Figure 4-4 shows that there were few differences in the estimates of disease before and after the nonresponse adjustment, and the direction of the differences was not consistent among diseases. The conclusions that can be drawn from this analysis are more limited because health care utilization characteristics were not made available to the committee, which prevented evaluations of the validity of self-reported disease. Moreover, it is likely that respondents are self-selected in ways that are unable to be observed or determined with certainty, such as experiences in theater, exposures, motivation to participate, and health concerns or outcomes. Without including the VA health care utilization factors in the adjustment, most of the differences between adjusted and unadjusted estimates are small.

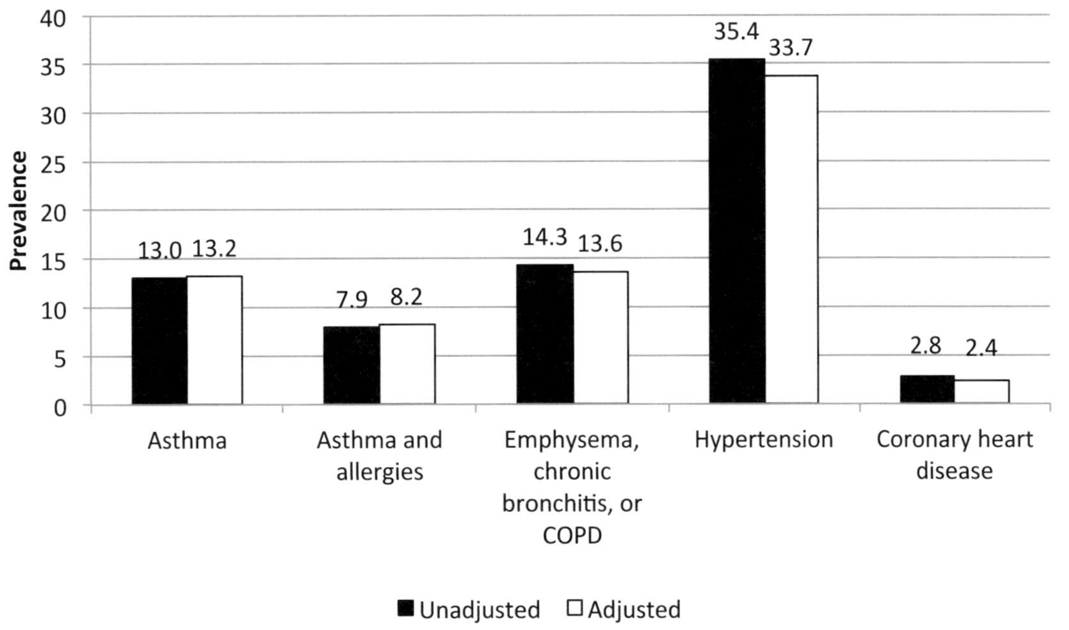

FIGURE 4-4 Comparison of unadjusted and nonresponse-adjusted estimates of selected disease categories for post-9/11 respondents.
NOTE: COPD = chronic obstructive pulmonary disease.

Based on both VA's and the committee's nonresponse bias analyses, the committee sought to determine what proportion of the variance in propensity is accounted for by the variables to which the committee had access. One method to distinguish between registry respondents and eligible non-respondents is to examine the distribution of propensity scores for the two groups from the logistic regression model predicting response status. Both respondents and nonrespondents had low mean propensity scores (0.024 and 0.017, respectively). A second method is to examine the cumulative percentage of non-respondents whose propensity score is below the 20th percentile of propensity scores for respondents. Most nonrespondents should have propensity scores that are lower than most respondents. If 80% of nonrespondents have propensity scores lower than the 20th percentile of respondent propensity scores, it suggests that the covariates included in the logistic regression strongly predict response status; however, in this analysis only 45% of nonrespondents had propensity scores that were less than the 20th percentile for respondents. The considerable overlap in the distribution of propensity scores for respondents and nonrespondents suggests that the demographic and military characteristics available to the committee do a modest to weak job of distinguishing between the two groups. Adding the VA health care utilization data would likely improve the proportion of variability that could be accounted for between respondents and eligible nonrespondents, but it still would not adjust for the greater motivation of people with presumed exposure and health effects to participate.

OVERVIEW OF QUESTIONNAIRE RESULTS

The committee examined the variability in responses to specific items on the questionnaire and the degree to which questions were not answered to further assess the degree to which biases are apparent in the registry data. To examine variability in responses, the committee looked at questions for which a large proportion of respondents endorsed an exposure or health condition. Because this registry is voluntary and some degree of selection bias is expected, the lack of variability in responses for exposure or health-related conditions may confirm or refute potential biases. An examination of items with high rates of nonresponse (missing answers, refusals, or "don't know") could indicate reporting or recall biases or evidence of poorly worded questions (as noted in Chapter 3). Reporting and recall biases would be evident as respondents are more likely to report or recall certain exposures or health conditions of concern than others. A lack of variability of responses also affects the ability to observe associations and draw conclusions.

Questions with Limited Variability in Responses

The committee examined variability in responses for six exposures of interest (burn pits, dust, construction duties, fuel exhaust, combat, and soot from oil-well fires) and all of the health outcomes, with a focus on respiratory and cardiovascular diseases. For questions that were part of a skip pattern, the variability in responses was examined only among those respondents who were eligible to answer the question. For questions about exposure that are asked on each deployment, the question was considered to have been "endorsed" if the respondent indicated exposure on any deployment. Responses that were not endorsed were either missing or considered as no exposure. The percentage of respondents who endorsed a question was calculated using all respondents eligible to answer the question. Items that the committee did not consider useful for assessing exposure or disease relationships were excluded, such as Question 8.1.B, which asks whether respondents use the Internet and was endorsed by more than 95% of respondents.

Table 4-6 shows the questions with the least amount of variability at the individual level, all of which were endorsed by at least 85% of respondents. Three exposure questions—near a burn pit, exposed to soot, and exposed to dust storms—were endorsed by the vast majority of respondents. Ever having been near a burn pit was endorsed on at least one deployment by 96% of respondents. Among the respondents who served in the 1990–1991 Gulf War, 89% reported that they were exposed to soot from oil-well fires. Exposure to dust storms during deployment, defined as at least one day per month, was reported by 85% of respondents.

Several other questions related to deployment experience or health outcomes had little variability. Question 3.1.H asked whether the respondent was concerned that in the future his or her health will be affected by something breathed during deployment, and it was endorsed by 92% of respondents. Question 1.4.D asked respondents whether, during deployment, they experienced wheezing, difficulty breathing, or itchy or irritated nose, eyes, or throat that they thought was the result of poor air quality; 88% of respondents answered yes to this question.

The lack of variability noted for these key exposure questions is important in that it limits the analysis that is possible using these items (see Chapters 5 and 6). Given that the registry was marketed with a focus on burn

TABLE 4-6 Questions with Limited Variability in Response

Question	Text	Number Eligible	Percent Endorsed
1.2.A	Were you exposed to soot, ash, smoke, or fumes from Gulf War oil well fires?	5,595[c]	89.3[a]
1.2.D	Were you near a burn pit during these dates (or on the base or close enough to the base for you to see the smoke)?	46,404	95.8[a]
1.2.H	On a typical day, how many hours were you outside or in an open tent or shelter?	46,404	96.3[b]
1.4.C	In a typical month, how many days were you exposed to dust storms?	46,404	85.3[b]
1.4.D	During your deployment(s), did you experience wheezing, difficulty breathing, an itchy or irritated nose, eyes, or throat that you thought was the result of poor air quality?	46,404	88.3[a]
3.1.H	Are you concerned that in the future that your health will be affected by something you breathed during deployment?	46,404	92.0[a]

[a] A response of "yes" is considered endorsed.
[b] A response of greater than 0 is considered endorsed.
[c] Question only asked to Gulf War service members.

pit exposures, it is not surprising that virtually all respondents endorsed this exposure. In effect, the high levels of endorsement for the items in Table 4-6 imply a very skewed sample. Many of the exposures that the questions sought to elucidate are relatively common so that nearly anyone who was deployed endorsed them. Moreover, there are no follow-up questions to allow for more granular detail. Questions that do not allow for variability are of little use for analyses examining associations.

Questions with High Rates of Nonresponse

For individual items on the questionnaire, nonresponse analyses were conducted, with missing, don't know, and refused responses grouped as a separate category. Respondents were required to answer every question—although "don't know" and "refused" options were provided for most questions. VA indicated that there was an "anomaly" in an earlier version of the questionnaire which allowed respondents to skip questions. This resulted in a third type of "missing" data. Item nonresponse rates were calculated for each item in the questionnaire. The item nonresponse rate was calculated as the number of nonresponses divided by the total number of respondents eligible to answer the question. Table 4-7 shows the percentage of nonresponses (don't know, refused, and missing responses) for each questionnaire item except for "check all that apply" questions (see Chapter 3 for a description of the types of questions contained in the questionnaire). Questions that were asked of each eligible deployment segment versus for the individual are indicated under the eligible column; the majority of questions were asked once at the individual level. Eligible refers to the number of deployment segments or individuals, after accounting for skip patterns, that were eligible to answer the question. For example, the question about the number of hours that smoke/fumes from the burn pit entered the work site/housing is restricted to deployment segments for which a participant reported having been near a burn pit.

Several items had nonresponse rates greater than 15%, and some had nonresponse rates of greater than 30%. "Don't know" responses contributed to the vast majority of the nonresponse rate. For example, Question 5.5.A, which asks respondents whether they have ever worked in a job with asbestos exposure, including military service, had a nonresponse rate (mostly attributed to don't know responses) of 35%. Such high levels of don't know and missing responses indicate that the data derived from these questions may not be valid or, therefore, may not be useful for analysis (as discussed in Chapter 3). The high rates of nonresponse, particularly "don't know" may be the result of poorly written questions that are difficult to answer, whereas questions with higher levels of refused or missing answers may be the result of survey fatigue. Notably, unlike the few questions with little variability in responses, item nonresponse was greatest for questions that were noted as being more nebulous, compound, and outside of the focus of the registry. As noted in Chapter 3, much could have been done to improve the questions and likely increase item response rates.

Table 4-8 presents the percentages of yes, no, and nonresponses for each exposure and health condition question. Among the exposure questions with high nonresponse rates was Question 1.2.D (whether a respondent was a near a burn pit on each deployment) which had a nonresponse rate of nearly 20%. In other words, for 20% of deployments, the response was don't know, refused, or missing for this question. Question 1.2.G (the number of hours per day during the deployment that smoke/fumes from a burn pit entered the work site or housing area) had a nonresponse rate of 18.8%. Question 1.2.I (the number of hours per day near sewage ponds) had a nonresponse rate of 39.1%. Question 1.3.B (the number of days exposed to smoke from heavy weapons) had a nonresponse rate of 22.5%.

A previous analysis of registry data found that nonresponse rates for questions asked by deployment segment were significantly higher on deployments that were asked about later in the questionnaire than those that were asked about earlier, a pattern that suggests that nonresponse may be due to respondent fatigue (Gasper and Katawa, 2015a). A further analysis found the nonresponse rate to Question 1.2.D (ever having been near a burn pit on deployment) was higher for deployments to bases where burn pits did not exist than on those where they did, suggesting that some of the "don't know" responses may actually be "not exposed" (Gasper and Katawa, 2015b). Because the committee was not given deployment dates or base information, these analyses could not be replicated.

In addition to the exposure questions, several questions on health conditions and concerns had high nonresponse rates. Most of those items were questions that the committee found to be poorly worded or otherwise prob-

TABLE 4-7 Number of Eligible Responses and Percentage of Nonresponses by Question

Question	Text	Eligible	Nonresponse N	Nonresponse %
Q1.2.A	Were you exposed to soot, ash, smoke, or fumes from the Gulf War oil well fires during any GWA deployment?	6,694[a]	585	8.7
Q1.2.B	Where did you spend most of your time during these dates?	206,373[b]	29,557	14.3
Q1.2.C	Where did you spend the second most amount of time?	206,373[b]	35,180	17.0
Q1.2.D	Were you near a burn pit during any deployment?	206,373[b]	40,648	19.7
Q1.2.F	Did you have any burn pit duties on any deployment?	129,192[b]	2,582	2.0
Q1.2.G	Did smoke or fumes from the burn pit enter your work site or housing on any deployment?	129,192[b]	24,321	18.8
Q1.2.H	Were you outside or in an open tent or shelter during any deployment?	206,373[b]	22,158	10.7
Q1.2.I	Did you smell or see sewage ponds on any deployment?	206,373[b]	80,743	39.1
Q1.3.A	Were you ever close enough to feel the blast from an IED (improvised explosive device) or other explosive device?	46,404	1,361	2.9
Q1.3.B	In a typical month, how many days were you near heavy smoke from weapons, signal smoke, markers or other combat items?	46,404	10,420	22.5
Q1.3.C	In a typical month, how many days were you in convoy or other vehicle operations?	46,404	2,677	5.8
Q1.3.D	In a typical month, how many days did you perform refueling operations?	46,404	3,575	7.7
Q1.3.E	In a typical month, how many days did you perform aircraft, generator, or other large engine maintenance?	46,404	2,786	6.0
Q1.3.F	In a typical month, how many days did you perform construction duties?	46,404	4,767	10.3
Q1.3.G	In a typical month, how many days did you perform pesticide duties for your unit?	46,404	5,008	10.8
Q1.4.A	Did you do anything differently during your deployment(s), when you thought or were informed air quality was bad (for example during dust storms or heavy pollution days)?	46,404	1,132	2.4
Q1.4.C	In a typical month during your deployment(s), how many days did you experience dust storms?	46,404	6,190	13.3
Q1.4.D	During your deployment(s), did you experience wheezing, difficulty breathing, an itchy or irritated nose, eyes or throat that you thought was the result of poor air quality?	46,404	3,701	8.0
Q1.4.E	How many days in an average month did you experience wheezing, difficulty breathing, an itchy or irritated eyes, nose, or throat that you thought was the result of poor air quality?	40,990	6,123	14.9
Q1.4.F	During your deployment(s), did you seek medical care for wheezing, difficulty breathing, an itchy or irritated nose, eyes, or throat that you thought was the result of poor air quality?	46,404	4,185	9.0
Q2.1.A	How difficult is it to run or jog 1 mile on a level surface?	46,404	693	1.5
Q2.1.B	How difficult is it to walk on a level surface for one mile?	46,404	566	1.2
Q2.1.C	How difficult is it to walk a 1/4 of a mile—about 3 city blocks?	46,404	565	1.2
Q2.1.D	How difficult is it to walk up a hill or incline?	46,404	495	1.1
Q2.1.E	How difficult is it to walk up 10 steps or climb a flight of stairs?	46,404	464	1.0
Q2.2.1.A	Have you been told by a doctor or other health professional that you had Hay fever or allergies to pollen, dust, or animals?	46,404	2,696	5.8
Q2.2.1.B	Have you ever been told by a doctor or other health care professional that you had asthma?	46,404	2,106	4.5

TABLE 4-7 Continued

Question	Text	Eligible	Nonresponse N	Nonresponse %
Q2.2.1.C	Have you ever been told by a doctor or other health care professional that you had emphysema?	46,404	1,595	3.4
Q2.2.1.D	Have you ever been told by a doctor or other health care professional that you had chronic bronchitis?	46,404	2,877	6.2
Q2.2.1.E	Have you ever been told by a doctor or other health care professional that you had chronic obstructive pulmonary disease also called COPD?	46,404	2,268	4.9
Q2.2.1.F	Have you ever been told by a doctor or other health care professional that you had some lung disease or condition other than asthma, emphysema, chronic bronchitis or COPD?	46,404	2,730	5.9
Q2.2.1.G	Have you ever been told by a doctor or other health care professional that you had constrictive bronchiolitis (CB)?	4,333	848	19.6
Q2.2.1.H	Have you ever been told by a doctor or other health care professional that you had idiopathic pulmonary fibrosis (IPF)?	4,333	650	15.0
Q2.2.1.J	Did this lung disease get better, worse, or about the same during deployment?	1,266	133	10.5
Q2.2.1.M	How would you rate your shortness of breath or breathlessness? (check the description/grade that applies to you.)	26,294	528	2.0
Q2.2.2.A	Have you ever been told by a doctor or other health care professional that you had hypertension, also called high blood pressure?	46,404	1,766	3.8
Q2.2.2.B	Have you ever been told by a doctor or other health care professional that you had coronary artery disease?	46,404	1,602	3.5
Q2.2.2.C	Have you ever been told by a doctor or other health care professional that you had angina pectoris?	46,404	2,932	6.3
Q2.2.2.D	Have you ever been told by a doctor or other health care professional that you had a heart attack, also called myocardial infarction?	46,404	996	2.1
Q2.2.2.E	Have you ever been told by a doctor or other health care professional that you had a heart condition other than coronary artery disease or angina or myocardial infarction?	46,404	1,783	3.8
Q2.2.3.A	During the past 12 months, have you regularly had insomnia or trouble sleeping?	46,404	993	2.1
Q2.2.3.B	During the past 12 months, have you had Neurological problems? (Some examples of neurological problems may include numbness, tingling, or weakness in your arms or legs or difficulties with thinking or memory.)	46,404	2,058	4.4
Q2.2.3.C	During the past 12 months, have you had problems of the immune system?	46,404	9,166	19.8
Q2.2.3.D	During the past 12 months, have you been told by a doctor or other health professional that you had any kind of liver condition?	46,404	1,656	3.6
Q2.2.3.E	During the past 12 months, have you been told by a doctor or other health professional that you had any chronic multi-symptom illness (examples include irritable bowel syndrome, chronic fatigue syndrome, and fibromyalgia)?	46,404	3,266	7.0
Q2.2.3.G	On average, how many hours of sleep do you get in a 24-hour period (round up 30 minutes or more to the next whole hour)?	46,404	1,505	3.2
Q2.2.3.H	Questions H and I are about snoring and breathing during sleep. To answer these questions, please consider both what others have told you and what you know about yourself. How often do you snore?	46,404	4,406	9.5
Q2.2.3.I	How often do you have times when you stop breathing during your sleep?	46,404	18,279	39.4
Q2.3.A	How tall are you without shoes?	46,404	216	0.5
Q2.3.B	How much do you weigh without shoes?	46,404	931	2.0
Q2.4.A	Have you ever been told by a doctor or other health professional that you had cancer or a malignancy (tumor) of any kind?	46,404	907	2.0

continued

TABLE 4-7 Continued

Question	Text	Eligible	Nonresponse N	Nonresponse %
Q2.4.B	What kind of cancer was it?	45,500	69	0.2
Q2.4.C	How old were you when this cancer was first diagnosed?	2,382	83	3.5
Q2.4.D	If you were diagnosed with a second cancer, what kind of cancer was it?	45,401	202	0.4
Q2.4.E	How old were you when this cancer was first diagnosed?	372	23	6.2
Q2.4.F	If you were diagnosed with a third cancer, what kind of cancer was it?	45,381	228	0.5
Q2.4.G	How old were you when this cancer was first diagnosed?	82	9	11.0
Q2.5.A	Have you smoked at least 100 cigarettes in your entire life?	46,404	1,208	2.6
Q2.5.B	How old were you when you first started to smoke fairly regularly?	18,984	738	3.9
Q2.5.C	Do you now smoke cigarettes every day, some days or not at all?	15,943	44	0.3
Q2.5.D	How long has it been since you quit smoking?	10,479	385	3.7
Q2.5.E	On the average, how many cigarettes do you now smoke a day?	5,420	201	3.7
Q2.5.F	Have you ever smoked tobacco products other than cigarettes even one time? (Such as cigars, pipes, water pipes or hookahs, small cigars that look like cigarettes, bidis, cigarillos, marijuana)?	46,404	898	1.9
Q2.5.G	Do you now smoke tobacco products other than cigarettes every day, some days, rarely, or not at all?	45,509	78	0.2
Q2.5.H	Have you ever used smokeless tobacco products even one time? (Such as chewing tobacco, snuff, dip, snus, or dissolvable tobacco)?	46,404	520	1.1
Q2.5.I	Do you now use smokeless tobacco products every day, some days, rarely, or not at all?	45,888	37	0.1
Q2.5.J	Are you exposed to second-hand smoke or environmental tobacco smoke every day, some days, rarely, or not at all?	46,404	646	1.4
Q2.6.A	Did you start smoking for the first time while being deployed?	18,984	161	0.8
Q2.6.B	How did deployment(s) change how much you smoked?	15,862	678	4.3
Q2.7.A	In the PAST YEAR, how often did you drink any type of alcoholic beverage. (Included are liquor such as whiskey or gin, beer, wine, wine coolers, and any other type of alcoholic beverage)? "On average, how many days per week did you drink?"	46,404	1,961	4.2
Q3.1.A	Compared to pre-deployment, would you say your overall health is better, worse, or about the same?	46,404	1,554	3.3
Q3.1.B	During your deployment(s), do you believe you were sick because of something you breathed?	46,404	13,804	29.7
Q3.1.C	Do you currently have a sickness or condition you think began or got worse because of something you breathed during deployment(s)?	46,404	14,709	31.7
Q3.1.D	When did the problem start?	25,703	18	0.1
Q3.1.E	Please rate your concern that something you breathed during deployment has already affected your health.	46,404	514	1.1
Q3.1.F	Please identify your biggest health concern that something you breathed during deployment has already affected your health.	43,868	1,435	3.3
Q3.1.G	Have you discussed this concern with your health care provider, medical professional or team?	43,868	—	—
Q3.1.H	Are you concerned that in the future that your health will be affected by something you breathed during deployment(s)?	46,404	3,002	6.5
Q3.1.I	Please rate your concern that something you breathed during deployment will affect your future health.	42,704	76	0.2
Q3.1.J	Please identify your biggest health concern that something you breathed during deployment will affect your future health.	42,602	645	1.5

TABLE 4-7 Continued

Question	Text	Eligible	Nonresponse N	Nonresponse %
Q3.1.K	Which exposure do you think has the biggest overall effect on your health?	45,614	2,937	6.4
Q5.1.A	Which of the following were you doing last week?	46,404	2,020	4.4
Q5.1.B	What is the main reason you did not work last week/have a job or business last week?	9,660	216	2.2
Q5.2.A	Select the occupational category that best describes your main occupation (the civilian job you've held the longest). Do not include your occupation during military service. If your occupation is not included, select other occupation.	46,404	4,803	10.4
Q5.2.B	Total years in this non-military job {0-99} years (enter 0 if less than one year).	46,404	6,847	14.8
Q5.3.A	Have you ever worked for a year or more in any dusty job outside the military?	46,404	1,219	2.6
Q5.3.B.1	For the job with the biggest dust exposure: Select the occupation category that best describes the job with the longest dust exposure. If your occupation is not included, select other occupation:	5,650	87	1.5
Q5.3.B.3	Total years in this job {0–99} years (enter 0 if less than one year).	5,650	186	3.3
Q5.3.B.4	Are you working in this dusty job now?	5,650	61	1.1
Q5.4.A	Have you ever been exposed to gas, smoke, chemical vapors or fumes in your work outside the military?	46,404	3,082	6.6
Q5.4.B.1	For the job with the biggest gas, smoke, vapor or fume exposure: Select the occupational category that best describes the job with the longest gas, smoke, chemical vapor, or fume exposures. If your occupation is not included, select other	8,060	76	0.9
Q5.4.B.3	Total years in this job {0–99} years (enter 0 if less than one year).	8,060	260	3.2
Q5.4.B.4	Are you working in this job with gas, smoke, chemical vapors, or fumes now?	8,060	81	1.0
Q5.5.A	Have you ever worked in a job with asbestos exposure, including military service?	46,404	16,259	35.0
Q5.5.B	Circle the type(s) of asbestos exposure that describe(s) how you were exposed to.	14,059	504	3.6
Q5.5.C	How many years did you work in a job with asbestos exposure? (enter 0 if less than one year)	14,059	2,045	14.5
Q5.5.D	Are you working in a job with asbestos exposure now?	14,059	1,777	12.6
Q6.1.A	Are there any traditional farm animals that live on your land or that you visit on a regular basis?	46,404	317	0.7
Q6.1.B	Have you ever removed mold in your home because of its effect on your health?	46,404	920	2.0
Q6.1.C	Have you ever lived in a home that had elevated radon levels?	46,404	8,868	19.1
Q6.1.E	How many total hours a week, on average, do you participate in all the above hobbies combined?	8,701	1,922	22.1
Q7.1.A	About how long has it been since you last saw or talked to a doctor or other health care professional about your own health? Include doctors seen while a patient in a hospital.	46,404	462	1.0
Q7.1.B	Do you wish to see a DoD or VA health care provider to discuss your health concerns related to airborne hazards during deployment?	46,404	9,452	20.4
Q8.1.A	How do you prefer to receive updated information on burn pits and other airborne exposures?	46,404	513	1.1
Q8.1.B	Do you use the Internet?	46,404	645	1.4
Q8.1.C	Do you send or receive emails?	46,404	32,348	69.7

NOTE: COPD = chronic obstructive pulmonary disease; GWA = Gulf War Area.

[a] Question only asked for Gulf War deployments.

[b] Question asked for each deployment.

lematic, and the high nonresponse rates further support the committee's evaluation of the questions, as discussed in Chapter 3. For example, Question 2.2.3.C asked about immune system problems and had an item nonresponse rate of 19.8%. This question is vaguely worded, does not ask about doctor diagnoses or provide examples, and therefore, respondents may not have known whether they had these problems. Question 2.2.3.I (whether the individual stops breathing during sleep) is also a poorly worded question (most people are unsure of their behaviors or the frequency of them while asleep) and, perhaps as a result, has high rates of nonresponse (39.0%). Other questions that asked respondents whether sickness experienced during deployment was because of something they breathed (Question 3.1.B) and whether they are currently sick because of something they breathed (Question 3.1.C) likewise had high nonresponse rates, approximately 30% and 32%, respectively.

SYNOPSIS AND CONCLUSIONS

Based on the material presented in this chapter, the committee has reached the following general observations and conclusions related to the availability and content of the AH&OBP Registry data. Subsequent chapters offer more detailed information on the exposure and health outcomes data provided for analysis and the methods used to analyze it to best effect.

Data made available by VA are limited in several respects that are important to the committee's ability to fully address its charge. Primary among these are the restrictions on the availability of some data because they were deemed to be potentially personally identifiable, including data on health care use or conditions as recorded in medical records through VA/DoD for either registry participants or the eligible population. The committee fully understands and appreciates VA's desire to ensure the security of personally identifiable information. However, VA's access restrictions limited the committee's ability to fully analyze participation in and the content of the registry and to construct and evaluate reliable measures of exposure potential and health outcomes. The restrictions also affected the confidence with which the committee can draw conclusions regarding the process of data acquisition and the validity of the information reported on exposure and health outcomes.

A second major limitation is that questionnaire and other data were only available for those who fully completed the questionnaire. A VA report (2015b) indicated that nearly 40% of those who initiated an AH&OBP Registry questionnaire did not complete it, an observation that should be followed up. **The committee recommends that VA evaluate whether and how registrants who did not complete the questionnaire differ from those who did, analyze the determinants of non-completion, and use this information to formulate strategies to encourage registrants to finish and submit their responses and improve the completion rate for future participants.**

The registry's title and the descriptions used by VA and DoD when communicating its availability were predominantly focused on environmental exposures experienced during deployment (as discussed in Chapter 3). That messaging likely resulted in higher levels of participation among those who feel that they experienced such exposures or whose health might have been affected by these or other military-related exposures. For example, 96% of all respondents reported being exposed to a burn pit on at least one deployment and 85.6% of Gulf War era respondents reported exposure to smoke from oil-well fires. Moreover, 85.2% of all respondents reported exposure to dust storms. The lack of data on those who were deployed and do not believe they were exposed to burn pits precludes using the registry to compare exposed with unexposed individuals. Therefore, the only means available for evaluating burn pit exposure is to examine gradations of exposure among the respondents using questions pertaining to being near a burn pit, having burn pit duties, and whether smoke or fumes from the burn pit entered the worksite or housing, as is described in detail in Chapter 5.

The makeup of the registry population differs in significant ways between the Gulf War and the post-9/11 respondent populations, as well as between each era of respondents and their respective eligible populations, for most demographic and military characteristics. When comparing registry participants either to nonrespondents only or to all eligibles it is important to recognize these differences, and examine outcomes separately by era of service as well as for the aggregated population. When a comparison group is necessary for analyses of registry data, the committee concludes that the eligible population stratified by era is the superior choice.

Although it is not possible to estimate the degree of reporting and response bias among those who choose to participate in the registry compared with the eligible population, comparing the eligible population with the

TABLE 4-8 Percentage of Endorsed and Nonresponses for Exposure and Health Condition Questions

Question	Text		Yes		No		Nonresponse	
		Eligible	N	%	N	%	N	%
Q1.2.A	Were you exposed to soot, ash, smoke, or fumes from the Gulf War oil well fires during any GWA deployment?	6,694[a]	5,729	85.6	380	5.7	585	8.7
Q1.2.D	Were you near a burn pit during any deployment?	206,373[b]	129,192	62.6	36,533	17.7	40,648	19.7
Q1.2.F	Did you have any burn pit duties on any deployment?	129,192[b]	66,087	51.2	60,523	46.8	2,582	2.0
Q1.2.G	Did smoke or fumes from the burn pit enter your work site or housing on any deployment?	129,192[b]	103,234	79.9	1,637	1.3	24,321	18.8
Q1.2.H	Were you outside or in an open tent or shelter during any deployment?	206,373[b]	177,880	86.2	6,335	3.1	22,158	10.7
Q1.2.I	Did you smell or see sewage ponds on any deployment?	206,373[b]	84,591	41.0	41,039	19.9	80,743	39.1
Q1.3.A	Were you ever close enough to feel the blast from an IED (improvised explosive device) or other explosive device?	46,404	33,517	72.2	11,526	24.8	1,361	2.9
Q1.3.B	In a typical month, how many days were you near heavy smoke from weapons, signal smoke, markers or other combat items?	46,404	28,845	62.2	7,139	15.4	10,420	22.5
Q1.3.C	In a typical month, how many days were you in convoy or other vehicle operations?	46,404	35,390	76.3	8,337	18.0	2,677	5.8
Q1.3.D	In a typical month, how many days did you perform refueling operations?	46,404	30,669	66.1	12,160	26.2	3,575	7.7
Q1.3.E	In a typical month, how many days did you perform aircraft, generator, or other large engine maintenance?	46,404	18,515	39.9	25,103	54.1	2,786	6.0
Q1.3.F	In a typical month, how many days did you perform construction duties?	46,404	16,754	36.1	24,883	53.6	4,767	10.3
Q1.3.G	In a typical month, how many days did you perform pesticide duties for your unit?	46,404	5,107	11.0	36,289	78.2	5,008	10.8
Q1.4.C	In a typical month during your deployment(s), how many days did you experience dust storms?	46,404	39,558	85.2	656	1.4	6,190	13.3
Q1.4.D	During your deployment(s), did you experience wheezing, difficulty breathing, an itchy or irritated nose, eyes or throat that you thought was the result of poor air quality?	46,404	40,990	88.3	1,713	3.7	3,701	8.0
Q1.4.E	How many days in an average month did you experience wheezing, difficulty breathing, an itchy or irritated eyes, nose, or throat that you thought was the result of poor air quality?	40,990	34,842	85.0	25	0.1	6,123	14.9
Q1.4.F	During your deployment(s), did you seek medical care for wheezing, difficulty breathing, an itchy or irritated nose, eyes, or throat that you thought was the result of poor air quality?	46,404	16,141	34.8	26,078	56.2	4,185	9.0
Q2.2.1.A	Have you been told by a doctor or other health professional that you had Hay fever or allergies to pollen, dust, or animals?	46,404	18,203	39.2	25,505	55.0	2,696	5.8

continued

TABLE 4-8 Continued

Question	Text	Eligible	Yes		No		Nonresponse	
			N	%	N	%	N	%
Q2.2.1.B	Have you ever been told by a doctor or other health care professional that you had asthma?	46,404	6,754	14.6	37,544	80.9	2,106	4.5
Q2.2.1.C	Have you ever been told by a doctor or other health care professional that you had emphysema?	46,404	588	1.3	44,221	95.3	1,595	3.4
Q2.2.1.D	Have you ever been told by a doctor or other health care professional that you had chronic bronchitis?	46,404	5,921	12.8	37,606	81.0	2,877	6.2
Q2.2.1.E	Have you ever been told by a doctor or other health care professional that you had chronic obstructive pulmonary disease also called COPD?	46,404	1,524	3.3	42,612	91.8	2,268	4.9
Q2.2.1.F	Have you ever been told by a doctor or other health care professional that you had some lung disease or condition other than asthma, emphysema, chronic bronchitis or COPD?	46,404	4,333	9.3	39,341	84.8	2,730	5.9
Q2.2.1.G	Have you ever been told by a doctor or other health care professional that you had constrictive bronchiolitis (CB)?	4,333	501	11.6	2,984	68.9	848	19.6
Q2.2.1.H	Have you ever been told by a doctor or other health care professional that you had idiopathic pulmonary fibrosis (IPF)?	4,333	99	2.3	3,584	82.7	650	15.0
Q2.2.2.A	Have you ever been told by a doctor or other health care professional that you had hypertension, also called high blood pressure?	46,404	16,502	35.6	28,136	60.6	1,766	3.8
Q2.2.2.B	Have you ever been told by a doctor or other health care professional that you had coronary artery disease?	46,404	858	1.8	43,944	94.7	1,602	3.5
Q2.2.2.C	Have you ever been told by a doctor or other health care professional that you had angina pectoris?	46,404	578	1.2	42,894	92.4	2,932	6.3
Q2.2.2.D	Have you ever been told by a doctor or other health care professional that you had a heart attack, also called myocardial infarction?	46,404	644	1.4	44,764	96.5	996	2.1
Q2.2.2.E	Have you ever been told by a doctor or other health care professional that you had a heart condition other than coronary artery disease or angina or myocardial infarction?	46,404	3,017	6.5	41,604	89.7	1,783	3.8
Q2.2.3.A	During the past 12 months, have you regularly had insomnia or trouble sleeping?	46,404	36,942	79.6	8,469	18.3	993	2.1
Q2.2.3.B	During the past 12 months, have you had neurological problems? (Some examples of neurological problems may include numbness, tingling, or weakness in your arms or legs or difficulties with thinking or memory.)	46,404	32,925	71.0	11,421	24.6	2,058	4.4
Q2.2.3.C	During the past 12 months, have you had problems of the immune system?	46,404	8,519	18.4	28,719	61.9	9,166	19.8

Q2.2.3.D	During the past 12 months, have you been told by a doctor or other health professional that you had any kind of liver condition?	46,404	3,107	6.7	41,641	89.7	1,656	3.6
Q2.2.3.E	During the past 12 months, have you been told by a doctor or other health professional that you had any chronic multi-symptom illness (examples include irritable bowel syndrome, chronic fatigue syndrome, and fibromyalgia)?	46,404	8,294	17.9	34,844	75.1	3,266	7.0
Q2.2.3.H	Questions H and I are about snoring and breathing during sleep. To answer these questions, please consider both what others have told you and what you know about yourself. How often do you snore?	46,404	40,291	86.8	1,707	3.7	4,406	9.5
Q2.2.3.I	How often do you have times when you stop breathing during your sleep?	46,404	20,454	44.1	7,671	16.5	18,279	39.4
Q2.4.A	Have you ever been told by a doctor or other health professional that you had Cancer or a malignancy (tumor) of any kind?	46,404	2,590	5.6	42,907	92.5	907	2.0

NOTE: COPD = chronic obstructive pulmonary disease; GWA = Gulf War Area.

[a] Question only asked for Gulf War deployments.

[b] Question asked for each deployment.

respondents on factors such as demographic characteristics may result in some insights that can be extrapolated to other areas of interest or included as effect modifiers in analyses of exposure and disease. However, these comparisons are weak because there are many unmeasured (or unavailable) variables, such as VA health care and utilization data and other factors that may influence the motivation to participate. Therefore, the committee was not able to draw firm conclusions from empirical evidence regarding the quality of the data. Applying propensity scores and weights designed to adjust for potential nonresponse bias using the available demographic and military characteristics met with limited success in reducing observed differences.

Based on the comparison of demographic and military characteristics between the respondents and eligible populations by era and the propensity analysis conducted, the committee concludes that the registry population is not representative of the eligible population. The committee further emphasizes that findings using the registry data are not generalizable to the broader, eligible population and should not be used for making inferences of that nature.

Several items have high rates of consistent responses (showing little variability), making them ultimately of little use for analyses. Items with little variability are evidence of selection bias (and perhaps other biases) which also affect their utility in analyses. Furthermore, several questions had nonresponse rates of greater than 15%, lending additional supporting evidence that many of the questions are poorly worded or otherwise problematic.

The AH&OBP Registry questionnaire data are amenable to analysis using standard statistical methods. However, the participants whose data were made available to the committee are a small (less than 2% of the eligible population), non-random, self-selected sample that is not broadly representative of the population of Gulf/OIF/OEF/OND and other eligible members of the military, and the results of analyses of them are not generalizable to the whole population of current or former service members covered by the registry. The committee therefore focused its analysis efforts on comparisons among registry participants rather than comparisons of registry participants to external populations. The distinctiveness of those who chose to participate precludes meaningful comparisons to the health experience of other populations. Internal analyses comparing registry participants to one another are limited by the quality of the exposure and disease information, but they have potential to generate hypotheses that would stimulate more rigorous epidemiologic evaluation through other approaches.

REFERENCES

AAPOR (American Association for Public Opinion Research). 2016. Standard definitions: Final dispositions of case codes and outcome rates for surveys. 9th edition. Oakbrook Terrace, IL: AAPOR.

Armed Forces Health Surveillance Center, Naval Health Research Center, and U.S. Army Public Health Command. 2010. *Epidemiological studies of health outcomes among troops deployed to burn pit sites.* Silver Spring, MD: Armed Forces Health Surveillance Center.

Baiocchi, D. 2013. *Measuring army deployments to Iraq and Afghanistan.* Santa Monica: RAND Corporation. http://www.rand.org/pubs/research_reports/RR145.html (accessed September 26, 2016).

Bonds, T. M., D. Baiocchi, and L. L. McDonald. 2010. *Army deployments of OIF and OEF.* Santa Monica, CA: RAND Corporation. http://www.rand.org/pubs/documented_briefings/DB587.html (accessed February 4, 2016)

Gasper, J., and J. Kawata. 2015a. Deployment segment analysis. Memo prepared for Paul Ciminera at the Department of Veterans Affairs Office of Public Health, Post-Deployment Health Group. April 27.

Gasper, J., and J. Kawata. 2015b. *An evaluation of selection bias and data quality: Open Burn Pit and Airborne Hazards Registry.* Prepared for the Department of Veterans Affairs Office of Public Health, Post-Deployment Health Group by Westat. September.

IOM (Institute of Medicine). 2013. *Returning home from Iraq and Afghanistan: Assessment of readjustment needs of veterans, service members, and their families.* Washington, DC: The National Academies Press.

Lezama, N.G. 2016. Responses from the Department of Veterans Affairs to questions from the Committee on the Review of the Department of Veterans Affairs Airborne Hazards and Open Burn Pit Registry. Received February 18.

Liu, J., N. Lezama, J. Gasper, J. Kawata, S. Morley, D. Helmer, and P. Ciminera. 2016. Burn pit emissions exposure and respiratory and cardiovascular conditions among airborne hazards and open burn pit registry participants. *Journal of Occupational and Environmental Medicine* 58(7):e249–e255.

Smith, B., C. A. Wong, E. J. Boyko, C. J. Phillips, G. D. Gackstetter, M. A. Ryan, and T. C. Smith. 2012. The effects of exposure to documented open-air burn pits on respiratory health among deployers of the Millennium Cohort Study. *Journal of Occupational and Environmental Medicine* 54(6):708–716.

Torreon, B. S. 2015. U.S. periods of war and dates of recent conflicts. Congressional Research Service. Report RS21405. https://www.fas.org/sgp/crs/natsec/RS21405.pdf (accessed October 3, 2016).

VA (Department of Veterans Affairs). 2015a. *Report on data from the Airborne Hazards and Open Burn Pit (AH&OBP) Registry*. April. http://www.publichealth.va.gov/docs/exposures/va-ahobp-registry-data-report-april2015.pdf (accessed September 10, 2016).

VA. 2015b. *Report on data from the Airborne Hazards and Open Burn Pit (AH&OBP) Registry, June 2015*. http://www.publichealth.va.gov/docs/exposures/va-ahobp-registry-data-report-june2015.pdf (accessed September 10, 2016).

Westat. 2016. Quality control procedures. Prepared by Westat in response to a request from the Committee on the Assessment of the Department of Veterans Affairs Airborne Hazards and Open Burn Pit Registry. Received October 3.

5

Analysis and Interpretation of Exposures Data

This is the second of three chapters that describe the methods and results of the committee's analysis of the initial months of Airborne Hazards and Open Burn Pit (AH&OBP) Registry data. It summarizes information on respondents' exposure to burn pits and other important service-related airborne exposures; discusses the registry's limited capacity to provide reliable estimates of the magnitude, duration, and frequency of the exposures; and explicates the committee's approach to using the exposure-related information that is available. An approach to analyzing the exposure information collected by the registry is then presented. The chapter closes with a summary and interpretation of this information. In presenting this, the committee wishes to emphasize the limitations of these data and of the results of analyses using it—the work described here is intended to respond to the Department of Veterans Affairs' (VA's) request for guidance on how to categorize the self-reported exposures collected by the AH&OBP Registry and not as an endorsement of the registry as a means for gathering exposure data.

INTRODUCTION

Human exposure is defined as contact between a chemical, physical, or biological agent and the outer boundary of a human organism. Exposure is quantified as the amount of an agent available at the exchange boundaries of the organism (for example, the skin, lungs, or gut) (EPA, 2011), and can include a time component as well. It is related to dose, which for a chemical agent is the amount taken into the body by a specific route of entry (such as dermal, inhalation, or ingestion). The pollutant concentration of a substance in a medium (that is, air, water, food, soil and the like) affects the dose taken into or in contact with the body. Substances can cause effects at the point of entry (the respiratory tract, gastrointestinal tract, skin, . . .) or lead to effects elsewhere in the body after absorption. Individuals differ in their absorption, metabolism to more or less toxic chemical intermediates, distribution in the body, and excretion. All of these factors give rise to a substantial degree of variability in the response, and potential health effects, of individuals' exposure to the same levels of an agent.

In health outcome studies, exposure is ideally measured at the individual level throughout the duration of exposure (by, say, a personal monitor). In absence of such data—which is typically the case—other measures can be used to approximate the relative degree of exposure, such as the amount of time spent in areas with elevated concentrations of airborne substances, the proximity of an individual's work or residence to an exposure source, or the type of job duties and their associated exposures. In characterizing exposure to burn pits, information on meteorological conditions, the substances and quantities burned at different times, satellite data, and self-reported information on exposure proximity and frequency might provide additional information on the relative magnitude and duration of exposure. The accuracy of the assessment of exposure, however, depends on the quality of the data and the extent to which surrogate information is representative of individual exposure. All of these measures have the potential for inaccuracies, bias, and confounding which affect the ability to detect or observe a true association between an exposure and resulting health effects.

Sources and Nature of Exposures

The duration and frequency of exposures to environmental contaminants vary greatly for service members, and the exposures occur in the presence of other physiological stressors. Because many U.S. personnel involved in the Southwest Asia theater of operations worked and lived at fixed sites, those who were in the presence of burn pits may have experienced exposures to burn pit emissions during the majority of their deployment to those sites, although the magnitude and nature of the exposures would have varied. Occupational duties, for example, may put some service members in close proximity to burn pits and other sources of airborne pollutants. Some exposures would be expected to be of relatively short duration: Joint Base Balad had a large burn pit, but the base functioned as a transit stop, resulting in a short-duration exposure for many service members passing through (IOM, 2011). Moreover, meteorological conditions varied, affecting whether pollutants were transported toward or away from individuals and affecting the rate of dissipation of the pollutant. Service members thus experienced large variations in the duration, frequency, and magnitude of exposures which are not easily characterized.

Burn Pits

From the beginning of the Southwest Asia conflicts, the uncontrolled open-air burning of waste in Iraq and Afghanistan has been the primary solid-waste management solution in those theaters of operations. Waste included that which would be expected in any community, such as food remains and latrine waste, but it also included plastic bottles, electronics, waste from medical facilities, weapons and munitions, paint, petroleum and lubricant products, plastics and Styrofoam, and rubber tires (VA, 2016). The usual waste management systems of land-filling, recycling, and incinerating these items were often not feasible, instead necessitating the use of open burning. However, open burning of the items generates more byproducts of incomplete combustion, thereby increasing potential health risks.

In 2009 the U.S. military's use of burn pits in Iraq and Afghanistan was restricted by law, and by the end of 2010 their use in Iraq had gradually been phased out, but it did continue in Afghanistan, where 197 burn pits were operating as of January 2011 (IOM, 2011). The use of burn pits varies depending on the size of the base, its activities, and its population. In Iraq, as of November 2009, burn pits were operating at 14 out of the 41 existing small military sites (defined as housing less than 100 U.S. service members), 30 of the 49 medium-size sites (between 100 and 1,000 service members), and 19 of the 25 large sites (more than 1,000 service members); however, data were not available for all sites (DoD, 2011). The number of burn pits used in Iraq declined in response to the 2009 law, and a 2010 Government Accountability Office (GAO) study of open-air pit burning in Iraq and Afghanistan listed only 22 in use in Iraq in August 2010 (GAO, 2010). Their use in Afghanistan continued, however, and in January 2011, 126 out of the 137 small sites, 64 of the 87 medium-size sites, and 7 of the 18 large military installation sites in Afghanistan still had operating burn pits (DoD, 2011).

In 2011, the Department of Defense (DoD) estimated that an average of 8 to 10 pounds of waste was generated each day by each person in theater (DoD, 2011). Joint Base Balad—the largest base, serving up to 25,000 people at a time—burned perhaps 100–200 tons of waste a day in 2007. By 2009, three incinerators were operational on the base, but the burn pit was still in use, burning approximately 10 tons of waste daily until it ceased operation in October 2009. A 2010 Army Institute of Public Health study of burn pits in Iraq and Afghanistan reported that large bases burned waste that consisted generally of 5–6% plastics, 6–7% wood, 3–4% miscellaneous noncombustibles, 1–2% metals, and 81–84% combustible materials (APHC, 2010; IOM, 2011).

Other Airborne Exposures

Air quality in the theaters of operation is affected by winds, temperature, humidity, meteorological events, and dust storms as well as by anthropogenic (civilian and military) sources such as power plants, industrial facilities, trash burning, agriculture, the Al-Mishraq (Iraq) sulfur plant fire (Baird et al., 2012), combat dusts from mortar fire and improvised explosive devices (IEDs), and—in the case of the 1990–1991 Gulf War—soot from oil-well fires. Additional emissions resulting from activities on military bases, including combustion products from vehicles, aircraft, and generators, and the evaporation of volatile compounds in fuel and occupational settings, such as vehicle maintenance, are also important considerations (IOM, 2011). Personal activities such as smoking also contribute to airborne exposures. Elevated exposures can thus occur on base, in the field, or in urban areas.

Composition of Air Pollutants

Not only are the emissions released by burn pits a complex mixture of various chemicals and particulates that depend on factors such as the composition of the trash burned, accelerant used, temperature, ventilation, and the burn rate (Woodall et al., 2012), but the composition and magnitude of air pollutants on military bases in theaters of operation are also affected by a variety of other anthropogenic and natural toxicants.

DoD conducted environmental monitoring and health studies at Joint Base Balad in Iraq starting in 2004. The base operated a large burn pit and was a central logistics hub for U.S. forces deployed in support of Operation Iraqi Freedom (OIF)/Operation Enduring Freedom (OEF)/Operation New Dawn (OND). In response to personnel complaints of odor, poor visibility, and health effects attributed to burn pit emissions, the U.S. Army Public Health Command and the Air Force Institute for Operational Health conducted a series of ambient-air sampling and screening health-risk assessments of burn pit exposures there in 2007 and again in 2009. The assessments were designed to measure the concentration of airborne pollutants released by burning at several sites on base and to detect potentially harmful inhalation exposures for personnel (APHC, 2010; CHPPM and AFIOH, 2009; IOM, 2011; Taylor et al., 2008). Even though these efforts were limited by their inability to contribute to individual exposure assessment as well as by their inability to distinguish the contributions from particular sources (combustion engines, burn pits, dust storms, and the like), they do yield some information about the constituents and ambient levels of airborne toxicants that may have been present on bases with burn pits.

A 2011 review of air monitoring efforts at Joint Base Balad conducted by a committee of the Institute of Medicine (IOM, 2011) found that

- Particulate matter (PM) concentrations in ambient air were on average higher than U.S. pollution standards. PM was most likely a result of local sources (vehicle traffic, aircraft emissions) and regional sources (long-range anthropogenic sources, dust storms), although the burn pit likely made some contribution.
- Polychlorinated dibenzo-*p*-dioxins and dibenzo-*p*-furans (PCDD/Fs) were detected at low concentrations. Although species associated with greater toxicity were higher than generally found in the United States or urban environments worldwide, they were lower than levels associated with some individual sources.
- Concentrations of volatile organic compounds (VOCs) and polycyclic aromatic hydrocarbons (PAHs) were similar to those reported in major urban areas outside the United States with major sources being regional background, ground transportation, stationary power generation, and the airport at Joint Base Balad (IOM, 2011).

Subsequent studies have also noted the contribution of the Joint Base Balad burn pit to PCDD/Fs on base as well as the important role of other sources of emissions including the airfield as the primary source of PAHs. Other important contributors to PAH levels were aircraft, vehicle emissions, space heaters, and diesel generators (Masiol et al., 2016a,b).

These observations were limited to the pollutants that were targeted by DoD and the conditions (meteorological, waste stream composition, operating conditions) present at the time of the measurements. Several criteria pollutants commonly monitored in the United States and likely released by burn pits were omitted by DoD's sampling, including sulfur dioxide, ozone, nitrogen dioxide, and carbon monoxide. Other pollutants not included in the sampling included those known to be associated with the burning of household waste (EPA, 1997, 2001; Lemieux et al., 2003, 2004), geologic material, carbon from combustion sources, metals from regional smelting activities, and other gaseous pollutants produces by combustion engines (Engelbrecht et al., 2009; IOM, 2011). Thus, the available monitoring data provide information on exposures to the major types of constituents from burn pit emissions, but they lack information on other chemicals that were likely present as well as exposure variability over time (IOM, 2011).

Other data collected as part of a monitoring program for a solid waste disposal facility at the Bagram Airfield in Afghanistan emphasize the variability of exposures associated with burn pits (Blasch et al., 2016). The facility operated a burn pit from 2005 to 2012. The investigators collected breathing zone samples, unlike the case with Joint Base Balad, but only PM and VOCs were studied. Sampling was conducted at four security locations (up to 125 meters from the burn pit) and a control location (4 km from the burn pit) during 30 12-hour shifts. Among the VOCs detected, only Acrolein exceeded the 1-year military exposure guideline but benzene was detected in all samples. The range of PM concentrations varied considerably in association with airfield activity (vegetation removal, demining, road construction, vehicle traffic, industrial activity, air traffic). The highest recorded concentrations of environmental $PM_{2.5}$[1] (0.615 mg/m^3) occurred at the solid waste disposal facility where the burn pit and incinerators were located. High $PM_{2.5}$ and PM_{10} concentrations were also noted at the bazaar, a highly populated site with unpaved roads and considerable vehicular traffic. The investigators concluded that "[t]he diversity of results support the concept of a complex environment with multiple polluting sources and changing meteorological and operational conditions" (Blasch et al., 2016, p. S38).

Limitations of Exposure Information

Emissions from burn pits were only one of the many potential exposures experienced by military personnel deployed to the Southwest Asia theater of operations. Other exposures included agents used as preventive measures (such as vaccines, pesticides, and insecticides), hazards of the ambient environment (such as sand, insects, air pollution, and endemic diseases), job-specific agents (such as paints, solvents, and diesel fumes), war-related agents (such as smoke from oil-well fires and depleted uranium), and hazards associated with cleanup operations in the 1990–1991 Gulf War (such as sarin and cyclosarin) (IOM, 2010). Neither DoD nor VA records nor other sources contain detailed information on all the agents to which military personnel might have been exposed, at what doses, or for what amount of time. The number and combination of these sources make it difficult to examine whether any agent or combination of agents may have caused or exacerbated health problems in a deployed population. Further complicating such research is the fact that other physical and psychological stressors that may have been experienced in addition to the airborne chemical and particulate exposures mentioned here could also have an effect on some health outcomes.

Determining whether veterans and service members deployed to the Southwest Asia theater of operations face an increased risk of illness because of their exposures during deployment requires information about specific agents, durations of exposure, routes of entry, internal dose, and documentation of adverse reactions. DoD initiated an environmental monitoring effort following the 1990–1991 Gulf War primarily because of concerns related to smoke from oil-well fires and possible exposure to sarin and cyclosarin, and it focused its information-gathering accordingly (IOM, 2010). When the military engaged in OEF in Afghanistan in 2001 and OIF in Iraq in 2003, it sampled the air, water, and soil to characterize the deployment environment. Using this baseline exposure information, DoD—through the U.S. Army's Public Health Command (formerly the Center for Health Promotion and Preventive Medicine)—designed and implemented its Enhanced Particulate Matter Surveillance Program to characterize and quantify particulate matter in the ambient environment at 15 sites in the Persian Gulf region (NRC, 2010). This was one of the first attempts to measure and characterize exposures to PM that could be used for studies of the health effects of service members and veterans deployed to those areas. The National Research Council (NRC) found, however, that this effort had several flaws in its methodology, and that its study design

---

[1] $PM_{2.5}$ is particulate matter with a diameter of 2.5 microns or smaller.

limited its usefulness. Accordingly, the NRC recommended that the methodology be improved before undertaking future monitoring efforts. Limitations included inappropriate design or analyses to address objectives that were not clearly defined, a lack of validation of the sampling equipment for high levels of PM concentrations recorded at the sites, the failure to collect sufficient particle mass and composition data on a consistent basis to be useful for quality assurance and for health-effects studies, and the use of different filter media, each analyzed with different techniques, which limited comparability among results and precluded source apportionment and mass-balance assessments (NRC, 2010).

Nevertheless, results from this PM exposure assessment effort clearly showed that service members deployed to the Southwest Asia theater of operations were exposed to high concentrations of PM and that the particle composition varied considerably over time and location. The NRC committee further concluded that to appropriately investigate health effects resulting from exposure to a complex mixture of pollutants, the monitoring strategy needs to be tailored to the specific goals and hypotheses that future health-surveillance and research studies are designed to address (NRC, 2010). In other words, study design should be based on a priori hypotheses that encompass basic analysis plans as opposed to attempting to design a study to fit the data after it has been collected, sometimes for a different purpose.

Other efforts to understand individual military personnel exposures are also fraught with challenges. One major issue is that reconstructing past exposure events is difficult and prone to problems. For example, one commonly used method to collect exposure information is to survey subjects about their perceived exposures to various agents, but this is limited by error in recollection and recall bias, which only increase with time. Individual responses have rarely been verified by in situ measurements or records (IOM, 2010).

Various exposure assessment tools and methods have been used in research to attempt to fill gaps in exposure information. Models specific to certain chemical exposures have been proposed (such as for in-theater exposure to sarin; GAO, 2004), but are of questionable reliability because of the difficulty in incorporating meteorological data, transport and dispersion data, and service member-unit location information—types of information that may not be recorded accurately, if at all, or easily available. For example, unit locations are rarely accurate, and are not reliable for reflecting the actual location of individual service members because members of the same unit may be deployed to different forward operating bases (GAO, 2004; IOM, 2006, 2013) These limitations on individual locations and experiences related to exposure greatly limit assessments of associations with, or the likelihood of increased risk for, diseases, symptoms, or other adverse health effects that are due specifically to airborne hazards for service personnel.

This committee concurs with the previous NRC (2010) committee that concluded that to appropriately investigate health effects resulting from exposure to a complex mixture of pollutants, the monitoring strategy needs to be tailored to the specific goals and hypotheses that future health-surveillance and research studies are designed to address (see Box 5-1). That includes matching the monitoring period with the deployment period of the military personnel being studied. In particular, different types of exposure monitoring may be required for the study of exposure events or conditions leading to potential persistent effects, such as asthma and chronic obstructive pulmonary disease, compared with the study of acute effects, such as day-to-day variability in respiratory or cardiac responses (NRC, 2010). Future monitoring studies should include other ambient pollutants that military personnel may be exposed to in the field and that may be relevant to human health, such as ozone, hazardous air pollutants, and other materials such as diesel exhaust. Consideration should also be given to the toxicity of mixtures. Although the health effects of mixtures are complex and incompletely known, the characterization of all relevant constituents is a necessary first step in understanding which combinations or concentrations might have combined effects for certain health endpoints. In addition, more repeated sampling with a consistent filter type would provide a greater library of gravimetric and chemical-specific data and thus increase statistical power. Finally, increasing the sampling frequency would make it possible to estimate more accurate annual-average concentrations of particle mass and chemical components (NRC, 2010).

BOX 5-1
Select Technical Recommendations from *Review of the Department of Defense Enhanced Particulate Matter Surveillance Program Report* (NRC, 2010)

In designing a comprehensive monitoring scheme, a set of study objectives should be developed that provides the rationale and selection of the samplers, filter media, sampling frequency, and data quality standards to be used.

- Future studies should use particle samplers that operate reliably on the basis of field testing in environmental conditions that are similar to the conditions in which they are likely to be used. For the EPMSP, such field testing was not conducted, and high PM concentrations may have led to overloading of the samplers, judging by prior results from Kuwait.
- The frequency of sampling and the types of analyses applied to the samples should be tailored to the study objectives. Such an approach maximizes the benefits of the resources expended on the study.
- In future monitoring studies, it is critical that quality assurance and control procedures be implemented and specified in writing to ensure the integrity of the samples collected and analyzed.
    - Replicate samples should be collected at selected sites during future monitoring efforts, where feasible, to assess sampler performance.
    - Measurement uncertainties should be reported for all PM components. That will make it possible to interpret, with caution, the concentration data on PM components whose concentrations are mostly below the detection limit of the analytic method, as in the case of the X-ray fluorescence data.
    - Mass closure (that is, comparison of particle mass with the sum of the individual-particle components) should be performed as part of the overall QA process.

Because this is likely to be a continuing effort, the military might consider developing real-time continuous particulate-matter monitoring equipment whose use is recommended in the EPMSP report. Such equipment could be based on commercially available models but adapted to withstand the theater environment, including extreme temperatures, moisture, and particle concentrations; rough handling; and minimal maintenance. The monitors should be battery-powered and should report particle-size mass concentrations.

NOTE: EPMSP = Enhanced Particulate Matter Surveillance Program; QA = quality assurance.

EXPOSURE INFORMATION COLLECTED BY THE AH&OBP REGISTRY

The exposure information collected by the AH&OBP Registry questionnaire consists of self-reports of deployment exposures, occupational activities, and personal habits. The result is qualitative information on individual-level exposures or exposure potential. This section briefly summarizes the questionnaire's collection of exposure information and discusses its limitations. Questions are discussed in the order in which they are asked in the questionnaire, which is reproduced in Appendix C: deployment-related exposures (Section 1.2); military occupational exposures (Section 1.3); and environmental exposures and regional air pollution (Section 1.4). Exposures not related to military service, including nonmilitary occupational exposures (Sections 5.1–5.5) and residential and hobby-related exposures (Section 6), are collected by the questionnaire but are not discussed here given the low response rates and quality issues with the questions themselves. Exposures to tobacco smoke (Sections 2.5 and 2.6) collected by the questionnaire are discussed as potential confounders of health effects in Chapter 6. Assessment of the questions themselves is found in Chapter 3.

Location-Specific Deployment Exposures

Section 1.2 asks respondents about exposures related to each deployment that the respondent verifies from the VA Defense Information Repository Database (referred to as a "deployment segment"). Thus, a person with multiple deployments will answer these questions multiple times. For each deployment, respondents are asked about three sources of emissions— oil-well fire smoke, burn pits, and sewage ponds—and about where the person spent the majority of his or her time with relation to these exposures.

The first of these questions (1.2.A), "Were you exposed to soot, ash, smoke, or fumes from the Gulf War oil well fires?" was asked only of individuals who had deployed during or directly after the 1990–1991 Gulf War. A yes or no response is required.[2] Two follow-up questions (1.2.B and 1.2.C) ask about deployment location(s) in the 1990–1992 time period. Respondents are asked to identify the base where they spent most of their time using a drop-down list or to type in an answer. They are then asked for the location where they spent the second most amount of time. Because the committee was only granted basic deployment information (country and year), these responses were not made available to the committee and thus were not part of the analysis.

The next four questions collect information about exposure to burn pits during each deployment. For service members who answer "yes," indicating that they were "near" a burn pit during a deployment (1.2.D), follow-up questions elicit who ran that burn pit (1.2.E), if their duties included direct work with the burn pit (1.2.F), and the numbers of hours they were exposed in a typical day (1.2.G). The committee focused on those later two questions, which characterized exposure to burn pits:

- Did your duties during these dates include the burn pit (examples include trash burning, hauling trash to the burn pits, burn pit security, trash sorting at the burn pit)? [Yes or No]
- On a typical day, how many hours did smoke or fumes from the burn pit enter your work site or housing? [Never or Enter 1–24 hours]

Two additional questions ask about the number of hours that a service member was outside or in an open tent or shelter (1.2.H) or near a sewage pond (1.2.I) on a typical day. For both questions, respondents could answer "never" or enter a number of hours. These questions had relatively high rates of nonresponse: 18.4% and 38.4% for Questions 1.2.G and 1.2.I, respectively. Additionally, the committee deemed sewage ponds to be a relatively small source of airborne exposure.

General Military Occupational Exposures

In Section 1.3, questions collect information about seven potential exposures related to occupational duties. These questions are asked only once of any respondent and apply to any deployment. The first question in the section asks the respondent to answer yes or no to "Were you ever close enough to feel the blast from an IED (improvised explosive device) or other explosive device?" All of the other questions are formatted in a similar manner, asking the number of days of exposure in a typical month. Those exposures include being near heavy smoke from weapons; being in a convoy; performing refueling duties; performing aircraft, generator, or other large engine maintenance; performing construction duties; or performing pesticide duties. All of these exposures were considered in the committee's analyses except for pesticide duties (Question 1.3.G). This question appears to have been included to generate information on confounding exposures, but it is too vague and open to interpretation to yield useful information.

Environmental Exposures and Regional Air Pollution

A series of questions in Section 1.4 collects information about air quality, dust storms, and their suspected impact on some symptoms. The first question asks "Did you do anything differently during your deployment(s)

[2] Almost all of the questions discussed in this chapter include options for "don't know" or "I don't wish to answer." These are characterized as nonresponses in the committee's analyses and discussed in Chapter 4.

when you thought or were informed air quality was bad (for example, during dust storms or heavy pollution days)?" Respondents could indicate "yes," "no," "never thought of this," or "I was not informed or aware of bad air quality" (1.4.A). Those who answered "yes" were asked what measures they took (1.4.B).

The next question (1.4.C) asks directly about dust storms as a specific exposure: "In a typical month during your deployment(s), how many days did you experience dust storms?" Respondents could answer "never" or enter a number of days from 1 to 31.

Three questions ask if respondents experienced "wheezing, difficulty breathing, an itchy or irritated nose, eyes or throat" that they "thought was the result of poor air quality" (1.4.D) and, if so, how many days in an average month were they affected (1.4.E). It also asks if they sought care for those symptoms (1.4.F).

Limitations of AH&OBP Registry Questionnaire Exposure Information

The information collected by the AH&OBP Registry questionnaire has a number of limitations, many of which are a consequence of it being self-reported. As noted in Chapter 2, self-reported data are by nature subject to recall and reporting biases. The ability of service members to accurately recall their exposures and activities—including duration and frequency, sometimes over multiple deployments as long as 25 years ago—is open to question. Another problem is the tendency for some individuals to over- or underreport exposures (reporting bias). This can result in exposure misclassification—that is, individuals who are actually exposed do not report relevant exposures, or individuals who were not or less exposed report that they were more exposed. Too much misclassification in one direction or the other can result in attenuated associations or spurious conclusions.

In considering the quality of the information provided for each deployment (Section 1.2), the potentially many responses required of respondents—particularly for those with multiple deployments (and thereby the potential for a longer duration of exposure)—may have resulted in less accurate information for those with more exposure because of the difficulty in remembering each individual deployment, or because of response fatigue (Bosnjak and Tuten, 2001). The amount of information or attention given to responding to each question may have decreased with each round of deployment-related queries in the section. Gasper and Kawata, in a 2015 analysis of respondents, reported that

> The order in which deployment segments are presented in the questionnaire is strongly related to nonresponse. For several of the items examined, nonresponse was higher on deployment segments that appeared to participants later than on those that appeared earlier. For example, for question 1.2.D about burn pit exposure, the nonresponse rate was 13 percent of the 1st segment, 19 percent for deployment segments 2 to 5, 25 percent for deployment segments 6 to 9, and 27 percent for deployment segment 10 or higher. This suggests that respondents may have recall problems when the number of deployment segments is larger, or that they may tire out and choose not to answer for later deployment segments. (pp. 3-3–3-4)

And if those with a greater potential for exposure were more motivated to complete the questionnaire if they had symptoms, this could result in reporting bias.

The accuracy of self-reported data can be improved if there is a way to confirm its validity. However, in the case of the exposure information collected by the AH&OBP Registry questionnaire, few data were available to verify self-reported exposures, although databases can confirm some deployment information, including dates, locations, and job categories. One way to validate self-reported exposure to burn pits would be to compare self-reported information to data on which bases had active burn pits and when incinerators went into use at various bases, thus defining periods of no or reduced burn pit exposure. The committee did not have access to data that would allow them to make such an evaluation, but Gasper and Kawata (2015) found

> Participants were more likely to report exposure to burn pits on deployment segments during which they were located at bases with documented burn pits compared to deployment segments at locations with no documented burn pits. Specifically, 97 percent of deployments to Joint Base Balad, 89 percent of deployments to Contingency Operating Base Speicher, and 88 percent of deployments to Camp Taji [which had burn pits] included reports of close proximity

> to a burn pit. In contrast, 37 percent and 30 percent of deployments at Camp Arifan and Camp Beuhring [which did not have burn pits], respectively, included reports of close proximity to a burn pit. (p. vii)

While some data were collected on particular predictors of exposure, such as job assignment and deployment locations, without additional information on the numbers of individuals who served in those jobs or at specific locations to serve as denominators in analyses, no inferences can be made about patterns in the prevalence of exposure. For example, if a large number of individuals who report a particular job duty participate in the registry and report a specific exposure, there is no way of differentiating among several possibilities—whether that reflects a large number of individuals who have those job duties, whether there was a greater motivation on the part of those with certain job duties to enroll in the registry, or whether a greater proportion of individuals with those job duties were exposed.

The IOM's 2011 report *Long-Term Health Consequences of Exposure to Burn Pits in Iraq and Afghanistan* highlights additional potential issues with the use of the registry data, beyond those identified earlier that are associated with using such data in general. While the levels of pollutants found at Joint Base Balad were elevated, urban areas around the globe have elevated levels of many of the pollutants observed as well, so individuals' exposures can be elevated not only when they are in the field, but also during duties and activities related to being in urban environments. Furthermore, it is important to note that similar exposures can be experienced at other sites that are not related to deployment. Such exposures may also contribute to, or exacerbate, the development of health conditions related to burn pit exposure.

The questions discussed above do not provide information on the intensity of exposure beyond a binary yes/no for exposure, and this is a major shortcoming. Intensity is a central component of exposure characterization. Information regarding it could be obtained from direct questions (a self-reported rating of the intensity of the smoke, for example), or indirect questions (How far from the burn pit did you live or work?), or by asking additional questions about burn pit duties or more frequent or longer exposure to burn pits. Other studies have used assessments of proximity to the source in studies of burn pit exposures (AFHSC et al., 2010; Jones et al., 2012; Powell et al., 2012; Smith et al., 2012), but such information on individual locations was not available to the committee.

The lack of good quantitative data for exposure assessment purposes also includes information that might allow one to characterize acute or chronic exposure or to identify the constituents and chemical species present, both of which are important to understanding potential health effects. For example, the health effects known to be associated with acute exposure to PM (such as increased mortality secondary to respiratory or cardiac diseases) are different from those associated with chronic PM exposure (such as lower respiratory symptoms and reduced lung function) (WHO, 2006). The AH&OBP Registry does not contain quantitative information on the level of exposure (beyond yes/no) and duration, so differentiating between acute or chronic exposures is not possible. Furthermore, there are several factors that may influence the toxicity of airborne PM, including bulk chemical composition, trace element content, strong acid or sulfate content, and particle size distribution (Harrison and Yin, 2000). Because there are no data available describing those aspects of exposure that can be linked to the registry, the ability of the data in the registry to be used to investigate health effects is further weakened.

THE COMMITTEE'S ANALYSIS OF AH&OBP EXPOSURE INFORMATION

The committee's analysis of the exposure data collected by the registry questionnaire included a careful inspection of descriptive data, consideration of how to create exposure variables, and a look at potential cumulative exposure. The examination in this chapter is limited to descriptive statistics for the questionnaire items related to exposures and the creation of independent variables describing exposure for use in the analysis of potential associations between exposures and health effects (discussed in Chapter 6). All analyses were carried out under the committee's direction by an external contractor using data supplied directly to the contractor by VA—neither committee members nor staff had access to the raw data.

The committee only examined exposure data as collected by the registry questionnaire. However, data on deployments (as described in Chapter 4) were extracted by VA from two DoD Defense Manpower Data Center datasets, the Gulf War Oil Well Fire Smoke Registry and the Contingency Tracking System (CTS) database. Deployment information was limited to country and year of deployment.

While other sources of data could potentially serve as exposure proxies (such as distance from the specific source of concern), the quality of much of that data is poor and is plagued by many of the same limitations as the registry data. Other data pertaining to exposures that could be linked to the registry data were not available to the committee.

During its deliberations, the committee recognized the variety of limitations to using the registry data as discussed above, with the objective of identifying potential methods to use the exposure data appropriately in its further analyses and execution of its charge. The committee's primary concerns included that exposures are multidimensional (time, intensity, chemical composition, and the like) and that data characterizing those dimensions were lacking, the high degree of correlation between responses to questionnaire questions, the potential for bias in those responses, and the high nonresponse rates for some questions. The text below describes the approaches the committee took to making the best use of the available data without over-interpreting it.

Descriptive Statistics

The committee focused its analysis on six potential sources or types of airborne emissions: burn pits, dust (including sandstorms and desert environment), diesel, exhaust, and fuel (including jet fuel, combat, construction, and soot [the last for 1990–1991 Gulf War respondents only]). All of these exposures are important contributors to PM, VOC, PAH, and PCDD/F exposures in theater, so it is inaccurate to solely focus on exposure to burn pits, and it is not reasonable to assume that individuals were exposed to these airborne substances as a result of burn pits alone.

Table 5-1 lists the exposures of interest, applicable questions in the questionnaire, and the type of response for those questions included in the committee's analyses.

Responses to questions related to exposures are summarized in Table 5-2. The majority of respondents indicated that they were near burn pits during their deployments (62.6% of deployments, but 96% of individuals overall; Question 1.2.D) but only about one-third of respondents indicated that their duties were related to burn pits (32.0% of deployments; Question 1.2.F). On average, respondents reported 7.5 hours per day of smoke or fumes from burn pits at their work site or housing, but 16% reported 24-hour exposure; about 7% reported 12 hours of exposure per day; and 15% reported 6 hours of exposure or less per day. A large proportion (72%) also reported having been near an IED blast or other explosion. For several of the duration questions (1.2.G, 1.2.I, 1.3.B, 1.3.C, 1.3.D, 1.3.E), about 15% of respondents indicated continuous exposure (24 hours per day or 31 days per month).

Not surprisingly, the reported exposure to dust storms was quite high, with a mean of 8.9 days per month. This, along with the relatively frequent convoy-related activities, would suggest that exposures to PM would be elevated. While it is true that PM resulting from dust storms and kicked up by convoys will be compositionally different than combustion-related PM (Cassee et al., 2013; Engelbrecht et al., 2008; Lyles et al. 2011), current standards do not differentiate risks based on PM composition (EPA, 2013).

Other than burn pit, dust storms, or Gulf War oil-well fire soot exposure, most respondents (68%) endorsed only one exposure related to their military service (Questions 1.3.A–G; heavy smoke from weapons, convoys, refueling duties, large engine maintenance, construction, and pesticides, Question 1.4.C not included). The numbers and percentages of respondents who reported up to 5 or more exposures are presented in Table 5-3.

The committee examined the distribution of responses to questions pertaining to the durations of exposures. Those distributions are depicted in Figures 5-1 and 5-2. In Figure 5-1 the duration of exposure to burn pits in hours per day, as reported for each deployment, is shown by bars that represent the number of deployments for which the duration of burn pit exposure was reported (1.2.G). Two-hour increments for exposure duration were chosen due to the tendency of respondents to endorse an even number of hours. Of the more than 206,000 eligible deployment segments, 37% skipped the question (reported no burn pit exposure on Question 1.2.D) and 12% responded "don't know," represented by the nonresponse bar to the right of the figure. Elevated frequencies for each duration occur at 24 hours (16% of deployments), 12 hours (7% of deployments), and 4 hours (4% of deployments).

TABLE 5-1 Exposures of Interest and Associated Questions

Question		Response[a]
Burn Pits		
1.2.F	Did your duties during these dates include the burn pit (examples include trash burning, hauling trash to the burn pit, burn pit security, trash sorting at the burn pit)?	Yes/No
1.2.G	On a typical day, how many hours did smoke or fumes from the burn pit enter your work site or housing?	1–24 hours
Dust		
1.4.C	In a typical month during your deployment(s), how many days did you experience dust storms?	1–31 days
1.3.C[b]	In a typical month, how many days were you in convoy or other vehicle operations?	1–31 days
1.3.F	In a typical month, how many days did you perform construction duties?	1–31 days
Diesel, Exhaust, and Fuel		
1.3.C[b]	In a typical month, how many days were you in convoy or other vehicle operations?	1–31 days
1.3.D	In a typical month, how many days did you perform refueling operations?	1–31 days
1.3.E	In a typical month, how many days did you perform aircraft, generator, or other large engine maintenance?	1–31 days
Combat		
1.3.A	Were you ever close enough to feel the blast from an IED (improvised explosive device) or other explosive device?	Yes/No
1.3.B	In a typical month, how many days were you near heavy smoke from weapons, signal smoke, markers or other combat items?	1–31 days
1.3.C[b]	In a typical month, how many days were you in convoy or other vehicle operations?	1–31 days
Construction		
1.3.F	In a typical month, how many days did you perform construction duties?	1–31 days
Soot (1990–1991 Gulf War only)		
1.2.A	Were you exposed to soot, ash, smoke, or fumes from the Gulf War oil fires?	Yes/No

[a] For questions that required input of time, "never" was the first response option and the range of hours or days was available as the second possible response to the question.
[b] Question 1.3.C was used to examine dust, diesel, exhaust, and fuel, and combat exposures.

Figure 5-2 shows the distribution of days per month of exposure to other hazards. The questionnaire did not ask respondents to estimate their burn pit exposure in days per month. Responses were grouped into 5-day categories because a review of the raw data suggested that those categories were natural breaking points, with greater numbers of respondents endorsing 5, 10, 20, 25, or 31 days of exposure per month. Most respondents indicated 0 to 5 days of exposure; however, for Questions 1.3.B, 1.3.C, 1.3.D, and 1.3.E, the second most commonly reported duration was 26 to 31 days.

Respondents with No Exposure

A small proportion of respondents indicated that they were *not* exposed to airborne hazards or open burn pit emissions. With regard to burn pit exposure, fewer than 18% of deployment segments represented in the registry were not near burn pits (1.2.D), and 29% of deployment segments represented in the registry did not have duties that involved the burn pit, (1.2.F; see Table 5-2). Pertaining to exposures to blasts, smoke from weapons, convoys, refueling, large engine maintenance, construction, and pesticides (Questions 1.3A–G), 24% of respondents indicated that they were not exposed to any of these hazards. Only 1% of respondents indicated that they had never experienced a dust storm (1.4.C). Among respondents who had been deployed in the 1990–1991 Gulf War, 6% reported that they were never exposed to soot from Gulf War oil-well fires (1.2.A; see Table 5-2).

TABLE 5-2 Descriptive Statistics for Questions Related to Deployment, Environmental, and Occupational Exposures

Question		Yes N (%)	No N (%)	Missing[a] N (%)	Total N
1.2.A	Were you exposed to soot, ash, smoke, or fumes from the Gulf War oil fires?	5,726 (85.6)	380 (5.7)	585 (8.7)	6,694[b]
1.2.D	Were you near a burn pit during these duties (on the base or close enough to the base for you to see the smoke)?	129,192 (62.6)	36,533 (17.7)	40,648 (19.7)	206,373[c]
1.2.F	Did your duties during these dates include the burn pit (examples include trash burning, hauling to the burn pit, burn pit security, trash sorting at the burn pit)?	66,087 (32.0)	60,523 (29.3)	79,763 (38.7)	206,373[c]
1.3.A	Were you ever close enough to feel the blast from an IED (improvised explosive device) or other explosive device?	33,517 (72.2)	11,526 (24.8)	1,321 (2.9)	46,404

		Mean	Med	SD	Missing[a] N (%)	Total N
1.2.G	On a typical day, how many hours did smoke or fumes from the burn pit enter your work site or housing?	7.5	3.0	9.2	101,502 (49.2)	206,373[c]
1.2.I	On a typical day, how many hours were you near (for example you could smell or see it) sewage ponds?	9.3	5.0	9.8	80,743 (39.1)	206,373[c]
1.3.B	In a typical month, how many days were you near heavy smoke from weapons, signal smoke, markers, or other combat items?	13.7	12.0	12.0	10,420 (22.5)	46,404
1.3.C	In a typical month, how many days were you in convoy or other vehicle operations?	15.3	15.0	12.0	2,677 (5.8)	46,404
1.3.D	In a typical month, how many days did you perform refueling operations?	12.6	10.0	12.1	3,575 (7.7)	46,404
1.3.E	In a typical month, how many days did you perform aircraft, generator, or other large engine maintenance?	8.2	0.0	12.2	2,786 (6.0)	46,404
1.3.F	In a typical month, how many days did you perform construction duties?	4.5	0.0	8.3	4,767 (10.3)	46,404
1.4.C	In a typical month during your deployment(s), how many days did you experience dust storms?	8.9	6.0	7.5	6,190 (13.3)	46,404

NOTE: Med = median; SD = standard deviation.

[a] Don't know, missing, refused, or skipped.

[b] Question asked only for Gulf War deployments.

[c] Question asked for each deployment.

TABLE 5-3 Number of Exposures Other Than Burn Pits, Gulf War Oil-Well Fire Soot, or Dust Storms Reported by Respondents*

Number of Exposures in Addition to Burn Pits, Gulf War Oil-Well Fire Soot, or Dust Storms Reported	Number of Respondents	Percentage of Respondents
0	10,963	23.8
1	31,248	67.7
2	3,311	7.2
3	524	1.1
4	88	0.2
5 or 6	11	0.0

*These exposures are blast, smoke from weapons, convoys, refueling operations, large engine maintenance, construction, or pesticides associated with military occupations (Questions 1.3.A–G).

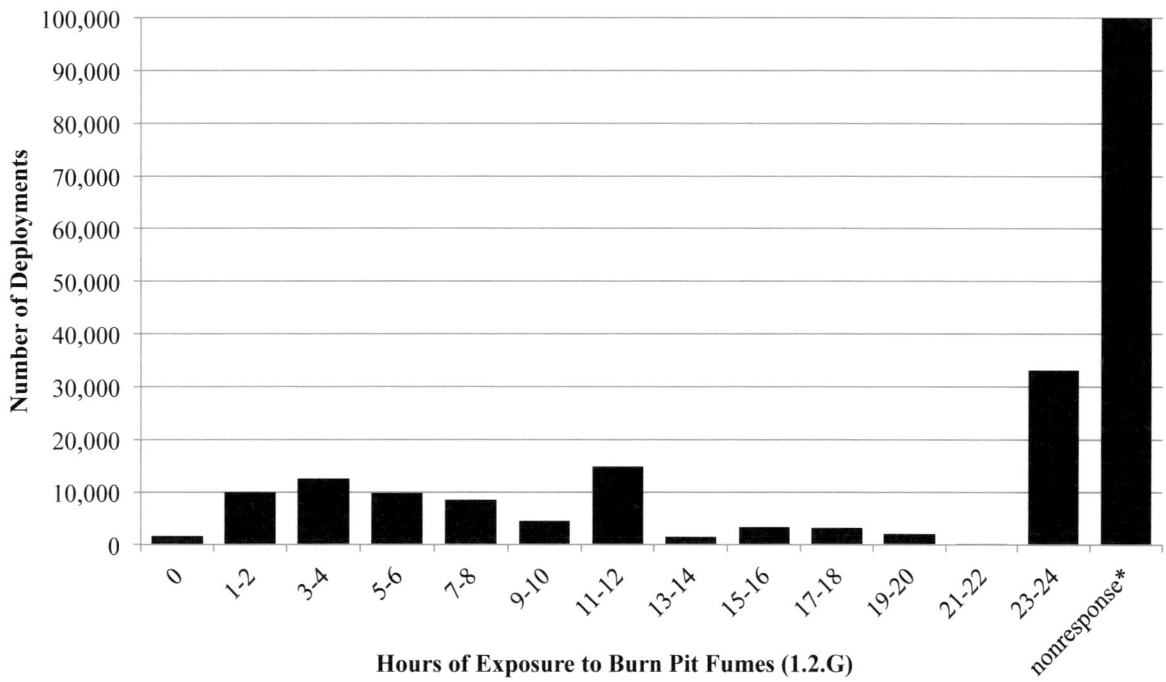

FIGURE 5-1 Distribution of reported hours of burn pit exposure per day.
* Nonresponse includes skipped, refused, missing, and "don't know."

All respondents reported exposure to at least two of the ten exposures that all respondents were asked about (soot was asked about only for Gulf War veterans): burn pits (1.2.D), sewage ponds (1.2.I), blasts (1.3.A), smoke from weapons (1.3.B), convoy (1.3.C), refueling operations (1.3.D), large engine maintenance (1.3.E), construction (1.3.F), pesticides (1.3.G), and dust storms (1.4.C). One individual reported exposure to just 3 of the 10 hazards, and only 4 reported exposure to only 4 of the 10 hazards. However, 2% of respondents did not answer all 10 questions (missing, don't know, or refused responses).

Because the proportion of respondents who were not exposed is small, a better reference group for analyses would be those with levels of exposure believed to be relatively low.

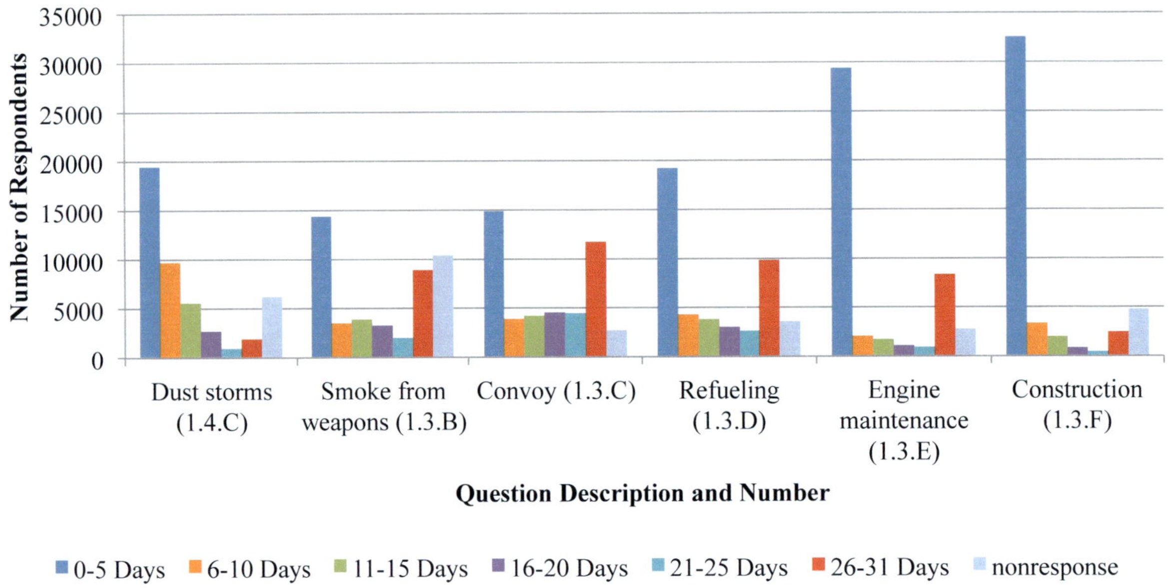

FIGURE 5-2 Distribution of reported duration (days per month) of exposures.
NOTE: Nonresponse includes refused, missing, and "don't know."

Committee-Created Exposure Variables

Several approaches were used to explore the data because each approach was viewed as having significant limitations by itself, given the underlying issues with the data. By using several approaches, the committee aimed for a more complete picture of the exposures. Thus, the committee examined burn pit exposure based on responses to the individual questions related to burn pit exposure as well as with a committee-created variable that combined those responses to create an exposure potential score. The committee also developed a cumulative metric for use in multivariate analyses.

One method the committee developed resulted in a reduced set of exposure metrics to characterize exposures to sources; the metric included information on whether an exposure may have occurred and, when possible, the duration of the exposure, with the recognition that any metric was potentially flawed, so undue precision would not be assumed. Self-reported information on exposures was interpreted as providing indicators of potential exposure to those agents (the committee uses the term "exposure potential" in its analysis to reflect this). Having a reduced set of metrics that more broadly characterizes exposures also serves in the interpretation of the exposure-related questions. These metrics can be used themselves or combined to provide a distribution of overall exposures to multiple potential agents.

Because of the qualitative nature of the information collected, ordinal variables expressing low, middle, or high levels were created for each of six exposures of interest (burn pits, dust, diesel/exhaust/fuel, combat, construction, and Gulf War oil-well fire soot) to express the gradation of exposures. Categories of exposure were assigned based on a score that incorporated responses to one, two, or three questions that, collectively, characterized the exposure potential of each respondent. (Gulf War oil-well fire soot exposure is based on one question, and that exposure is specific to soot exposure in Kuwait during the Gulf War, as opposed to soot exposure to burn pits or other combustion sources. This exposure was viewed as being particularly unique, given the timing and source differences.) The magnitude of the score is used as an indicator of the magnitude of exposure, ranging from 0 to indicate never exposed or no exposure to 6 to indicate greater exposure. The integration of both the binary and duration aspects of the questions captures both potential acute and chronic exposures, recognizing that individuals have varying levels of response and reflecting the limited information on exposure–response relationships for the agents of concern. A representation of how the exposure potential score was derived for each exposure (burn pits, dust, diesel/exhaust/fuel, combat, construction, and Gulf War oil-well fire soot) is presented in Box 5-2.

For burn pit exposure potential, the possible responses to Questions 1.2.F and 1.2.G (see Table 5-1) were combined. Because burn pit exposure was assessed for each deployment and the committee sought a representation of exposure for each individual, the created variable used responses to Question 1.2.F to indicate *ever exposed* (3) or *never exposed* (0) and the cumulative number of hours of burn pit exposure collected by Question 1.2.G across deployments. For persons with a cumulative number of hours of burn pit exposure in the highest tertile, 3 points were assigned, for the middle tertile, 2 points were assigned, and for the lowest tertile, 1 point was assigned. By adding the scores for each part (ever/never exposed + tertile of cumulative hours of burn pit exposure), the resulting exposure potential scores ranged from 0 (never/no exposure) to 6 (ever exposed/highest tertile of cumulative hours).

The dust exposure potential scores were based on responses to Questions 1.4C, 1.3C, and 1.3F, which asked the respondent to report the number of days per month that the respondent experienced dust storms, was in a convoy, and performed construction duties. Scores were derived by adding the points assigned for each questions (highest tertile, 2 points; middle tertile 1 point; and lowest tertile 0 points). The resulting dust exposure potential score ranged from 0–6.

The diesel/exhaust/fuel exposure potential scores were based on responses to Questions 1.3C, 1.3.D, and 1.3.E, which elicited the number of days per month the respondent was in a convoy, performed refueling duties, and performed large engine maintenance. Scores were derived by adding the points assigned for each questions (highest tertile, 2 points; middle tertile, 1 point; and lowest tertile, 0 points). The resulting diesel/exhaust/fuel exposure potential score ranged from 0–6.

The combat exposure potential score was based on responses to three questions (1.3.A, 1.3.B, and 1.3.C). The possible responses to 1.3.B and 1.3.C were combined so that respondents with a high number of days exposed to combat-related smoke and a high number of days in a convoy (and thus were potentially exposed to combat-related stressors), were assigned a score of 4, whereas respondents reporting low exposure to both were assigned a score of 0. Responses indicating exposure to blast were incorporated by adding two points for a positive (yes) response to question 1.3.A. This resulted in a range of scores from 0 to 6 for combat exposure.

The construction exposure potential was based on one question that asked the respondent to report the number of days per month that he or she performed construction duties (1.3.F). Again, points were awarded based on tertiles, the lowest tertile was assigned a score of 0, the middle tertile was assigned a score of 3, and the highest tertile was assigned a score of 6 so that scores ranged from 0 to 6.

The Gulf War oil-well fire soot exposure potential score was based on the response to one question (1.2.A) with a yes or no response. A response endorsing exposure to Gulf War oil-well fire soot was assigned a score of 6, whereas 0 was assigned to responses reporting no exposure.

The exposure potential scores were transformed into categories of exposure potential reflecting assumed low, medium, and high potential for exposure so that each service member had one estimate of exposure potential for each exposure variable. Low, medium, or high exposure was meant to reflect the gradation of exposures but not in a quantitative way. Table 5-4 shows how scores for each variable were assigned to low, medium, or high categories of exposure potential.

The distributions for the exposure variables that the committee created for the six exposures of interest are shown in Figure 5-3. A medium level of exposure potential is most predominant for burn pits, dust, diesel/exhaust/fuel, and combat exposure. However, there were nearly as many respondents with a high exposure potential to burn pits as medium. Combat exposure also had a relatively high proportion of respondents in the high and medium exposure potential groups. Low-level exposure potential is dominant for construction. Gulf War oil-well fire soot exposure is different in that nearly all who had eligible Gulf War deployments indicated that they had exposure, which falls into the high exposure potential category. Only 146 Gulf War respondents indicated that they did not have exposure to oil-well fire soot.

BOX 5-2
Scores for Potential Exposures

Burn Pit exposure potential based on responses to Questions 1.2.F and 1.2.G:

1.2.F (ever exposed)		**1.2.G** (cumulative number of hours of exposure)		
No = 0		No exposure = 0		
Yes = 3	+	Lowest tertile = 1	=	0 to 6
		Middle tertile = 2		
		Highest tertile = 3		

Dust exposure potential based on responses to Questions 1.4C, 1.3C, and 1.3F:

1.4.C		**1.3.C**		**1.3.F**		
Lowest tertile = 0		Lowest tertile = 0		Lowest tertile = 0		
Middle tertile = 1	+	Middle tertile = 1	+	Middle tertile = 1	=	0 to 6
Highest tertile = 2		Highest tertile = 2		Highest tertile = 2		

Diesel, Exhaust, and Fuel exposure potential based on responses to Questions 1.3C, 1.3.D, and 1.3E:

1.3.C		**1.3.D**		**1.3.E**		
Lowest tertile = 0		Lowest tertile = 0		Lowest tertile = 0		
Middle tertile = 1	+	Middle tertile = 1	+	Middle tertile = 1	=	0 to 6
Highest tertile = 2		Highest tertile = 2		Highest tertile = 2		

Combat exposure potential based on responses to Questions 1.3.A, 1.3.B, and 1.3.C:

1.3.A		**1.3.B**		**1.3.C**		
No = 0		Lowest tertile = 0		Lowest tertile = 0		
Yes = 2	+	Middle tertile = 1	+	Middle tertile = 1	=	0 to 6
		Highest tertile = 2		Highest tertile = 2		

Construction exposure potential based on the response to Question 1.3.F:

1.3.F		
Lowest tertile = 0		
Middle tertile = 3	=	0 to 6
Highest tertile = 6		

Gulf War Oil-Well Fire Soot exposure potential based on response to Question 1.2.A:

1.2.A		
No = 0	=	0 to 6
Yes = 6		

TABLE 5-4 Categories of Exposure Potential Based on Exposure Potential Scores for Each Exposure Variable

Categories of Exposure Potential	Variables: Burn Pits	Dust	Diesel	Combat	Construction	Gulf War Oil-Well Fire Soot
Low	0	0	0	0	0	0
	1	1	1	1		
Medium	2	2	2	2	3	
	3	3	3	3		
	4	4	4	4		
High	5	5	5	5	6	6
	6	6	6	6		

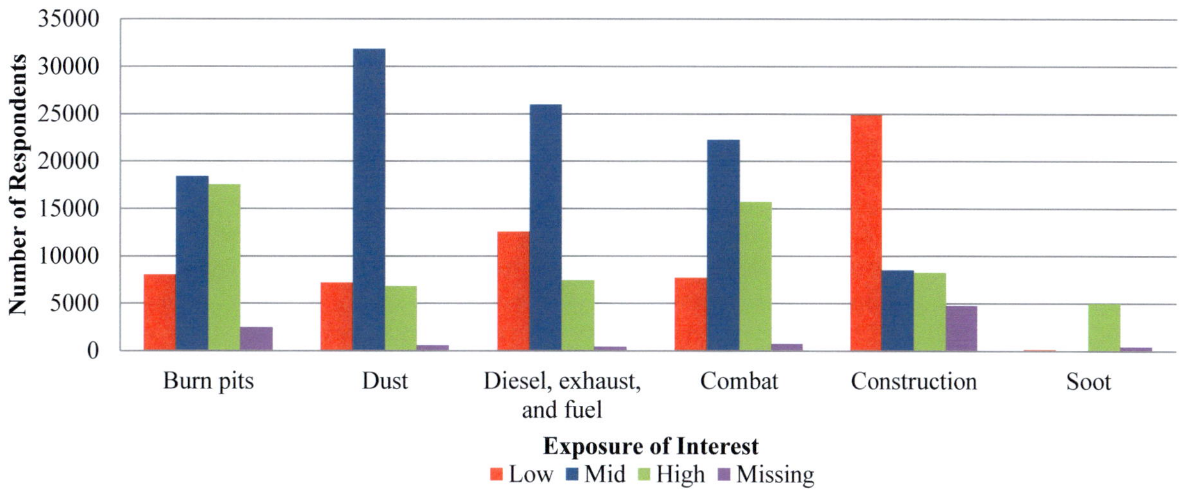

FIGURE 5-3 Number of respondents in each exposure category for each exposure of interest.

Collinearity Among Exposure Categories

Potential collinearity among exposure potential scores (0 to 6) was examined using Pearson correlation coefficients. An assessment of collinearity is informative in interpreting the resulting health analyses, and it also reflects the tendencies of persons to report being consistently highly exposed to multiple agents, as discussed previously.

All six exposure potential variables were statistically significantly correlated, with one exception. Construction and Gulf War soot from oil fires were the only two not correlated ($R = 0.02$, $p = 0.15$). All other correlations ranged from 0.04 (Gulf War oil-well fire soot and burn pits, $p = 0.01$) to 0.71 (construction and dust, $p < 0.01$). This indicates that respondents who reported one exposure of interest were likely to report other exposures. However, correlation between dust, diesel/exhaust/fuel, and combat exposure variables was high due to the fact that all three incorporate Question 1.3.C (days per month being in a convoy or other vehicle operations). The same occurs for dust and construction, which both incorporate Question 1.3.F (days performing construction duties).

Exposures Among Gulf War Service Members

Given the different context and activities of the 1990–1991 Gulf War compared with the later conflicts, the committee examined the exposure potential categories (low, medium, high) reported among the 5,595 service members who had been deployed during the Gulf War time period.[3] The proportions of Gulf War service members with low, medium, or high exposure potential to burn pits, dust, diesel/exhaust/fuel, combat, and construction were very similar to those of the complete cohort of respondents (see Figure 5-3). Generally, the difference between Gulf War service members and the larger group was less than 5% for each level and each variable. The greatest difference occurred between the percentage of Gulf War service members with medium exposure potential to burn pits (49.7%) and those of all respondents (41.8%).

Composite Exposure Potential

Because there were many sources of airborne emissions that contributed to a service member's exposures to PM (dust storms, convoys, construction), and PAHs, VOCs, and PCDD/Fs (refueling operations, convoys, large engine maintenance) in addition to burn pits, and because there were insufficient data[4] with which to determine which sources contributed the most or posed the most harm, the committee chose to weigh each potential exposure equally and to create a metric that places emphasis on the totality of exposures. This is similar to the U.S. National Ambient Air Quality Standards PM standard which is defined for all particles in a certain size range and does not discriminate by source (EPA, 2013). Given the construction of the variables developed, this approach acknowledges that multiple physiologic insults weigh on the health of the service member.

To examine the experience and potential health effects of highly exposed individuals, the committee devised an approach to identify composite exposure potential levels across all six exposure potential variables. For each individual, the number of low and high exposure potentials, up to six each, was determined. For example, a respondent who had a low exposure potential for two of the exposure potential variables, a medium exposure potential for two variables, and a high exposure potential for the last two variables. A composite exposure potential score to all variables was then created by assigning a 0 to the lowest exposure tier, 1 to the middle tier, and 2 to the most exposed tier from the self-reported exposures. The respondent in the example above would have a score of 6 $[(2\times0)+(2\times1)+(2\times2) = 6]$. The possible scores range from 0 (low exposure potential to all six variables) to 12 (high exposure potential to all six variables). Respondents with equal numbers of high and low exposures, or many medium-level exposures would score about 6.

The committee-created composite exposure potential score generally followed a normal distribution (see Figure 5-4). The majority of scores (16.6%) were at the mid-point (6), and very few individuals reached either extreme (0.3% on the high end, and none on the low end). This is not unexpected given how the exposure metrics were created (e.g., using tertiles of the distributed variables), though it indicates that few individuals were at one extreme (low or high) for the majority of the exposure metrics. However, the proportion of respondents in the two most exposed groups (composite exposure potential scores 11 and 12) is higher compared with the two least exposed groups (composite exposure potential scores 0 and 1), indicating some skewness. As noted previously, there is a significant degree of correlation among the exposure metrics, and respondents who reported on one exposure of interest were likely to report other exposures.

EXPOSURE METRICS FOR MULTIVARIATE ANALYSES

A set of "cumulative" exposure metrics were developed to characterize burn pit exposure. The multivariate analyses were conducted only among post-9/11 respondents to control for different latency periods and because the types of exposures encountered were likely different for Gulf War and post-9/11 service members. These measures exploit the fact that burn pit exposure is asked separately for each deployment, allowing the construction of cumulative exposure measures by multiplying responses by the length of each deployment. These metrics focused on burn pit exposures and were calculated in three different ways:

[3] This group includes service members who also deployed in the theater of operations in later time periods.

[4] There are a few exceptions where environmental monitoring data are available for some sort of quantitative exposure assessment, such as at Joint Base Balad, but those data are quite limited and were not viewed as being sufficient for a meaningful exposure analysis (IOM, 2011). Furthermore, the committee did not have access to base-specific deployment information.

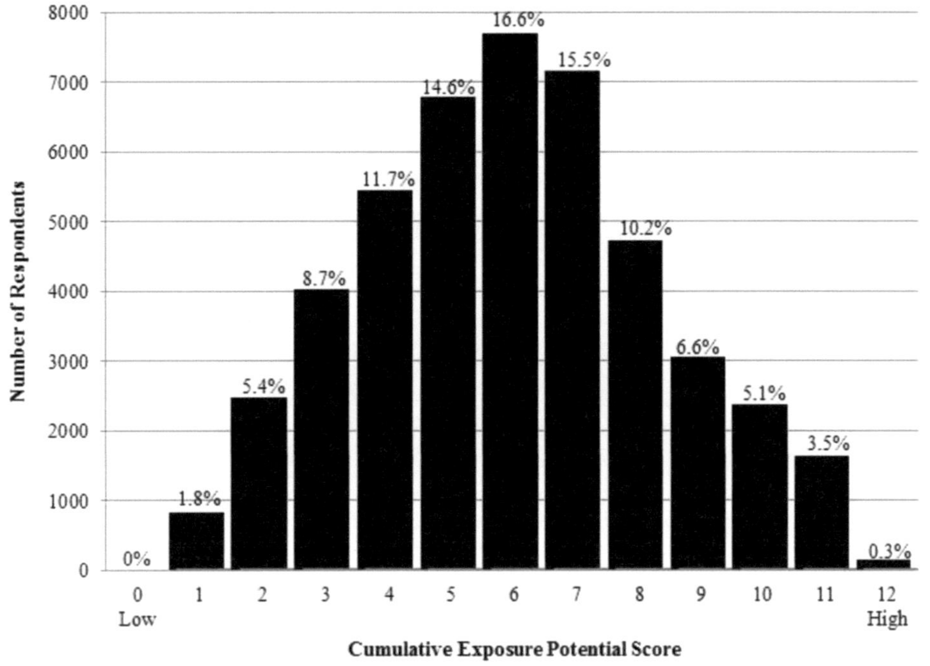

FIGURE 5-4 Distribution of composite exposure potential score.

1. Cumulative days deployed near a burn pit, derived by summing the number of days of each deployment for which a respondent indicated that he or she was near a burn pit (based on Question 1.2.D) expressed as quartiles;
2. Cumulative days deployed with burn pit duty, derived by summing the number of days of each deployment for which a respondent indicated that he or she had duties that included the burn pit (based on Question 1.2.F) expressed as quartiles; and
3. Cumulative hours of exposure to burn pit smoke, derived as the product of the number of days of deployed times the hours per day that smoke or fumes from burn pits entered the work site or housing (based on Question 1.2.G), summed over all deployments and expressed as quartiles.

Given the importance of the other sources of the main pollutants released by the burn pits (PM, PAHs, VOCs, PCDD/Fs), the committee felt it was appropriate to create a variable that would characterize exposure to these pollutants. To do so, it used the composite exposure potential variable created to qualitatively express the potential exposure to all six main sources (burn pits, dust, diesel/exhaust/fuel, combat, construction, and Gulf War oil-well fire soot). In the multivariate analyses, the composite exposure potential score (0 to 12) was further consolidated and expressed as quartiles. For the three cumulative metrics of burn pit exposure and the composite exposure potential metric expressed as quartiles, the values that define each quartile are presented in Table 5-5.

The association between health outcomes and exposure potential was also examined using the committee's exposure potential metric (0 to 6) for dust, diesel/exhaust/fuel, combat exposure, and construction. Soot from Gulf War oil-well fires was not examined for the multivariate analyses because the question about soot was only asked of the Gulf War respondents. Burn pit exposure as expressed by the committee's exposure potential variable was included as a qualitative fourth proxy of burn pit exposure to be consistent with the presentation of the other exposures.

TABLE 5-5 Quartiles of Burn Pit and Composite Exposure Potential Metrics Used in Multivariate Analyses

Variable	Q1	Q2	Q3	Q4
Composite exposure potential (score 0 to 12)	0–4	5–5	6–7	8–12
Cumulative deployment days near a burn pit	0–208	209–333	334–518	519–4,513
Cumulative deployment days with burn pit duties	0–0	1–183	184–348	34–4,144
Cumulative smoke hours	0–729	730–2,726	2,727–6,116	6,120–97,692

Changes in Burn Pit Exposure Over Time

The committee examined changes in the patterns of reporting burn pit exposure over time by location. The analyses were limited to country and year of deployment, although others have been able to access and conduct similar analyses with more detailed data, including base location, compared with knowledge about when and where incinerators were in use (Gasper and Kawata, 2015). The number of deployments for which high exposure to burn pits as defined by the committee's exposure potential variable was reported (rather than by the individual) by the country of deployment (Iraq, Afghanistan, or Kuwait) were examined by year of deployment, shown in Figure 5-5. The use of burn pits in Iraq and Afghanistan but not in Kuwait is well described (Liu et al., 2016).

As shown in Figure 5-5, reported burn pit exposure by year of deployment could be interpreted as an indicator of registry or deployment data validity because there were fewer reports of high burn pit exposures in Kuwait than

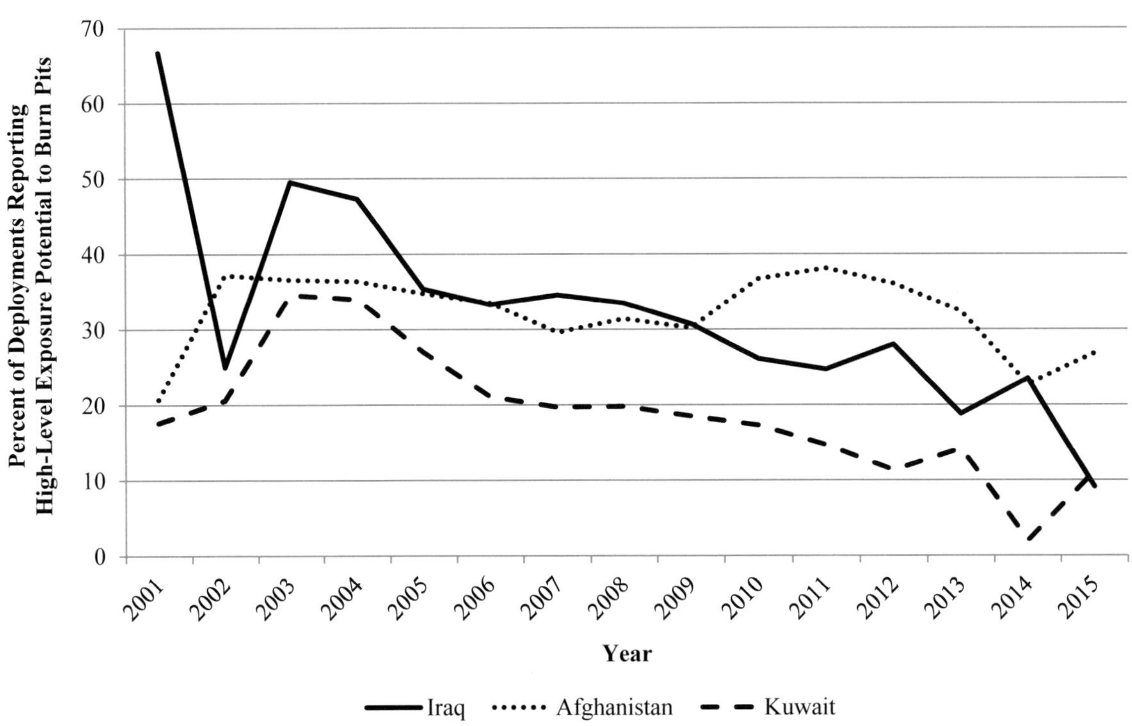

FIGURE 5-5 Percentage of deployments reporting high exposure potential to burn pits by country and year of deployment.

for Iraq and Afghanistan, as would be expected. On the other hand, the figure also raises some uncertainty about responses since if there were no burn pits in Kuwait, the exposures should presumably have been even lower than reported and shown. Perhaps the Kuwait data in part reflects the issue of multiple deployments and uncertainty as to where and when the burn pit exposures occurred (for example, military versus civilian trash-burning activities). Smaller-scale trash burning at bases without large burn pits may have been interpreted as burn pits.

As noted early in this chapter, the use of burn pits in Iraq and Afghanistan began to decline in 2009, and one would thus expect to see a dip in the number of deployments for which high burn pit exposure was reported after 2009. While the percentage of deployments to Iraq with high burn pit exposure potential shows a relatively steady decline since 2003, there is no notable deviation around 2009. Deployments to Afghanistan, however, show a rise in the percent of deployments with reported burn pit exposure in 2010 and a decrease after 2012.

SYNOPSIS AND CONCLUSIONS

Based on the information presented in this chapter, the committee has reached the following findings, conclusions, and recommendations regarding the analysis and interpretation of AH&OBP Registry exposure data.

Assessing exposure using self-reported registry data has a number of inherent limitations—even if the registry was well designed and there is a careful selection of items for analysis. The information collected by the AH&OBP Registry questionnaire consists of self-reports of

- location-specific deployment-related exposures (oil-well fires, burn pits, and sewage ponds);
- general military occupational exposures (being near heavy smoke from weapons; being in a convoy; performing refueling duties; performing aircraft, generator, or other large engine maintenance; performing construction duties; or performing pesticide duties);
- environmental exposures and regional air pollution (air quality, dust storms);
- exposures not related to military service, including nonmilitary occupational exposures (such as working as a fire fighter or in a dusty job);
- residential and hobby-related exposures (living near a farm or recreational woodworking, for example); and
- exposures to tobacco smoke and consumption of alcohol.

Other potentially problematic exposures such as endemic diseases, insects, depleted uranium, and hazards associated with cleanup operations in the 1990–1991 Gulf War were not included.

As detailed in Chapters 3 and 4, the committee identified several flaws with the way these data were collected and found that there was a particular problem with deployment-related exposure questions, which asked for specific information for each separate segment of a respondent's time in theater. In addition, the information collected has a number of limitations, including the fact that self-reported data are subject to recall and reporting biases. The questions do not provide information on the intensity of exposure beyond a binary yes/no for exposure even though intensity is a central component of exposure characterization. While potential surrogates for the intensity of exposure to sources such as distance from a source are often used in analyses like this, such information was not available to the committee.

The data on burn pit exposures are limited by the lack of details on the chemicals and PM that comprised that exposure; other occupational and environmental sources of airborne pollutants, troop location, meteorological, satellite, or other data by which to conduct exposure assessments; and the absence of information that would allow for the consideration of acute versus chronic exposures. The analysis of the exposure data is complicated by the high fraction of registry participants reporting potential exposures to both burn pit emissions and dust, particularly dust storms but also convoys and construction. For many of the questions, there was a very high percentage of respondents indicating exposures, and there was a tendency for individuals reporting exposures to one type of source to report exposures to other sources as well. Some questions had high rates of non-response.

Thus, given the charge, and given the concern for over-interpreting the data at hand, the committee developed a reduced set of metrics to express exposure potential. Because there were many sources of airborne emissions that contributed to a service member's exposures to PM (dust storms, convoys, construction), and PAHs, VOCs, and PCDD/Fs (refueling operations, convoys, large engine maintenance) in addition to burn pits, and because there are insufficient data by which to determine which sources contributed the most or posed the most harm, the committee chose to weigh each potential exposure equally and to focus on the totality of exposures. Specifically, the metrics combined responses to multiple questions into single indicators of potential exposure for each of six exposures of interest: burn pits, dust, diesel/exhaust/fuel, construction, combat, and Gulf War oil-well fire soot. Combining responses that are related to a similar exposure can reduce the resulting number of variables (dimensionality) to be considered in health studies and can be used to construct an overall exposure to multiple stressors. The chapter text contains descriptive statistics related to the metrics.

On the basis of its evaluation, the committee concludes that the exposure data are of insufficient quality or reliability to make them useful in anything other than the most general evaluations of exposure potential. Under these limited circumstances, it believes that there may be some circumstances where supplementing these data with information from on-site environmental monitoring or meteorological, satellite, or other relevant measurements or observations might yield results that would better quantify the variation in exposures to specific constituents, thereby allowing more detailed assessments of health outcomes in particular populations.

The exposure potential metrics are also used in the health outcome assessment described in Chapter 6, individually and in combination, to indicate cumulative exposure potential in order to account for both the exposures to sources individually and multiple exposures to potentially harmful agents or stressors. The committee wishes to emphasize that this was done in the service of fulfilling the statement-of-task directive to address associations of self-reported exposures with self-reported health conditions and is not an endorsement of the data's suitability for this task.

REFERENCES

AFHSC (U.S. Armed Forces Health Surveillance Center), Naval Health Research Center, and U.S. Army Public Health Command. 2010. *Epidemiological studies of health outcomes among troops deployed to burn pit sites*. Silver Spring, MD: Defense Technical Information Center. http://fhpr.dhhq.health.mil/Libraries/NEWS_DOCUMENTS/100604_FINAL_Burn_Pit_Epi_Studies.sflb.ashx (accessed October 1, 2016).

APHC (U.S. Army Public Health Command). 2010. *Screening health risk assessments, Joint Base Balad, Iraq, 11 May–19 June 2009*. Aberdeen Proving Ground, MD: U.S. Army Center for Health Promotion and Preventive Medicine. July.

Baird, C. P., S. DeBakey, L. Reid, V. D. Hauschild, B. Petruccelli, and J. H. Abraham. 2012. Respiratory health status of U.S. Army personnel potentially exposed to smoke from 2003 Al-Mishraq sulfur plant fire. *Journal of Occupational and Environmental Medicine* 54(6):717–723.

Blasch, K. W., J. E. Kolivosky, and J. M. Heller. 2016. Environmental air sampling near burn pit and incinerator operations at Bagram Airfield, Afghanistan. *Journal of Occupational and Environmental Medicine* 58(8 Suppl 1):S38–S43.

Bosnjak, M., and T. L. Tuten. 2001. Classifying response behaviors in Web-based surveys. *Journal of Computer-Mediated Communication* 6(3). doi:10.1111/j.1083-6101.2001.tb00124.x.

Cassee, F. R., M. E. Héroux, M. E. Gerlofs-Nijland, and F. J. Kelly. 2013. Particulate matter beyond mass: Recent health evidence on the role of fractions, chemical constituents and sources of emission. *Inhalation Toxicology* 25(14):802–812.

CHPPM and AFIOH (U.S. Army Center for Health Promotion and Preventive Medicine and U.S. Air Force Institute for Operational Health). 2008. *Screening health risk assessment burn pit exposures, Balad Air Base, Iraq, and Addendum, May 2008*. USACHPPM Report No. 47-MA-08PV-08/AFIOH Report No. IOH-RS-BR-TR-2008-0001. Aberdeen Proving Ground, MD: U.S. Army Center for Health Promotion and Preventive Medicine.

DoD (Department of Defense). 2011. Exposure to toxins produced by burn pits: Congressional data request and studies. In Memorandum for the Assistant Secretary of Defense for Health Affairs. Washington, DC: Department of Defense.

Engelbrecht, J. P., E. V. McDonald, J. A. Gillies, and A. W. Gertler. 2008. *Department of Defense Enhanced Particulate Matter Surveillance Program (EPMSP)*. Final report. Reno, NV: Desert Research Institute. https://phc.amedd.army.mil/PHC%20Resource%20Library/Final%20EPMSP%20Report%20without%20appx%20Feb08.pdf (accessed September 4, 2016).

Engelbrecht, J. P., E. V. McDonald, J. A. Gillies, R. K. M. Jayanty, G. Casuccio, and A. W. Gertler. 2009. Characterizing mineral dusts and other aerosols from the Middle East—Part 1: Ambient sampling. *Inhalation Toxicology* 21(4):297–326.

EPA (Environmental Protection Agency). 1997. *Evaluation of emissions from the open burning of household waste in barrels, Vol 1*. Technical Report EPA/600/R-97-134a. Research Triangle Park, NC: Environmental Protection Agency.

EPA. 2001. Database of sources of environmental releases of dioxin-like compounds in the United States. EPA/600/C-01/012. http://cfpub.epa.gov/ncea/cfm/recordisplay.cfm?deid=20797 (accessed September 23, 2010).

EPA. 2011. Community-based air pollution projects glossary. https://iaspub.epa.gov/sor_internet/registry/termreg/searchandretrieve/glossariesandkeywordlists/search.do?details=&vocabName=Community-Based%20Air%20Pollution&uid=1821538 (accessed September 15, 2016).

EPA. 2013. National ambient air quality standards for particulate matter. *Federal Register* 78 (10):3086–3287. https://www.gpo.gov/fdsys/pkg/FR-2013-01-15/pdf/2012-30946.pdf (accessed September 15, 2016).

GAO (Government Accountability Office). 2004. *Gulf War illnesses: DOD's conclusions about U.S. troops' exposure cannot be adequately supported*. Testimony before the Subcommittee on National Security, Emerging Threats, and International Relations, Committee on Government Reform, House of Representatives. GAO-04-821T. Washington, DC: GAO.

GAO. 2010. DOD should improve adherence to its guidance on open pit burning and solid waste management. http://www.gao.gov/assets/320/311365.pdf (accessed September 15, 2016).

Gasper, J., and J. Kawata. 2015. An evaluation of selection bias and data quality: Open burn pit and airborne hazards registry. Prepared for the Department of Veterans Affairs Office of Public Health, Post-Deployment Health Group by Westat. September.

Harrison, R. M., and J. Yin. 2000. Particulate matter in the atmosphere: Which particle properties are important for its effects on health? *Science of the Total Environment* 249(1–3):85–101.

IOM (Institute of Medicine). 2006. *Gulf War and health, volume 4: Health effects of serving in the Gulf War.* Washington, DC: The National Academies Press.

IOM. 2010. *Gulf War and health, volume 8: Update of health effects of serving in the Gulf War.* Washington, DC: The National Academies Press.

IOM. 2011. *Long-term health consequences of exposure to burn pits in Iraq and Afghanistan*. Washington, DC: The National Academies Press.

IOM. 2013. *Returning home from Iraq and Afghanistan: Assessment of readjustment needs of veterans, service members, and their families*. Washington, DC: The National Academies Press.

Jones, K. A., B. Smith, N. S. Granado, E. J. Boyko, G. D. Gackstetter, M. A. K. Ryan, C. J. Phillips, T. C. Smith, and the Millennium Cohort Study Team. 2012. Newly reported lupus and rheumatoid arthritis in relation to deployment within proximity to a documented open-air burn pit in Iraq. *Journal of Occupational and Environmental Medicine* 54(6):698–707.

Lemieux, P. M., B. K. Gullett, C. C. Lutes, C. K. Winterrowd, and D. L. Winters. 2003. Variables affecting emissions of PCDD/Fs from uncontrolled combustion of household waste in barrels. *Journal of the Air & Waste Management Association* 53(5):523–531.

Lemieux, P. M., C. C. Lutes, and D. A. Santoianni. 2004. Emissions of organic air toxics from open burning: A comprehensive review. *Progress in Energy and Combustion Science* 30(1):1–32.

Liu, J., N. Lezama, J. Gasper, J. Kawata, S. Morley, D. Helmer, and P. Ciminera. 2016. Burn pit emissions exposure and respiratory and cardiovascular conditions among airborne hazards and open burn pit registry participants. *Journal of Occupational and Environmental Medicine* 58(7):e249–e255.

Lyles, M. B., H. L. Fredrickson, A. J. Bednar, H. B. Fannin, D. W. Griffin, and T. M. Sobecki. 2011. Medical geology: Dust exposure and potential health risks in the Middle East. 34th International Symposium on Remote Sensing of Environment. Sydney, Australia. April 10–15, 2011. http://www.isprs.org/proceedings/2011/ISRSE-34 (accessed September 15, 2016).

Masiol, M., C. T. M. Mallon, K. M. Haines Jr, M. J. Utell, and P. K. Hopke. 2016a. Airborne dioxins, furans, and polycyclic aromatic hydrocarbons exposure to military personnel in Iraq. *Journal of Occupational and Environmental Medicine* 58(8S):S22–S30.

Masiol, M., C. T. M. Mallon, K. M. Haines Jr, M. J. Utell, and P. K. Hopke. 2016b. Source apportionment of airborne dioxins, furans, and polycyclic aromatic hydrocarbons at a United States forward operating air base during the Iraq war. *Journal of Occupational and Environmental Medicine* 58(8S):S31–S37.

NRC (National Research Council). 2010. *Review of the Department of Defense Enhanced Particulate Matter Surveillance Program report*. Washington, DC: The National Academies Press.

Powell, T. M., T. C. Smith, I. G. Jacobson, E. J. Boyko, T. I. Hooper, G. D. Gackstetter, C. J. Phillips, B. Smith, and the Millennium Cohort Study Team. 2012. Prospective assessment of chronic multisymptom illness reporting possibly associated with open-air burn pit smoke exposure in Iraq. *Journal of Occupational and Environmental Medicine* 54(6):682–688.

Smith, B., C. A. Wong, E. J. Boyko, C. J. Phillips, G. D. Gackstetter, M. A. K. Ryan, T. C. Smith, and the Millennium Cohort Study Team. 2012. The effects of exposure to documented open-air burn pits on respiratory health among deployers of the Millennium Cohort Study. *Journal of Occupational and Environmental Medicine* 54:708–716.

Taylor, G., V. Rush, A. Deck, and J. A. Vietas. 2008. *Screening health risk assessment burn pit exposures, Balad Air Base, Iraq and addendum report*. IOH-RS-BR-TR-2008-0001/USACHPPM 47-MA-08PV-08. Brooks City-Base, TX: Air Force Institute for Operational Health and U.S. Army Center for Health Promotion and Preventative Medicine. May.

VA (Department of Veterans Affairs). 2016. U.S. Department Of Veterans Affairs: 10 things to know about burn pits. http://www.publichealth.va.gov/docs/exposures/ten-things-to-know-fact-sheet.pdf (accessed September 15, 2016).

WHO (World Health Organization). 2006. *Health risks of particulate matter from long-range transboundary air pollution*. Bonn, Germany: European Centre for Environment and Health. http://www.euro.who.int/__data/assets/pdf_file/0006/78657/E88189.pdf (accessed September 16, 2016).

Woodall, B. D., D. P. Yamamoto, B. K. Gullett, and A. Touati. 2012. Emissions from small-scale burns of simulated deployed U.S. military waste. *Environmental Science and Technology* 46:10997–11003.

6

Analysis and Interpretation of Registry Health Outcome Data

As discussed in the previous chapters, the data collected by the Airborne Hazards and Open Burn Pit (AH&OBP) Registry questionnaire have important limitations. Nonetheless, this was the information available to the committee. This chapter uses the health outcomes data collected by the questionnaire to examine overall prevalence rates of selected conditions among registry participants and to assess associations between those health outcomes and the exposures of interest.

The chapter begins with a description of the health outcomes data collected from the registry questionnaire and an assessment of the scientific validity of the collected information. Using that information and the results of previous reports and epidemiological studies on burn pit exposures and health effects, the committee addresses the conditions and diseases that are the most plausibly affected by exposures to burn pits and other airborne hazards that might have been experienced during deployment to the Southwest Asia theater of operations. Selected demographic and military characteristics are then presented and described for those health conditions the committee considered to be of most interest: respiratory and cardiovascular conditions. The final part of the chapter describes the methods and results of multivariate analyses of exposures and the selected health outcomes, a mandated part of the committee's charge that it has addressed with a full awareness of the serious inadequacies of using the data for such purposes. The chapter concludes with a discussion of how to interpret the associations of health outcome data and exposure data, but due to limitations noted, the committee does not draw any conclusions based on its analyses.

HEALTH OUTCOMES COLLECTED BY THE QUESTIONNAIRE

Box 6-1 lists all health outcomes identified in the questionnaire. All outcomes are self-reported. Some questions asked respondents about symptoms they have experienced or are currently experiencing. Others sought to capture the presence of specific diseases and were phrased as "Has a doctor or other health professional ever told you . . . ?" A few questions throughout the questionnaire included long lists of symptoms and conditions and asked respondents to indicate all that were applicable. However, many of the conditions in those lists were not included in more detail as separate questions elsewhere in the questionnaire.

The majority of health-related outcomes in the questionnaire focused on the respiratory and cardiovascular systems. While a few outcomes related to other organ systems were included, the committee regarded these questions as being too general to be of value in its analyses. Exposure to particulate matter from burn pits, dust, and other potential sources has been shown to acutely affect organ systems in addition to the respiratory and cardiovascular

BOX 6-1
Health Outcomes Examined in Questionnaire

Symptoms

- Wheezing or whistling in the chest
- Shortness of breath; breathlessness
- Decreased ability to exercise
- Hay fever or other respiratory allergy
- Sore throat, hoarseness, change in voice
- Chest pain, chest discomfort or chest tightness
- Chronic sinus infection/sinusitis
- Sputum or phlegm production for more than 3 weeks
- Neurological problems (numbness, tingling, or weakness in your arms or legs or difficulties with thinking or memory)
- Immune system problem
- Snoring and sleep apnea

Doctor-Diagnosed Conditions and Diseases

- Hay fever or allergies to pollen, dust, or animals
- Asthma
- Emphysema
- Chronic bronchitis
- Chronic obstructive pulmonary disease (COPD)
- Other lung disease or condition
- Constrictive bronchiolitis
- Idiopathic pulmonary fibrosis
- Hypertension
- Coronary artery disease
- Angina pectoris
- Myocardial infarction (heart attack)
- Liver condition
- Chronic multisymptom illness (examples include irritable bowel syndrome, chronic fatigue syndrome, and fibromyalgia)
- Cancer or a malignancy (tumor) (up to three types)

systems, and thus these other systems are potential candidates for further research on acute and long-term effects as well (IOM, 2011; see Tables 5-1 and 6-2).

The questionnaire does not elicit information on many of the conditions that may be possible confounders—such as a history of specific lung infections—or that have been found to be associated with similar exposures, such as reduced lung function. The committee cautions that the limited information—both in number of conditions asked about and the failure to elicit specific details—precluded it from performing detailed analyses of many conditions.

Responses to questions on cancer diagnoses and history were not included in the committee's analysis. This decision was based on several factors. First, the questions included in the cancer section were very general and limited to type and age at diagnosis for a maximum of three cancer diagnoses. No information was collected on stage, timing of related symptoms or additional diagnoses (that is, before, during, or after deployment), or treatment. Second, the committee considers the time since exposure to have been too short and the mean age of both the respondents and eligible populations of Gulf War and post-9/11 veterans to be too young for the development of most cancers that could be due to airborne hazards or open burn pit–related pollutants.

The committee limited its analysis of health outcomes to focus on symptoms, conditions, and diseases with latencies between the exposure and the likely disease onset of more than 6 months and less than 10 years; the health outcomes were also limited to those associated with the respiratory and the cardiovascular systems because these are the most plausible and well-documented potential health effects of the exposures of concern and have the most information collected from the registry. However, the registry's data on health outcomes were not captured with the specificity that would be necessary to draw conclusions about the presence or absence of specific diagnoses in the registry population.

Strategies for Assessing Self-Reported Health Information

The committee applied multiple methods to both directly and indirectly examine the characteristics, attributes, and quality of the registry health data. Direct methods included analyses to check for internal consistency of the questionnaire; indirect methods included reviewing published comparisons of analyses using registry data and the Department of Veterans Affairs (VA) medical records.

This section reports the results of both the analyses that the committee directed and those that VA provided. Although all the analyses were carried out by the same contractor, they differ in their use of respondent data from different points in time, thereby consisting of different numbers of registry participants. VA provided data to the contractor that were not made available for the committee's analyses because of VA-imposed restrictions on the use of personally identifiable information, and these data were used by the contractor to perform analyses that were not possible for the committee to conduct.

Comparisons with VA Health Record Diagnoses

VA evaluated the validity of self-reported conditions from registry respondents by comparing them with information contained in the VA health records for these same respondentswho used VA health care. The sample contained 7,078 registry participants who had completed the registry questionnaire as of March 31, 2015, and had at least one inpatient or outpatient visit with VA health care in each of the four years prior to registry participation (19.0% of all registry participants). *International Classification of Diseases, Ninth Revision* (ICD-9) codes for primary diagnosis and up to nine secondary diagnoses were used to confirm diagnoses. Prevalence was calculated using the proportion of participants who reported the condition on the questionnaire or who had the diagnostic code in their VA medical records. Sensitivity (the proportion of respondents who had the condition in the medical record and also affirmed it in the questionnaire), specificity (the proportion of respondents who did not have the condition in the medical record and did not report having it on the questionnaire), and positive predictive value (the proportion of respondents who reported the condition in the questionnaire and had it recorded in the VA medical record) were computed (see Table 6-1).

For most conditions, a higher prevalence was found on responses to the questionnaire than in VA medical records. The exceptions were for chronic obstructive pulmonary disease (COPD), coronary artery disease, myocardial infarction, and cancer; the largest differences in prevalence were for allergies and chronic bronchitis. The reasons for these differences are unclear, but the differences may be due to the conditions being diagnosed more than 4 years prior to participation in the registry, diagnosis by a non-VA provider, or, most likely for allergies, people having reported the symptoms after self-diagnosing it but never having sought medical treatment for them. Similarly, people who have been diagnosed with acute bronchitis may not know that it is distinct from chronic bronchitis. In addition, a diagnosis in the medical record (for example, coronary artery disease) may be self reported based on the clinical event (such as myocardial infarction) or symptom (such as angina pectoris) that was experienced by the respondent. Similarly, a diagnosis in the medical record with a broad scope, such as COPD, may be self-reported as a more specific condition (e.g., emphysema or chronic bronchitis). The highest sensitivity was 77.6% for hypertension. Sensitivity was moderate or low for most conditions, which suggests either that people diagnosed with these conditions are often not reporting them on the questionnaire or that people who report having the condition do not have it documented. The latter would explain the higher prevalence of most conditions based on the questionnaire. Specificity was high (greater than 90%) for most conditions, and lowest for

TABLE 6-1 Disease Prevalence Comparing VA Medical Records and Questionnaire Responses for 7,078 Registry Respondents

	Medical Record Diagnosis		Prevalence (%)		Questionnaire as a Screen for Disease from Codes (%)		
Disease	Yes	No	Medical Record	Questionnaire	Sensitivity	Specificity	Positive Predictive Value
Allergies	1,895	4,699	28.7	45.1	64.1	62.6	45.1
Asthma	1,202	5,477	18.0	20.7	60.3	87.9	20.7
Emphysema	31	6,703	0.5	2.6	61.3	97.7	10.8
Chronic bronchitis	91	6,425	1.4	18.6	62.6	81.7	4.6
Chronic obstructive pulmonary disease	694	5,916	10.5	7.2	29.7	95.4	43.3
Constrictive bronchiolitis	8	6,323	0.1	1.9	62.5	98.2	4.1
Idiopathic pulmonary fibrosis	3	6,335	<.1	0.4	0.0	99.6	0.0
High blood pressure	3,045	3,758	44.8	47.2	77.6	77.4	73.5
Coronary artery dissease	339	6,350	5.1	3.8	37.2	98.0	50.2
Myocardial infarction	317	6,536	4.6	2.8	31.2	98.6	52.4
Angina pectoris	61	6,318	1.0	2.5	32.8	97.8	12.5
Cancer	660	6,230	9.6	7.8	40.6	95.6	49.6

SOURCE: Gasper and Kawata, 2015.

hypertension (77.4%) and allergies (62.6%). The high specificity for most conditions indicates that respondents who do not have the conditions (according to their VA medical records) also do not report having the conditions on the questionnaire. Positive predictive value varied considerably, from 0% for idiopathic pulmonary fibrosis to 73.5% for hypertension (Gasper and Kawata, 2015).

In a subsequent analysis, Liu et al. (2016) used a subpopulation of registry respondents (N = 4,343) for whom deployment dates and specific location information (military bases with and without documented burn pits) were available to examine the associations between geographic and self-reported burn pits emissions exposure and self-reported respiratory and cardiovascular conditions. Self-reported diagnoses were compared with VA medical record information for 2,857 respondents who used VA health care at least once between January 2007 and November 2015. Participants who reported having been diagnosed with a condition before deployment were excluded from both the analyses of self-report and the comparisons with VA medical records for that condition. Models were adjusted for demographic, lifestyle, and military service characteristics. The burn pit exposure measures were restricted to a population who were deployed to locations during periods when burn pits were known to operate. A strong dose–response association was found between cumulative days deployed within a 2-mile radius of a burn pit and self-reported emphysema, chronic bronchitis, or COPD. The same association was weaker when VA medical record diagnoses were used in place of self-report. In both the self-report and VA medical record analyses, no dose–response associations were found between cumulative days of deployment near a documented burn pit and the incidence of asthma, hypertension, or cardiovascular disease. When the number of self-reported hours per day of burn pit smoke was used as the exposure measure in place of days deployed near a burn pit, the associations with self-reported hypertension and the combined respiratory outcome of emphysema, chronic bronchitis, or COPD were strong and larger. On the other hand, neither of the burn pit emissions exposure measures variables (days or hours) showed a dose–response association with any of the respiratory or cardiovascular diagnoses in VA health records. The limited correlation between self-reported diagnoses and the diagnoses recorded in VA medical records may indicate a misidentification of self-reported health conditions, which in turn implies that analyses performed using self-reported diagnoses may likewise be affected.

Internal Consistency of the Questionnaire

As a way to check the internal consistency of the questionnaire, the committee performed a cross-tabulation to determine the agreement between respondents who endorsed functional limitations due to hypertension (2.1.F) and who also endorsed hypertension in 2.2.2.A. A similar cross-tabulation was made for those who endorsed functional limitations due to a lung/breathing problem (2.1.F) and those who endorsed at least one respiratory condition in 2.2.1.B–H. Table 6-2 shows the cross-tabulation between these health conditions and their corresponding functional limitations. Among those respondents who did not report hypertension, 0.5% reported functional limitations due to hypertension; in addition, among those who did not report any respiratory conditions, 13.6% reported functional limitations due to a lung/breathing problem. The disagreement for hypertension is likely due to erroneous self-report. The discordance between a lung/breathing problem and for any respiratory condition is likely due to both erroneous self-report and because functional limitations due to lung/breathing problems may result from undiagnosed respiratory conditions or perhaps other factors such as a lack of fitness. This analysis of comparing correspondence between reported functional limitations due to hypertension or lung problems and the reporting of the conditions themselves is by no means a perfect measure, especially because the wording of the questions for functional limitations and diagnosed respiratory and cardiovascular conditions differs, but it nonetheless raises questions. While there was high agreement with functional limitation due to hypertension and self-reported diagnosis of hypertension, a better correspondence would be expected between self-reported lung/breathing problem and any respiratory diagnoses. This suggests that there was reasonable internal consistency in the questionnaire, but, based on the lung and respiratory indicators, a closer correspondence would have generated more confidence in the reliability of the self-reported questionnaire data.

A second method that the committee used to check internal consistency was to examine the correlation of strong predictors, such as sex and smoking status and smoking status and respiratory and cardiovascular health outcomes. A higher proportion of women than men were nonsmokers, for example, which is consistent with the general population (CDC, 2014a). Using nonsmokers as the reference group, the associations between smoking status and each health outcome were considered. Former smokers and current smokers both had lower odds of asthma than nonsmokers. For the other conditions, former and current smokers had higher odds of each health outcome compared with nonsmokers. Nonsmokers were more likely to report no respiratory or cardiovascular conditions than were former and current smokers, but by contrast, nonsmokers were more likely to report having cardiovascular conditions alone than persons in the other smoking categories. Current smokers were more likely than nonsmokers and former smokers to report diagnoses of both respiratory and cardiovascular conditions and also respiratory conditions alone (data not shown). There were too few nonsmokers who reported no exposures to burn pits or other airborne hazards to use that group as a comparison to isolate the effect of airborne exposures on respiratory or cardiovascular outcomes.

TABLE 6-2 Endorsement of Health Conditions and Their Associated Functional Limitations

	Functional Limitations Due to Hypertension (Q 2.1.F.16)	
	Yes	No
Hypertension (Q 2.2.2.A)		
Yes	5,324 (32.4%)	11,109 (67.6%)
No	134 (0.5%)	27,856 (99.5%)
	Functional Limitations Due to Lung/Breathing Problem (Q 2.1.F.19)	
	Yes	No
Any respiratory condition (Q 2.2.1.B–H)		
Yes	7,019 (53.6%)	6,088 (46.4%)
No	4,017 (13.6%)	25,589 (86.4%)

The committee also examined body mass index (BMI) in relation to respiratory and cardiovascular outcomes. As expected, the respondents who were obese reported more respiratory and cardiovascular diagnoses and diagnoses of cardiovascular conditions alone than respondents with lower BMI. Normal BMI respondents were most likely to report neither respiratory nor cardiovascular condition diagnoses. Underweight responents had the highest percentage who reported a respiratory condition only. The strong correlation of smoking and BMI individually with higher odds of most outcomes is consistent with expected associations of smoking and BMI with respiratory and cardiovascular disease. This likewise suggests reasonable internal consistency of the questionnaire.

HEALTH CONDITIONS USED IN ANALYSES

The committee focused on the following health outcomes for its analysis:

- Asthma
- Functional limitation due to lung or breathing problem
- Emphysema/chronic bronchitis/COPD
- Any respiratory symptom
- Hypertension
- Coronary artery disease (CAD), myocardial infarction (MI), angina

The committee chose to group some of these conditions for analysis purposes even though separate questions were asked for each because there is overlap in the symptoms of many of these conditions, and the committee was concerned about the ability of respondents to distinguish among them. All except one of the questions used (functional limitation due to lung or breathing problem) were limited to respiratory and cardiovascular health outcomes that were diagnosed by a medical professional ("Have you ever been told by a doctor or health care provider that you have . . . ?"). Asthma and respiratory symptoms represent reactive airway disease. Respiratory symptoms included cough for more than 3 weeks, wheeze, sputum for more than 3 weeks, and non-angina chest pain in the past 12 months. Emphysema/chronic bronchitis/COPD represent chronic obstructive lung diseases, while CAD/MI/angina represent ischemic cardiac disease. In particular, both emphysema/chronic bronchitis/COPD and CAD/MI/angina represent groupings of common diagnoses whose meaning and usage by laypersons often overlap.

Functional limitation due to a lung or breathing problem was also included in the health outcomes examined by the committee because this represents the primary outcome of concern for registry participants with burn pit and other deployment-related inhalational exposures and provides information on the severity of the health condition. Studies of deployed service members have shown respiratory symptoms to be associated with substantial functional limitation and an inability to perform the military physical fitness testing required to maintain active duty status (King et al., 2011). The questionnaire section on functional limitations asked respondents to rate on a five-point scale how difficult specific activities (e.g., jogging a mile on level ground, walking a mile on level ground, walking up an incline, climbing a flight of stairs) were to perform. For any activity that was reported as "difficult," respondents were then asked to indicate what condition or health problem caused the difficulty or difficulties. "Lung/breathing problem (for example, asthma and emphysema)" was one of the 36 possible choices that respondents could choose. It is not possible to determine which conditions cause the difficulty in activities if more than one functional limitation or condition is indicated. Of the 41,350 respondents who reported that at least one activity was "difficult," more than two-thirds (67.8%) of respondents indicated more than one cause, and 12.4% indicated seven or more causes. If the cause of a functional limitation is a lung/breathing problem, this series of questions is designed to assess how severely it affects the respondent.

Table 6-3 shows the overall responses by health condition for all 46,404 participants for which the committee was given data. No adjustments have been made. The highest prevalence for a respiratory condition was for a functional limitation due to lung/breathing problem (25.7%) followed by allergies (23.0%), and the lowest prevalence was reported for emphysema (1.3%). Cardiovascular outcomes were less common, but more than one-third of registry respondents reported hypertension (35.6%). Few respondents reported other cardiovascular outcomes.

TABLE 6-3 Overall Respiratory and Cardiovascular Disease Frequencies Reported by All Respondents

	Yes	No	Refused	Missing	Don't Know
Respiratory Conditions					
Allergies	18,203 (23.0%)	25,505 (55.0%)	254 (0.6%)	7 (0.02%)	2,435 (5.3%)
Asthma	6,754 (14.6%)	37,544 (81.0%)	342 (0.7%)	7 (0.02%)	1,757 (3.8%)
Emphysema	588 (1.3%)	44,221 (95.3%)	178 (0.4%)	7 (0.02%)	1,410 (3.0%)
Chronic bronchitis	5,921 (12.8%)	37,606 (81.0%)	216 (0.5%)	7 (0.02%)	2,654 (5.7%)
COPD	1,524 (3.3%)	42,612 (91.8%)	201 (0.4%)	7 (0.02%)	2,060 (4.4%)
Lung disease other than asthma, emphysema, chronic bronchitis, COPD	4,333 (9.3%)	39,341 (84.8%)	219 (0.5%)	7 (0.02%)	2,504 (5.4%)
Functional limitation due to lung/breathing problem*	11,899 (25.7%)	34,194 (73.7%)	199 (0.4%)	4 (0.01%)	78 (0.2%)
Cardiovascular Conditions					
Hypertension	16,502 (35.6%)	28,136 (60.6%)	229 (0.5%)	7 (0.02%)	1,530 (3.3%)
Coronary artery disease	858 (1.9%)	43,944 (94.7%)	165 (0.4%)	7 (0.02%)	1,430 (3.1%)
Angina pectoris	578 (1.3%)	42,894 (92.4%)	156 (0.3%)	8 (0.02%)	2,768 (6.0%)
Myocardial infarction	644 (1.4%)	44,764 (96.5%)	136 (0.3%)	7 (0.02%)	853 (1.8%)
Heart condition other than CAD, angina, or MI	3,017 (6.5%)	41,604 (89.7%)	167 (0.4%)	7 (0.02%)	1,609 (3.5%)

NOTE: CAD = coronary artery disease; COPD = chronic obstructive pulmonary disease; MI = myocardial infarction.
* Excludes 30 respondents who skipped this question because they did not report having a functional limitation.

The committee finds that the overall prevalence of reported respiratory and cardiovascular conditions (higher prevalence of asthma, allergies, and hypertension and lower prevalence of emphysema, COPD, angina pectoris, coronary artery disease, and myocardial infarction) is not unexpected for a population that is predominantly male, aged 25–60 (average age 38.7 years), and for whom about one-third report a current or former history of smoking (CDC, 2014b; Mozaffarian et al., 2015; Yoon et al., 2015).

The distribution of the number of respiratory and cardiovascular conditions reported by respondents based on their endorsement of one or more the 11 possible conditions listed in the questionnaire is shown in Table 6-4. Having a respiratory condition was defined as a yes response to any question from 2.2.1.A–2.2.1.F. Similarly, having a cardiovascular condition was defined as a yes response to any of the questions from 2.2.2.A–E. Just more than one-third (34.7%) of all registry respondents did not report a diagnosis of any respiratory or cardiovascular condition (see Table 6-5). For respondents who reported a diagnosis, 33.6% indicated a single condition, and 31.7% reported two or more conditions.

Table 6-6 shows the number and percent of respondents reporting both respiratory (answered yes to at least one question in 2.2.1.B–H or 2.1.F) and cardiovascular outcomes (answered yes to at least one question in 2.2.2.A–E), although the orgin of the conditions is unknown. The numbers in the cells are not mutually exclusive, as one person could have reported multiple respiratory and cardiovascular outcomes. For example, among respondents to the asthma and hypertension questions, 6.4% answered yes to having both health conditions.

Descriptive Statistics for Health Outcomes

The stratified analysis excludes respondents who reported that they were diagnosed before deployment (Question 2.2.1.I for respiratory conditions or Question 2.2.2.F for cardiovascular conditions) because it is not plausible that deployment caused the condition. However, because of the lack of specificity for these two questions, someone who, for example, self-reported post-deployment coronary heart disease (i.e., coronary artery disease, myocardial

TABLE 6-4 Distribution of Self-Reported Respiratory and Cardiovascular Conditions

Number of Health Conditions	Frequency	Percent
0	16,116	34.7
1	15,577	33.6
2	8,444	18.2
3	3,690	8.0
4	1,581	3.4
5	633	1.4
6	219	0.5
7	94	0.2
8	33	0.1
9	7	0.0
10	7	0.0
11	3	0.0

TABLE 6-5 Cross-Tabulation of Self-Reported Respiratory and Cardiovascular Conditions

	Hypertension		Coronary Artery Disease		Angina Pectoris		Myocardial Infarction		Other Heart Conditions	
	N	%	N	%	N	%	N	%	N	%
Asthma	2,762	6.4	194	0.5	154	0.4	150	0.3	608	1.4
Emphysema	318	0.7	67	0.2	40	0.1	47	0.1	82	0.2
Chronic bronchitis	2,748	6.5	229	0.5	174	0.4	193	0.5	630	1.5
COPD	805	1.9	144	0.3	85	0.2	101	0.2	222	0.5
Other lung conditions	1,990	4.7	200	0.5	132	0.3	125	0.3	560	1.3
Constrictive bronchiolitis	259	0.6	26	0.1	24	0.1	19	0.0	71	0.2
Any pulmonary fibrosis	63	0.2	14	0.0	8	0.0	7	0.0	28	0.1
Functional limitation due to lung/breathing problem	4,648	10.5	324	0.7	234	0.5	255	0.6	987	2.2

NOTE: COPD = chronic obstructive pulmonary disease.

TABLE 6-6 Frequencies of Self-Reported Diagnoses for Any Respiratory or Cardiovascular Conditions Before Versus During or After Deployment

	Deployment Status	
	Before Deployment	During/After Deployment
Any respiratory conditions (N = 12,563)	1,266 (10.1%)	11,297 (89.9%)
Any cardiovascular conditions (N = 17,168)	1,208 (7.0%)	15,960 (93.0%)

infarction, or angina pectoris) could have been excluded if the person had pre-deployment hypertension rather than pre-deployment coronary heart disease, but there is no way to address this other than through exclusions. The committee examined the distribution of the responses with regard to deployment status, in which respiratory or cardiovascular diagnoses before deployment were grouped separately from diagnoses that occurred during and after deployment. No respondents selected all three time periods (before, during, or after deployment), and none chose both before and during deployment. About 12% (1,630) of respondents selected both during and after deployment. As shown in Table 6-6, of the 12,563 participants who were diagnosed with any respiratory condition and had information on the timing of diagnosis with regard to deployment status, 10.1% were diagnosed before deployment. Of the 17,168 participants who were diagnosed with any cardiovascular condition and had information on the timing of diagnosis with regard to deployment status, 7.0% were diagnosed before deployment.

Tables 6-7 and 6-8 show each of the health outcome category diagnoses limited to those reported during and after deployment and stratified by select demographic and lifestyle characteristics that were available to the committee. The questionnaire does not contain questions on demographic characteristics; that information was obtained by linking the registry data to the Department of Defense (DoD) administrative data (Contingency Tracking System [CTS] and Gulf War Oil Well Fire Smoke Registry files). Service members deployed during the stabilization era (1992–2001) only are not included in those files, and thus their demographic characteristics are missing and not available for comparison. The administrative data do not contain updated information on demographic covariates, such as marital status or education; however, those factors are not considered central to the analyses. Gender and race/ethnicity information was derived from the most recent deployment in the CTS or Gulf War Oil Well Fire Smoke Registry file for which the data were not missing; age was taken from the AH&OBP Registry at the time of questionnaire completion. Smoking status was derived using Questions 2.5.A and 2.5.C from the questionnaire. Table 6-7 shows the number and percentage of respondents by demographic and other attributes who did and did not report a diagnosis of each health outcome category. Table 6-8 shows the odds ratios, upper and lower confidence limits, and p-values for each demographic category by health outcome. For both tables, the associations with sex, age, or race as the main independent variable are unadjusted. Associations where smoking status is the main independent variable are adjusted for sex, age, and race. Associations where BMI is the main independent variable are adjusted for sex, age, race, and smoking status. The finding that the proportion of respondents with a functional limitation due to a lung/breathing problem was nearly equal for all age groups was unexpected (see Table 6-7). However, the committee was unable to further assess this finding because national prevalence data for lung/breathing problems due to functional limitation as elicited in the questionnaire are not available.

Potential confounders such as age, smoking, and BMI were examined as they are often associated with respiratory diseases (i.e., emphysema, chronic bronchitis, or COPD) and cardiovascular diseases (i.e., CAD, MI, or angina pectoris). Table 6-8 provides the adjusted associations of age, smoking, and BMI with selected groups of respiratory and cardiovascular conditions, better reflecting an independent association with these predictors. As expected, older age, being a current smoker, and being overweight or obese were associated with a higher prevalence of self-reported respiratory or cardiovascular diseases. These results provide some assurance of the quality of self-reported information on potential confounders and health conditions.

Other covariates that were used in the committee's analyses include military occupation, country of deployment, service branch, rank, unit component, and total length of deployment. The distribution of military characteristics by era for respondents and eligible nonrespondents is shown in Table 4-2 in Chapter 4. Although all military factors were statistically significant predictors because of the size of the respondent population, they are not included in the health outcomes tables because they are not informative for assessing the accuracy of self-reported health outcomes.

Women report a higher prevalence of asthma and chronic respiratory conditions (diagnoses of emphysema, chronic bronchitis, and COPD combined) than men. Men reported a higher prevalence of all cardiovascular conditions. Using younger than 30 years old as the reference group, the odds of each health outcome category increased with each increasing age group. Similarly, using white race as the reference group, the odds of each health outcome were increased for black, Hispanic, and other races. Compared with a BMI classified as underweight or normal weight, overweight and obese respondents had higher odds of each of the health outcomes.

TABLE 6-7 Number and Percent of Health Outcomes Reported During or After Deployment by Demographic Categories

	Asthma						Emphysema, Chronic Bronchitis, or COPD						Any Respiratory Symptoms				Functional Limitation Due to Lung Problem					
	Yes		No		Missing		Yes		No		Missing		Yes		No		Yes		No		Missing	
	N	%	N	%	N	%	N	%	N	%	N	%	N	%	N	%	N	%	N	%	N	%
Total	5,861	13.0	37,189	82.4	2,088	4.6	6,470	14.3	35,286	78.2	3,382	7.5	34,533	74.4	11,871	25.6	11,899	25.6	34,194	73.7	311	0.7
Sex																						
Male	4,873	12.1	33,487	83.2	1,871	4.7	5,435	13.5	31,749	78.9	3,047	7.6	30,723	74.3	10,622	25.7	10,487	25.4	30,582	74.0	276	0.7
Female	912	20.1	3,425	75.6	192	4.2	939	20.7	3,292	72.7	298	6.6	3,498	74.8	1,180	25.2	1,295	27.7	3,351	71.6	32	0.7
Missing	76	20.1	277	73.3	25	6.6	96	25.4	245	64.8	37	9.8	312	81.9	69	18.1	117	30.7	261	68.5	3	0.8
Age at questionnaire completion (years)																						
<30	914	11.9	6,364	83.2	373	4.9	985	12.9	6,117	80.0	549	7.2	5,838	73.8	2,068	26.2	2,006	25.4	5,842	73.9	58	0.7
30–39	2,159	12.4	14,505	83.0	816	4.7	2,162	12.4	14,045	80.3	1,273	7.3	13,254	73.7	4,729	26.3	4,460	24.8	13,392	74.5	131	0.7
40–49	1,922	13.7	11,478	81.9	619	4.4	2,202	15.7	10,693	76.3	1,124	8.0	10,685	74.4	3,682	25.6	3,724	25.9	10,551	73.4	92	0.6
50–59	749	14.4	4,210	80.9	246	4.7	979	18.8	3,861	74.2	365	72.0	4,134	77.3	1,212	22.7	1,474	27.6	3,847	72.0	25	0.5
≥60	117	14.9	632	80.7	34	4.3	142	18.1	570	72.8	71	9.1	622	77.6	180	22.4	235	29.3	562	70.1	5	0.6
Race																						
White	3,447	11.7	24,666	83.9	1,269	4.3	3,865	13.2	23,567	80.2	1,950	6.6	22,248	73.6	7,966	26.4	7,762	25.7	22,231	73.6	221	0.7
Black	640	17.5	2,818	77.2	190	5.2	710	19.5	2,590	71.0	348	9.5	2,858	75.7	915	24.3	980	26.0	2,774	73.5	19	0.5
Hispanic	490	14.9	2,599	79.0	200	6.1	443	13.5	2,491	75.7	355	10.8	2,632	78.1	740	21.9	801	23.8	2,548	75.6	23	0.7
Other	173	12.1	1,161	81.3	94	6.6	183	12.8	1,105	77.4	140	9.8	1,064	71.9	416	28.1	329	22.2	1,138	76.9	13	0.9
Missing	1,111	15.0	5,945	80.4	335	4.5	1,269	17.2	5,533	74.9	589	8.0	5,731	75.8	1,834	24.2	2,027	26.8	5,503	72.7	35	0.5
Smoking																						
Nonsmoker	3,449	13.5	20,848	81.6	1,242	4.9	3,218	12.6	20,357	79.7	1,964	7.7	18,823	71.8	7,389	28.2	6,535	24.9	19,485	74.3	192	0.7
Former smoker	1,297	12.8	8,510	83.8	352	3.5	1,588	15.6	7,944	78.2	627	6.2	7,904	75.4	2,575	24.6	2,794	26.7	7,663	73.1	22	0.2
Current smoker	567	10.8	4,493	85.7	182	3.5	1,005	19.2	3,931	75.0	306	5.8	4,501	83.0	919	17.0	1,463	27.0	3,948	72.8	9	0.2
Missing	548	13.1	3,338	79.5	312	7.4	659	15.7	3,054	72.7	485	11.6	3,305	77.0	988	23.0	1,107	25.8	3,098		88	2.0
Body mass index																						
Underweight/Normal	784	10.4	6,407	85.3	316	4.2	859	11.4	6,135	81.7	513	6.8	5,268	68.1	2,467	31.9	1,669	21.6	6,003	77.6	63	0.8
Overweight	2,478	11.4	18,300	84.3	943	4.3	2,746	12.6	17,425	80.2	1,550	7.1	16,035	71.9	6,272	28.1	5,351	24.0	16,818	75.4	138	0.6
Obese	2,455	16.5	11,758	78.9	686	4.6	2,688	18.0	11,081	74.4	1,130	7.6	12,470	81.3	2,863	18.7	4,618	30.1	10,666	69.6	49	0.3
Missing	144	14.2	724	71.6	143	14.1	177	17.5	645	63.8	189	18.7	760	73.9	269	26.1	261	25.4	707	68.7	61	5.9

TABLE 6-7 Continued

	Hypertension						CAD, MI, or Angina Pectoris					
	Yes		No		Missing		Yes		No		Missing	
	N	%	N	%	N	%	N	%	N	%	N	%
Total	15,531	34.4	27,908	61.7	1,757	3.9	1,513	3.3	40,485	89.6	3,198	7.1
Sex												
Male	14,440	35.8	24,219	60.1	1,631	4.0	1,376	3.4	35,975	89.3	2,939	7.3
Female	888	19.6	3,530	77.9	114	2.5	100	2.2	4,216	93.0	216	4.8
Missing	203	54.3	159	42.5	12	3.2	37	9.9	294	78.6	43	11.5
Age at questionnaire completion (years)												
<30	2,375	30.9	4,978	64.7	336	4.4	178	2.3	7,034	91.5	477	6.2
30–39	5,463	31.3	11,294	64.6	718	4.1	404	2.3	15,947	91.3	1,124	6.4
40–49	5,214	37.2	8,308	59.3	498	3.6	595	4.2	12,341	88.0	1,084	7.7
50–59	2,128	40.7	2,921	55.9	181	3.5	280	5.4	4,491	85.9	459	8.8
≥60	351	44.9	407	52.0	24	3.1	56	7.2	672	85.9	54	6.9
Race												
White	9,206	31.3	19,095	64.9	1,115	3.8	826	2.8	26,849	91.3	1,741	5.9
Black	1,650	45.5	1,846	50.9	130	3.6	208	5.7	3,030	83.6	388	10.7
Hispanic	1,077	32.6	2,032	61.5	195	5.9	97	2.9	2,922	88.4	285	8.6
Other	503	34.7	863	59.6	83	5.7	40	2.8	1,278	88.2	131	9.0
Missing	3,095	41.8	4,072	55.0	234	3.2	342	4.6	6,406	86.6	653	8.8
Smoking												
Nonsmoker	8,612	33.8	15,917	62.5	948	3.7	779	3.1	22,939	90.0	1,759	6.9
Former Smoker	3,673	36.0	6,188	60.7	336	3.3	386	3.8	9,227	90.5	584	5.7
Current Smoker	1,828	34.4	3,317	62.4	174	3.3	219	4.1	4,748	89.3	352	6.6
Missing	1,418	33.7	2,486	59.1	299	7.1	129	3.1	3,571	85.0	503	12.0
Body mass index												
Underweight/Normal	1,344	17.7	5,965	78.7	270	3.6	136	1.8	7,028	92.7	415	5.5
Overweight	6,679	30.8	14,148	65.2	874	4.0	620	2.9	19,716	90.9	1,365	6.3
Obese	7,122	47.8	7,292	48.9	494	3.3	725	4.9	12,949	86.9	1,234	8.3
Missing	386	38.3	503	49.9	119	11.8	32	3.2	792	78.6	184	18.3

NOTES: There are no missing data for any respiratory symptoms because the data were derived from a check-all-that-apply item and coded 1 if the respondent checked any symptom and 0 otherwise. CAD = coronary artery disease; COPD = chronic obstructive pulmonary disease; MI = myocardial infarction.

TABLE 6-8 Odds Ratios and Confidence Intervals for Health Outcomes Reported During or After Deployment by Demographic Categories

	Asthma				Emphysema, Chronic Bronchitis, or COPD				Any Respiratory Symptoms				Functional Limitation Due to Lung Problem			
	OR	LCL	UCL	P-value	OR	LCL	UCL	P-value	OR	LCL	UCL	P-value	OR	LCL	UCL	P-value
Sex																
Male	(ref)															
Female	1.8	1.7	2.0	*	1.7	1.5	1.8	*	1.0	1.0	1.1	0.49	1.1	1.1	1.2	*
Missing	1.9	1.5	2.4	*	2.3	1.8	2.9	*	1.6	1.2	2.0	*	1.3	1.1	1.6	0.02
Age (years)																
<30	(ref)															
30–39	1.0	1.0	1.1	0.40	1.0	0.9	1.0	0.28	1.0	0.9	1.1	0.81	1.0	0.9	1.0	0.33
40–49	1.2	1.1	1.3	*	1.3	1.2	1.4	*	1.0	1.0	1.1	0.39	1.0	1.0	1.1	0.39
50–59	1.2	1.1	1.4	*	1.6	1.4	1.7	*	1.2	1.1	1.3	*	1.1	1.0	1.2	*
≥60	1.3	1.0	1.6	0.02	1.5	1.3	1.9	*	1.2	1.0	1.5	0.02	1.2	1.0	1.4	0.02
Race																
White	(ref)															
Black	1.6	1.5	1.8	*	1.7	1.5	1.8	*	1.1	1.0	1.2	*	1.0	0.9	1.1	0.77
Hispanic	1.3	1.2	1.5	*	1.1	1.0	1.2	0.14	1.3	1.2	1.4	*	0.9	0.8	1.0	0.01
Other	1.1	0.9	1.3	0.44	1.0	0.9	1.2	0.90	1.1	1.1	1.2	*	0.8	0.7	0.9	*
Missing	1.3	1.2	1.4	*	1.4	1.3	1.5	*	0.9	0.8	1.0	0.14	1.1	1.0	1.1	0.07
Smoking																
Nonsmoker	(ref)															
Former smoker	1.0	0.9	1.0	0.26	1.3	1.2	1.4	*	1.9	1.8	2.1	*	1.1	1.0	1.1	*
Current smoker	0.8	0.7	0.9	*	1.7	1.5	1.8	*	1.2	1.2	1.3	*	1.1	1.0	1.2	*
Missing	1.0	0.9	1.1	0.49	1.4	1.3	1.6	*	1.3	1.2	1.4	*	1.1	1.0	1.2	0.07
Body mass index																
Underweight/Normal	(ref)															
Overweight	1.2	1.1	1.3	*	1.2	1.1	1.3	*	1.3	1.1	1.5	*	1.2	1.1	1.3	*
Obese	1.8	1.7	2.0	*	1.8	1.7	2.0	*	2.1	2.0	2.2	*	1.6	1.5	1.7	*
Missing	1.5	1.2	1.8	*	1.9	1.6	2.3	*	1.2	1.2	1.3	*	1.3	1.1	1.5	*

TABLE 6-8 Continued

	Hypertension				Coronary Artery Disease, Angina Pectoris, or Myocardial Infarction			
	OR	LCL	UCL	P-value	OR	LCL	UCL	P-value
Sex								
Male	(ref)							
Female	0.4	0.4	0.5	*	0.6	0.5	0.8	*
Missing	2.1	1.7	2.6	*	3.3	2.3	4.7	*
Age (years)								
<30	(ref)							
30–39	1.0	1.0	1.1	0.65	1.0	0.8	1.2	0.99
40–49	1.3	1.2	1.4	*	1.9	1.6	2.3	*
50–59	1.5	1.4	1.6	*	2.5	2.0	3.0	*
≥60	1.8	1.6	2.1	*	3.3	2.4	4.5	*
Race								
White	(ref)							
Black	1.9	1.7	2.0	*	2.2	1.9	2.6	*
Hispanic	1.1	1.0	1.2	0.02	1.1	0.9	1.3	0.49
Other	1.2	1.1	1.4	*	1.0	0.7	1.4	0.92
Missing	1.6	1.5	1.7	*	1.7	1.5	2.0	*
Smoking								
Nonsmoker	(ref)							
Former smoker	1.1	1.1	1.2	*	1.3	1.1	1.4	*
Current smoker	1.0	1.0	1.1	0.33	1.4	1.2	1.6	*
Missing	1.1	1.0	1.1	0.08	1.1	0.9	1.3	0.30
Body mass index								
Underweight/Normal	(ref)							
Overweight	1.9	1.8	2.0	*	1.5	1.3	1.8	*
Obese	3.8	3.6	4.1	*	2.5	2.1	3.0	*
Missing	3.4	2.9	3.9	*	2.0	1.4	3.0	*

NOTE: LCL = lower confidence level; OR = odds ratio; ref = reference group; UCL = upper confidence level.
* Denotes p-value <0.01.

ANALYSES OF EXPOSURES AND HEALTH OUTCOMES

Methods

For the committee's multivariate analysis of airborne hazard exposure potentials and selected respiratory and cardiovascular health outcomes, the committee limited the population to post-9/11 respondents and considered several metrics of exposure. First, the committee examined exposure to burn pits using three measures that combine duration and presumed intensity of exposure, as first described in Chapter 5: (1) cumulative days deployed near a burn pit, derived by summing the number of days of each deployment for which a respondent indicated that he or she was near a burn pit (based on Question 1.2.D); (2) cumulative days of burn pit duty (based on Question 1.2.F); and (3) cumulative hours of exposure to smoke from burn pits, created by multiplying the average number of hours each day that smoke or fumes from burn pits entered the worksite or housing by the number of days of deployment (based on Question 1.2.G). Those measures are presented as quartiles with the first (lowest exposure) quartile serving as the reference group. Other exposures contribute pollutants that may lead to respiratory distress, and therefore dust, diesel/exhaust/fuel, combat, and construction exposure potential scores were considered in the analysis and are presented as categories ranging from 0 to 6, where 0 is the lowest level of exposure potential and also serves as the reference group for comparisons. The soot exposure measure was not included because the question was asked only of respondents who served in the 1990–1991 Gulf War. Finally, the committee used a composite exposure potential measure, which was calculated based on the levels of exposure to the six individual exposure categories (burn pits, dust, diesel, combat, construction, soot; described in Chapter 5) to examine the association between the totality of airborne exposures of concern and each health outcome. Those results are presented as quartiles, with the first quartile serving as the reference.

Outcome measures were limited to selected respiratory and cardiovascular diseases. Respiratory outcomes included asthma; emphysema, chronic bronchitis, or COPD as a composite variable; any functional limitation due to a lung or breathing problem; and respiratory symptoms as a composite variable. Cardiovascular disease included hypertension and cardiovascular disease (CAD, MI, or angina pectoris) as a composite variable. The models excluded respondents diagnosed with a respiratory or cardiovascular disease before deployment. Respondents with missing exposure or disease were excluded from the analysis. Sample sizes differed across models and ranged from 32,178 for construction exposure and emphysema, chronic bronchitis, or COPD to 39,271 for each of the three burn pit metrics (cumulative days near a burn pit, cumulative days of burn pit duties, and cumulative hours of exposure to smoke from burn pits) and any respiratory symptoms.

All of the models adjusted for the following covariates: sex; age at questionnaire completion in approximate quartiles (19–30, 31–37, 38–44, or ≥45 years); education level (less than college or "some college or more"); race/ethnicity (white or "other"); BMI (underweight/normal [<25 kg/m^2], overweight [25–30 kg/m^2], or obese [>30 kg/m^2]); smoking status (current smoker, former smoker, or never smoker); unit component (active duty, National Guard, or reserve); rank (enlisted, warrant officer, or commissioned officer); service branch (Army, Navy, Marine Corps, Air Force, or Coast Guard); and primary duty occupational specialty (10 broad groups based on the Millennium Cohort Study[1]). For covariates with less than 5% missing data, the missing data were imputed with the modal category. For covariates with greater than 5% missing data, a separate category for missing was entered into the regression.

Results

Analyses involving the three specific burn pit exposure variables (amount of time near burn pits; amount of time deployed with burn pit duties; and number of hours exposed to burn pit smoke) yielded associations that were consistent with those found by using the burn pit exposure potential variable constructed by the committee. The results of the associations with health outcomes using the three derived burn pit exposure variables are shown in

[1] Personal communication from Joseph Gasper, senior study director at Westat. Emailed responses to a request from the Committee on the Assessment of the Department of Veterans Affairs Airborne Hazards and Open Burn Pit Registry by Cynthia LeardMann, senior epidemiologist at the Deployment Health Research Department, Naval Health Research Center. San Diego, California, September 2, 2015.

Figures 6-1a–c. Tables of multivariable model result estimates from which the figures were derived are shown in Appendix E. Quartile 1 (the lowest exposure potential) is used as the referent group. All three burn pit exposure variables are associated in a dose–response relationship with all health outcomes of interest (p <0.01; test for p trend data not shown). The magnitude of association is substantial, there are generally monotonic gradients across the quartiles of exposure, and given the sample size, almost all associations are statistically significant. Those registry participants who reported being more proximal to burn pits, having more hours of smoke exposure, or having burn pit duties, in general, reported a greater prevalence of respiratory and cardiovascular disease.

To gain further insight into the quality of the data, the three burn pit exposure variables were each partitioned based on the country of deployment (information on the location of deployment, including base names, was not provided to the committee). Separate burn pit exposure measures were created for exposure that was reported in Iraq or Afghanistan, where burn pits were most likely to be located, considered "plausible exposure," and exposure that was reported in other countries, such as Kuwait, where burn pits were least likely to be located, considered "implausible exposure." The results are shown in Figures 6-2a–e. Respondents were divided into quartiles, with the first quartile serving as the reference group (see Chapter 5, Table 5-4, for quartile cut points).

Using the measure of days deployed near a burn pit, all quartiles of plausible exposure were statistically significant for any respiratory symptoms only. Tests for p trend showed statistical significance (p <0.01) for any respiratory symptoms, hypertension, and CAD/MI/angina pectoris. For the measure of implausible exposure of near a burn pit, not all of the quartiles of exposure were statistically significantly associated with any one health outcome; however, the test for trend was statistically significant (p <0.01) for emphysema, chronic bronchitis, or COPD; any respiratory symptoms; and both categories of cardiovascular outcomes.

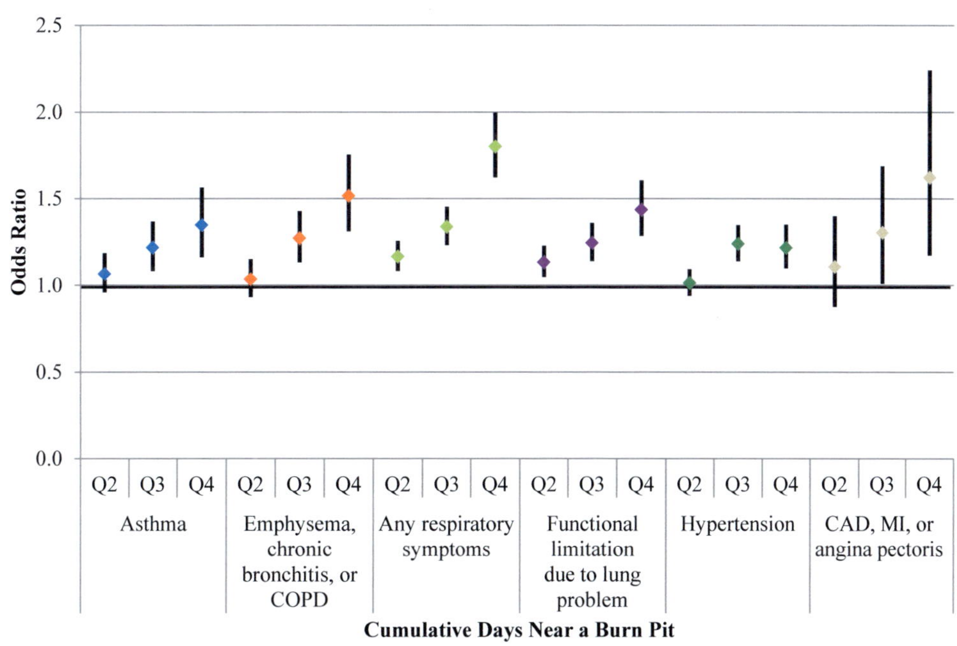

FIGURE 6-1a Adjusted odds ratios for health outcomes by quartile levels of exposure for cumulative days near a burn pit.
NOTE: CAD = coronary artery disease; COPD = chronic obstructive pulmonary disease; MI = myocardial infarction.

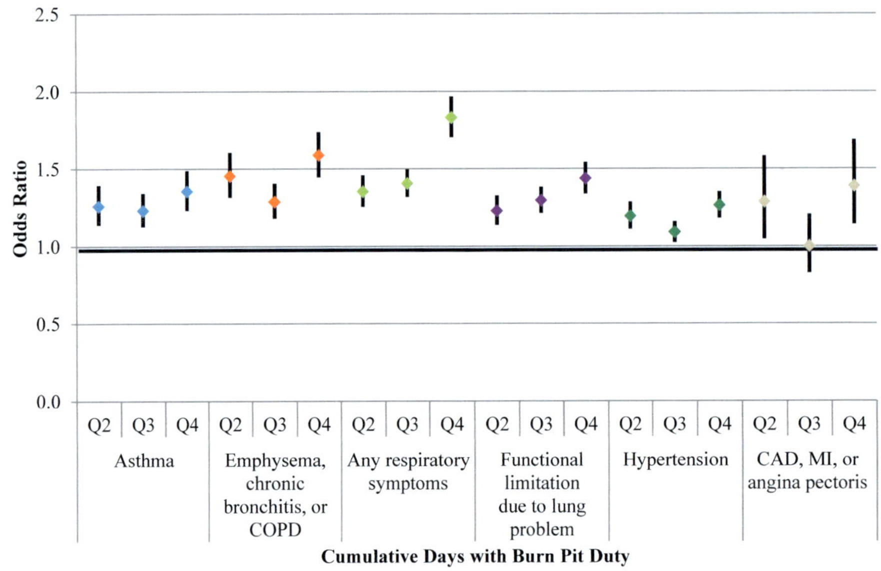

FIGURE 6-1b Adjusted odds ratios for health outcomes by quartile levels of exposure for cumulative days with burn pit duties.
NOTE: CAD = coronary artery disease; COPD = chronic obstructive pulmonary disease; MI = myocardial infarction.

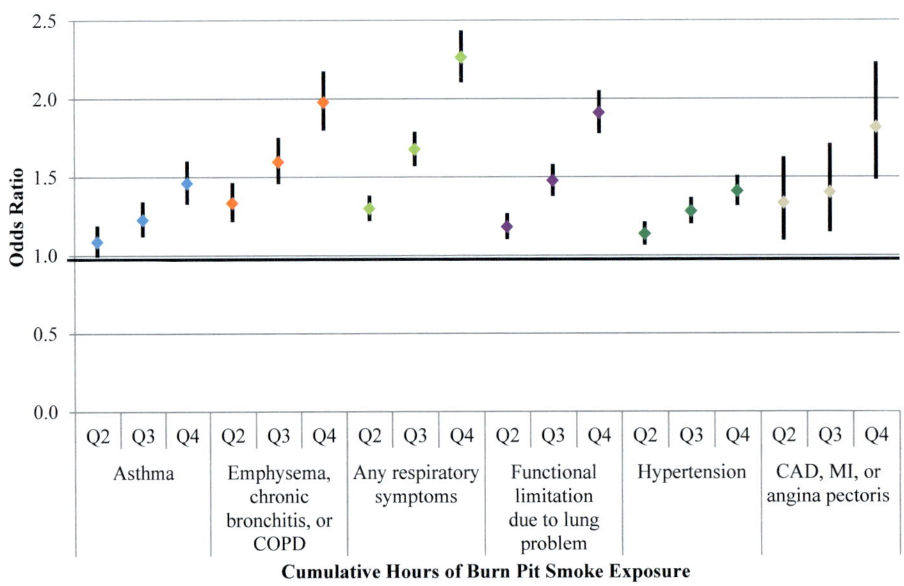

FIGURE 6-1c Adjusted odds ratios for health outcomes by quartile levels of exposure for cumulative hours exposed to smoke from burn pits.
NOTE: CAD = coronary artery disease; COPD = chronic obstructive pulmonary disease; MI = myocardial infarction.

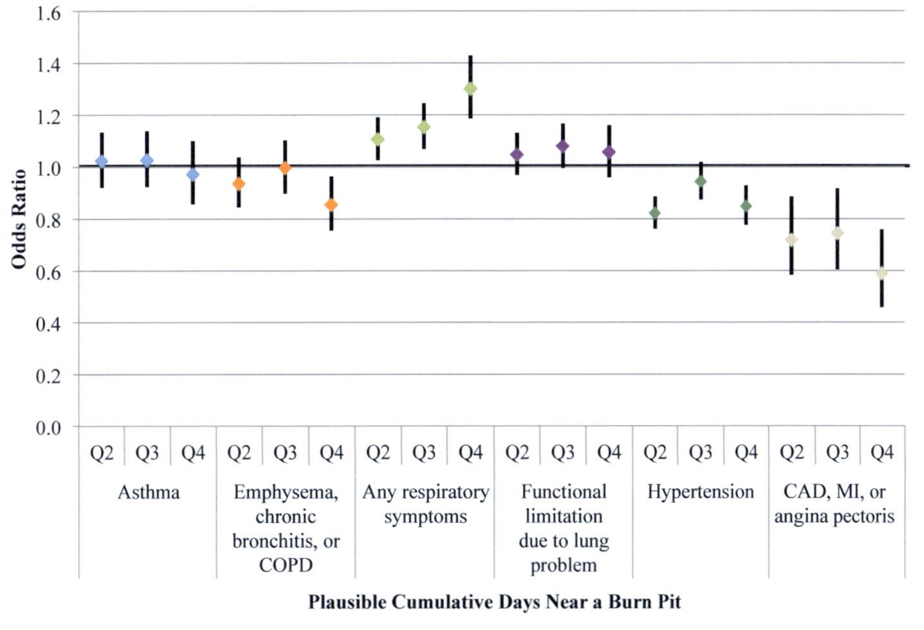

FIGURE 6-2a Adjusted odds ratios for health outcomes by quartile levels of exposure for plausible cumulative days near a burn pit.
NOTE: CAD = coronary artery disease; COPD = chronic obstructive pulmonary disease; MI = myocardial infarction.

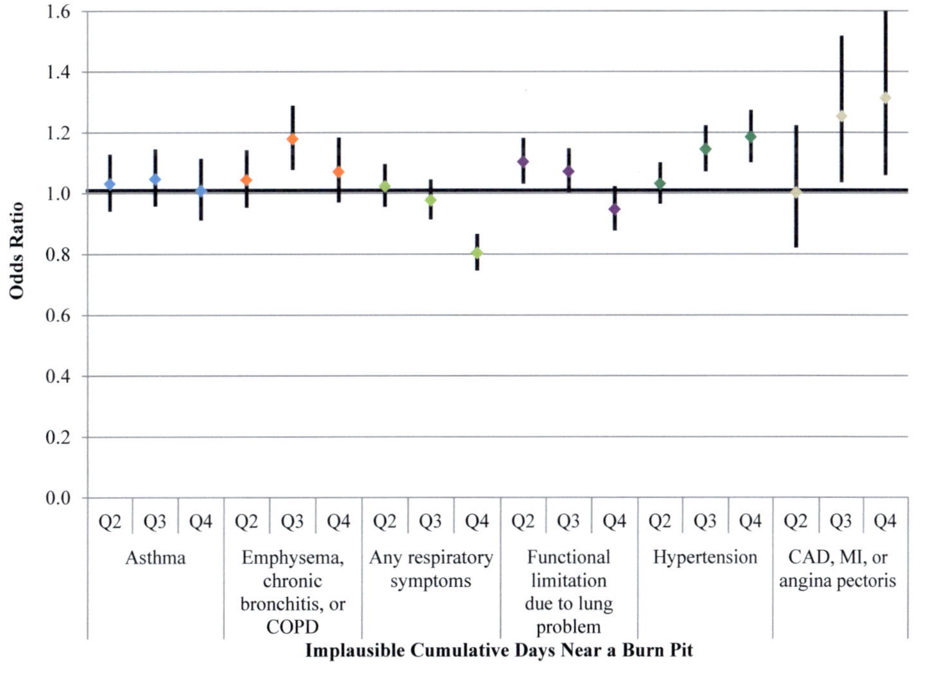

FIGURE 6-2b Adjusted odds ratios for health outcomes by quartile levels of exposure for implausible cumulative days near a burn pit.
NOTE: CAD = coronary artery disease; COPD = chronic obstructive pulmonary disease; MI = myocardial infarction.

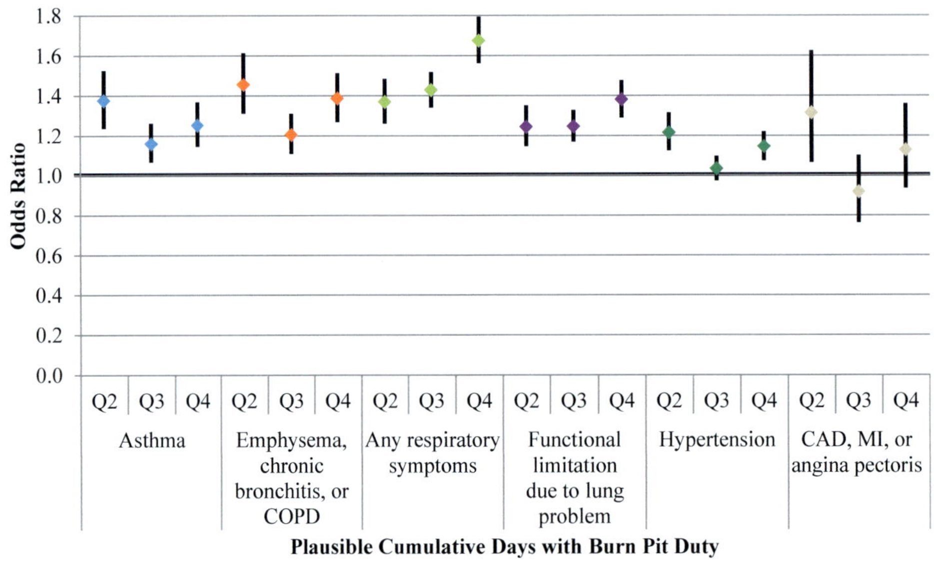

FIGURE 6-2c Adjusted odds ratios for health outcomes by quartile levels of exposure for plausible cumulative days of burn pit duty.
NOTE: CAD = coronary artery disease; COPD = chronic obstructive pulmonary disease; MI = myocardial infarction.

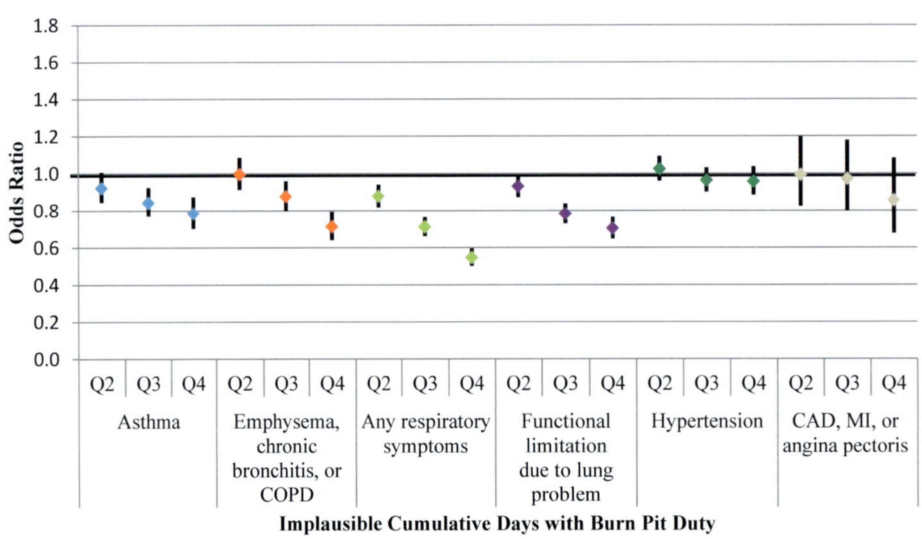

FIGURE 6-2d Adjusted odds ratios for health outcomes by quartile levels of exposure for implausible cumulative days of burn pit duty.
NOTE: CAD = coronary artery disease; COPD = chronic obstructive pulmonary disease; MI = myocardial infarction.

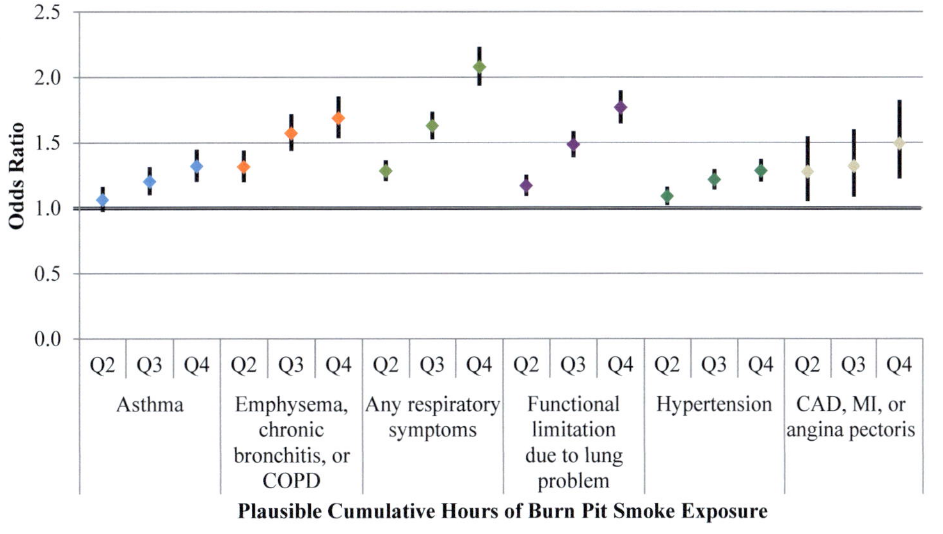

FIGURE 6-2e Adjusted odds ratios for health outcomes by quartile levels of exposure for plausible cumulative smoke hours. NOTE: CAD = coronary artery disease; COPD = chronic obstructive pulmonary disease; MI = myocardial infarction.

For burn pit duties, all plausible exposure quartiles were associated with all four respiratory outcomes, and the tests for trend were all statistically significant. Associations with the cardiovascular outcome categories were not as consistent, although the test for trend for hypertension was statistically significant ($p < 0.01$). Implausible exposures followed the same general pattern as the plausible exposure, and were associated with all the respiratory outcomes (especially quartiles 3 and 4), but no quartile of exposure was associated with either of the cardiovascular condition categories. Tests for p-trend for implausible exposure of burn pit duties were statistically significant for all respiratory categories, but not for either cardiovascular category.

The measure of smoke hours is only presented for plausible exposures because it indicates self-report burn pit exposure that occurred outside of Iraq and Afghanistan, and 90% of respondents had reported 0 hours for that variable. In effect, applying the definitions to create the quartiles would make the first three quartiles zero. Plausible smoke hours showed showed positive, strong associations with all the respiratory and cardiovascular conditions categories, and the tests for trend were all also statistically significant.

Associations between the five exposure category variables (burn pits, diesel/exhaust/fuel, construction, dust, and combat) and respiratory and cardiovascular health outcomes were examined and adjusted for potential confounding as described above. Figures 6-3a–e show the results of the regression models using the exposure potential scores, with the lowest level (0) as the reference group. Higher levels of each of the exposure category variables were strongly and uniformly associated with an increased prevalence of each of the health outcomes. The magnitude of the association was substantial, with point estimates often above 2.0 in the highest category, and they generally followed a monotonic dose–response gradient of stronger associations with increasing exposure score. The pattern of exposures is similar to those found for the three burn pit metrics in Figures 6-1a–c. The consistency across diverse and such fundamentally different exposures and across distinctly different health outcomes is notable and will be discussed in more detail under Interpretation of Registry Data. The tables show a clear trend that those who reported greater exposure also reported greater prevalence of self-reported respiratory and cardiovascular disease.

Figure 6-4 shows the results of the regression models that used the composite exposure potential measure. Quartile 1 (having the lowest exposure potential) was used as the reference group. Higher levels of the composite exposure were strongly and uniformly associated with increased odds for each of the respiratory and cardiovascular outcome categories. The p-test for trend for all health outcome categories was statistically significant ($p < 0.01$).

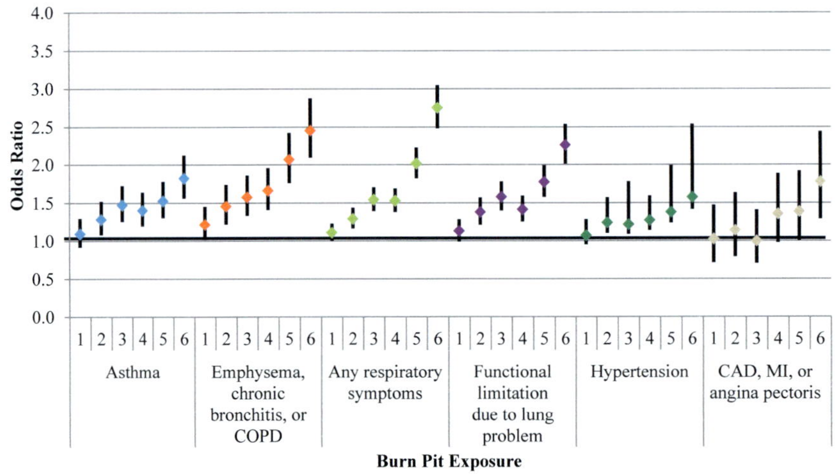

FIGURE 6-3a Adjusted odds ratios for health outcomes by levels of burn pits exposure.
NOTE: CAD = coronary artery disease; COPD = chronic obstructive pulmonary disease; MI = myocardial infarction.

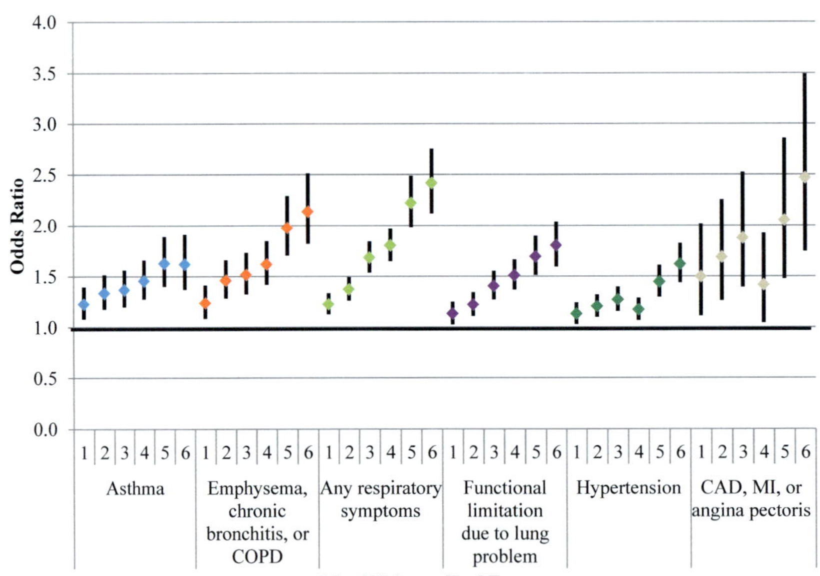

FIGURE 6-3b Adjusted odds ratios for health outcomes by levels of diesel/exhaust/fuel exposure.
NOTE: CAD = coronary artery disease; COPD = chronic obstructive pulmonary disease; MI = myocardial infarction.

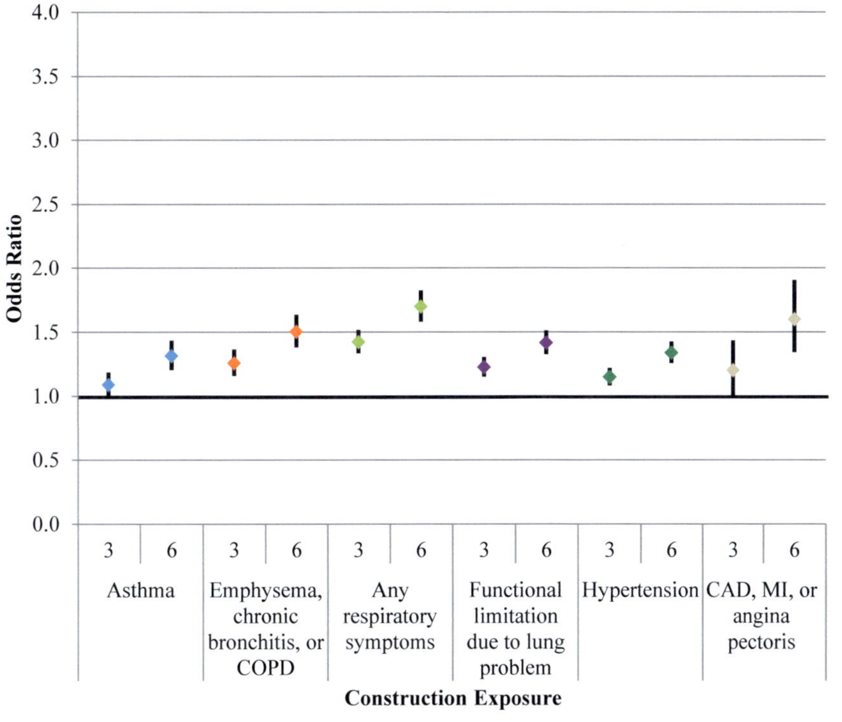

FIGURE 6-3c Adjusted odds ratios for health outcomes by levels of construction exposure.
NOTE: CAD = coronary artery disease; COPD = chronic obstructive pulmonary disease; MI = myocardial infarction.

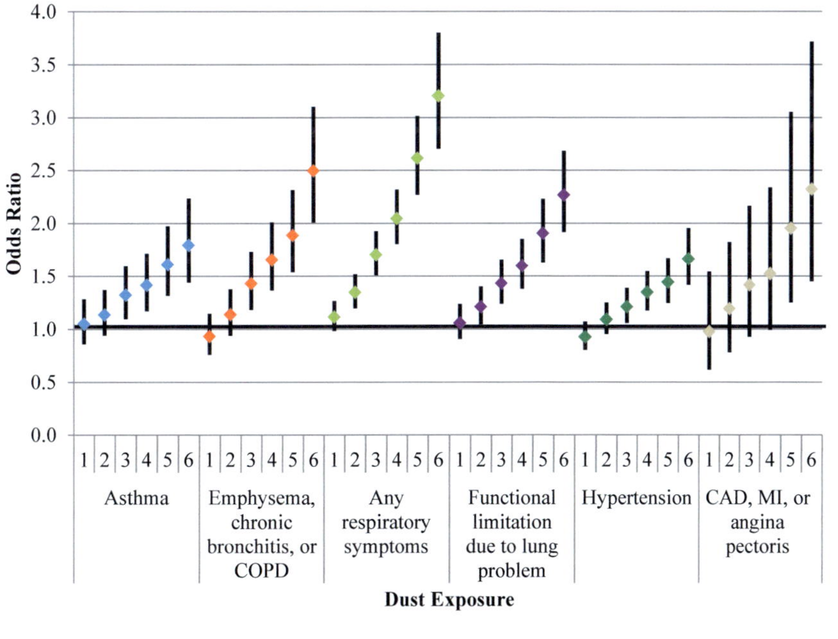

FIGURE 6-3d Adjusted odds ratios for health outcomes by levels of dust exposure.
NOTE: CAD = coronary artery disease; COPD = chronic obstructive pulmonary disease; MI = myocardial infarction.

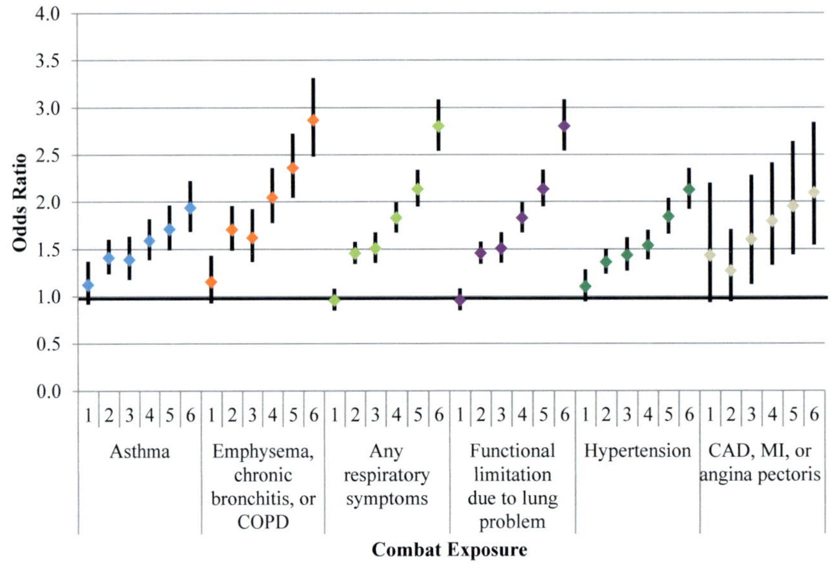

FIGURE 6-3e Adjusted odds ratios for health outcomes by levels of combat exposure.
NOTE: CAD = coronary artery disease; COPD = chronic obstructive pulmonary disease; MI = myocardial infarction.

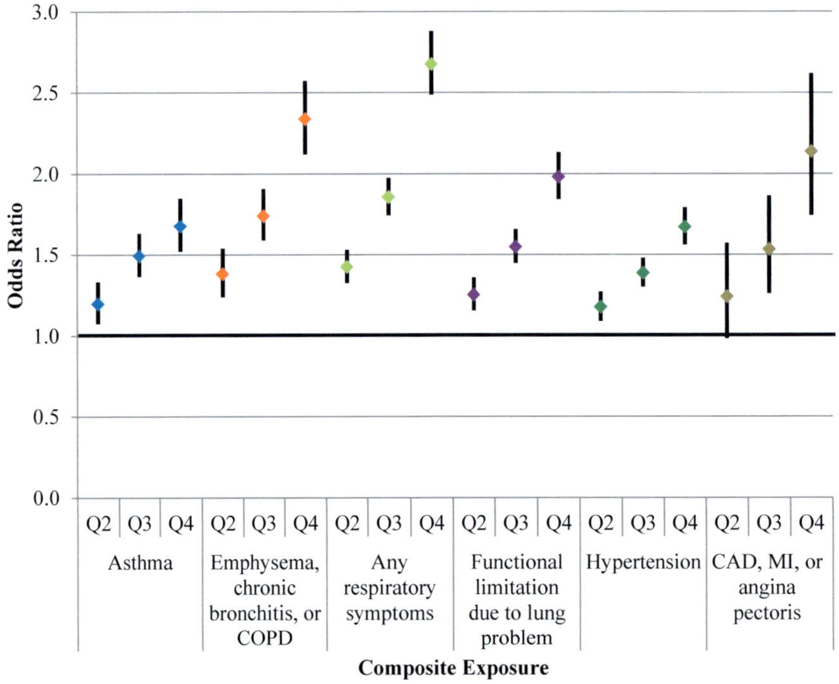

FIGURE 6-4 Adjusted odds ratios for health outcomes by levels of composite exposure potential measure.
NOTE: CAD = coronary artery disease; COPD = chronic obstructive pulmonary disease; MI = myocardial infarction.

Unexpected Patterns of Association

The committee used an approach motivated by Hill's criteria for causation (Hill, 1965) to analyze the potential for a relationship of exposure to airborne hazards and open burn pits emissions with health outcomes. In addition to the strength of association, consistency, specificity, temporality, and biologic gradient, the committee felt that it was important to explore the plausibility of the findings of its analysis of registry data. In so doing, it uncovered some unexpected findings that are not consistent with the currently understood scientific mechanisms of exposure and outcome.

The committee's approach was an analysis of associations between exposures and health conditions collected by the registry questionnaire that, based on current scientific knowledge, do not have a recognized or known linkage or plausible biologic mechanism by which the exposure could influence the health condition. The goal of this analysis was to inform the committee's assessment of the potential for reporting bias by using data (albeit, of admittedly poor quality) on less likely or unexpected associations.

The committee originally planned to examine such outcomes as varicose veins, hernia, and knee problems collected in Question 2.1.F, but VA did not provide those data for analysis. Instead, the committee used non-military exposure to asbestos (5.5.A) as an exposure considered unlikely to be related to neurological conditions (2.2.3.B), immune problems (2.2.3.C), and liver conditions (2.2.3.D) to examine those associations. All models were adjusted for sex, age at questionnaire completion, education level, race/ethnicity, BMI, smoking status, unit component, rank, service branch, primary duty occupational specialty, and total deployment time. While it is remotely possible for there to be some unknown connection between the exposure and the health conditions, such a connection is far less plausible than for respiratory and cardiovascular conditions (for which literatures exist). To prevent confounding, respondents who self-reported respiratory and cardiovascular conditions were excluded.

For this analysis, the asbestos exposure variable was categorized in five groups: no exposure and four quartiles of self-reported exposure duration (from lowest to highest). Those with no exposure were used as the reference group. The "no exposure" category consisted of those who answered "no" to Question 5.5.A; the quartile categories were created from those who provided a response to the number of years of asbestos exposure in 5.5.C. Table 6-9 shows that higher exposures to asbestos were statistically significantly associated with increased odds of each of the examined health outcomes, and the tests for trend were also all statistically significant (p-trend <0.01).

Thus, although there is little known scientific evidence that asbestos exposure is associated with a higher prevalence of neurologic, immune, or liver conditions, the analyses found a statistically significant association between higher levels of exposure and a higher prevalence of each of these health conditions. The relationship was relatively monotonic. These findings suggest that the statistical relationships between self-reported exposure to airborne hazards and the prevalence of self-reported health outcomes in the registry responses are biased toward

TABLE 6-9 Associations Between Asbestos and Selected Health Outcomes

		Neurological Conditions N = 36,399				Immune Problems N = 30,759				Liver Conditions N = 36,883			
Asbestos (5.5.A and 5.5.C)	Years [Median (min, max)]	OR	LCL	UCL	P	OR	LCL	UCL	P	OR	LCL	UCL	P
Q1	1 (1,2)	1.27	1.15	1.40	*	1.31	1.18	1.45	*	1.25	1.07	1.47	*
Q2	3 (3,4)	1.42	1.27	1.58	*	1.24	1.10	1.39	*	1.04	0.86	1.25	0.71
Q3	6 (5,8)	1.20	1.07	1.34	*	1.27	1.13	1.44	*	1.13	0.93	1.37	0.22
Q4	14 (9,41)	1.53	1.36	1.71	*	1.42	1.26	1.59	*	1.30	1.10	1.55	*
			P-trend*				P-trend*				P-trend*		

NOTE: LCL = lower confidence level; OR = odds ratio; UCL = upper confidence level.
* Denotes p-value <0.01.

a positive association, although the committee does not have direct information on what may be driving that pattern. The associations might arise because of selection bias and self-reporting or because of circumstances that are unique to the exposures and conditions experienced by the questionnaire respondents. The source of or reasons for any associations, however, cannot be determined from the available data.

OVERALL INTERPRETATION OF REGISTRY DATA ANALYSIS

Registry respondents represent a very small (less than 2%) and self-selected portion of the U.S. service members who deployed to Iraq, Afghanistan, or other countries in Southwest Asia. Those individuals who enrolled in the registry tended to report a range of exposures and respiratory and cardiovascular health problems, and participants who reported higher levels of exposure tended to also report a higher prevalence of adverse health outcomes. The respiratory and cardiovascular conditions reported by registry participants are relatively common in the U.S. population and are not unexpected, based on the respondents' age range and other demographic factors. Empirically, among registry respondents, the overall and stratified prevalence of self-reported respiratory and cardiovascular conditions differed by health outcome, but all were statistically significantly associated with exposure to burn pits and other potentially harmful deployment exposures. The analysis indicated that each exposure category had strong and consistent associations with the self-reported health outcomes of interest: asthma; emphysema, chronic bronchitis, or COPD; any respiratory symptoms; functional limitations due to lung or breathing problems; cardiovascular disease; and hypertension. These associations were observed for several indicators of burn pit exposure as well as for a range of other deployment exposures, such as exposures to diesel/exhaust/fuel, construction, dust, and combat. All of these exposures could potentially affect any of the health outcomes considered.

The combined overall prevalence among registry respondents was lowest for CAD at 3.6% and greatest for hypertension at 35.8%. Asthma, allergies, emphysema, chronic bronchitis, COPD, and hypertension are all relatively common chronic diseases, even among young and middle age adults (CDC, 2014b; Mozaffarian et al., 2015; Yoon et al., 2015). Because many of these conditions do not result in hospitalizations, it is difficult to produce accurate prevalence estimates.

The registry respondents represent a unique population with a high proportion of self-reported exposure to burn pits or the smoke from burn pits; 96% of respondents reported being near a burn pit on at least one deployment. Within this unique population, the prevalence of self-reported health outcomes appears to increase, with few exceptions, as the intensity of the self-reported exposure to burn pits or the smoke from burn pits increases, even after adjusting for multiple possible confounders. The same trend is found for the other deployment-related exposures (dust, construction, diesel/exhaust/fuel, and combat). Similar strong and consistent associations were seen for several outcomes (neurologic, immune, and liver conditions) that the committee did not consider likely to be affected by exposures to asbestos. The magnitude of association is substantial, with adjusted odds ratios of 2.0 and higher in the uppermost exposure categories for nearly all of the associations examined. Generally monotonic dose–response gradients are seen, with more exposure associated with a higher prevalence of disease in registry respondents. However, it seems unlikely that these statistical trends reflect a biological link between exposure and a health outcome, given that associations that do not seem biologically plausible show patterns of association that are similar to those seen for exposures and conditions that could be directly related.

Given these strong and consistent findings for both biologically plausible and also biologically less plausible associations, and despite substantial and highly statistically significant indications of association, these associations may reflect possible positive bias in the dataset or indicate a need for appropriately designed studies to follow up on the many potential health effects reported for more exposed individuals. The current registry data have limitations for use in quantifying the effects of any or all deployment exposures on the subsequent risk or prevalence of disease within the registry population, let alone suggesting that the associations are generalizable to the broader deployed veteran population. However, these statistical associations do warrant closer scrutiny and evaluation so as to allow for a better understanding of what is generating the findings and trend patterns.

First, as previously discussed in Chapter 2, there are substantial limitations to using registry data to draw conclusions about the strength and plausibility of associations. The questions that elicited the self-reported exposures are nonspecific, were often poorly worded, and can only be used to make the most general of associations.

The questions and the data that result from them cannot be used to quantify the duration or intensity of those exposures, which would be more useful and accurate information for determining the strengths of association with the health outcomes of interest.

Second, because participation in the AH&OBP Registry is voluntary, responses from self-selected participants that enrolled in the registry may result in an inaccurate measure of the true association between exposure and health outcomes. The very small proportion of individuals who successfully enrolled in the registry may have selectively included those who were more likely to have a disease and perhaps were more likely to have been heavily exposed as well, resulting in higher exposure potential. Very few respondents reported no exposures, but approximately one-third did not report having been diagnosed with a respiratory or cardiovascular condition (see Table 6-4). However, to generate the observed patterns, there would need to be selective participation among those with both exposure and disease (jointly) or absence of (or less) exposure and absence of disease (jointly). While that is possible, given the very limited enrollment, the observations are not readily explained. An alternative explanation is that among participants, those who reported more exposure were also more likely to report disease; even modest correlations of this sort can generate substantial associations (Brenner et al., 1993). Without knowing the true exposure or disease status of respondents or the eligible population, it is difficult to evaluate the plausibility of reporting error and its effect on the observed associations.

Third, limitations in the registry data constrain the analyses to suggest, at most, where epidemiologic studies are needed to quantify, confirm, or refute the associations between exposures and health outcomes. It is clear from the registry responses and comments from veterans service organizations that some service members and veterans with burn pit and other airborne exposures from deployment to Southwest Asia are experiencing serious health effects. Nonetheless, a valid assessment of the relationship between measured exposures and health outcomes can only be accomplished through well-designed epidemiologic studies, such as that recommended in *Long-Term Health Consequences of Exposure to Burn Pits in Iraq and Afghanistan* (IOM, 2011).

SYNOPSIS AND CONCLUSIONS

Based on the information presented in this chapter, the committee has reached the following findings, conclusions, and recommendations regarding the analysis and interpretation of AH&OBP Registry data. These observations address four major areas included in the committee's charge.

(1) What is the best approach to analyzing the registry data in order to gain insights into the potential effects of burn pit and airborne exposures on health outcomes?

As delineated in Chapter 5, the available exposure data are of poor quality, but it is possible to construct a set of indicators of exposure potential from them that while being of uncertain validity, are worthy of examination. The committee believes it is possible to perform some informative analyses of registry health information by using these indicators along with data on a circumscribed set of health outcomes, delineated below. So long as the results are not overinterpreted with respect to the association of exposures with health outcomes, there is value in conducting such analyses. The most useful approach to carrying them out is to generate variables using multiple indicators of exposures and multiple grouped indicators of health outcomes. Descriptive information on registry participants also has value in documenting the experience of those who chose to complete the questionnaire.

The population of registry respondents is sufficiently large to permit analyses of exposure and disease with adjustment for some potential confounders. Such analyses are subject to severe limitations, though, and must be presented with strong caveats regarding their implications. Given all the uncertainties, it is prudent to take multiple approaches to analyzing the data in order to assess the robustness of results. Undertaking additional methodologic work to better understand the sources of bias that resulted in the committee's counterintuitive finding that all exposures were statistically associated with all outcomes would, additionally, be productive.

(2) Which health outcomes in the registry are most amenable to analysis?

The committee believes that the limitations of the AH&OBP Registry questionnaire and the data collected by it are too great to allow any firm conclusions to be drawn from its analysis. Within that constraint, it believes that the health outcomes data related to the symptoms, conditions, and diseases associated with the respiratory and the cardiovascular systems are the best candidates for analysis because these constitute the most plausible and well-documented potential health effects of the exposures of concern. Cancer outcomes data from the questionnaire are unlikely to be informative because the questions are too general, insufficient time has elapsed since the exposures for effects to manifest, and the population is too young for most cancers to have developed.

Many respondents reported respiratory and cardiovascular conditions, as would be expected given that those with health problems are likely to be motivated to participate in the registry. Well-established predictors of these conditions, such as age and smoking, followed the expected patterns, offering some evidence for the accuracy of the self-reported data. Less than 10% of respondents indicated a pre-deployment diagnosis of a respiratory or cardiovascular condition. Excluding persons who reported being diagnosed with a condition before deployment had little effect on the prevalence rates of the health conditions examined.

Over half of the all respondents (55.5%) reported having either a respiratory or cardiovascular diagnosis, 15.3% of all respondents reported at least one respiratory and cardiovascular diagnosis, 14.8% reported only a respiratory diagnosis, and 25.4% reported only a cardiovascular diagnosis. Generally speaking, the observed higher prevalence of asthma, allergies, and hypertension and the lower prevalence of emphysema, COPD, angina pectoris, CAD, and MI appear consistent with what would be expected in a population that is predominantly male, aged 25–60 (average age 38.7 years), and for whom about one-third report a current or former history of smoking.

While it would not be informative to compare the absolute rates of disease among registry participants—a small and highly self-selected population—to other more general populations, the health data may be of sufficient quality to justify internal comparisons in which subsets of registry participants with varying levels of potential exposure are compared with one another. Analyses of organ systems other than respiratory and cardiovascular would be less valuable as the questions were worded too broadly to allow interpretation of any results using those variables.

Analyses of plausible and implausible exposure to burn pits based on the country of deployment were conducted using three burn pit exposure metrics. For cumulative days deployed near a burn pit, many of the point estimates for quartiles of plausible exposure were lower than the point estimates for the same disease category for implausible exposure. All quartiles of plausible exposure of cumulative days near a burn pit were statistically significantly associated with any respiratory symptoms only. For the measure of implausible exposure of near a burn pit, not all of the quartiles of exposure were statistically significantly associated with any one health outcome. For cumulative days of burn pit duties, all plausible exposure quartiles were associated with all four respiratory outcomes, but only select quartiles showed statistically significant associations with the cardiovascular outcome categories. Implausible exposures showed a general decreasing trend with increasing exposure potential for all categories of conditions. Point estimates of health outcome categories for plausible exposure by quartile were higher than nearly all corresponding estimates for implausible exposure. The measure of smoke hours could only be analyzed for plausible exposures and showed positive, strong associations with all the respiratory and cardiovascular conditons categories. These results are supportive evidence for the accuracy of self-reported exposure to burn pits.

While the committee recognizes serious limitations in the quality of exposure and disease information and in the self-selected nature of registry participants, examinations of the relationship between exposure and health outcomes among participants may provide a foundation for hypothesis generation for more rigorous research investigations.

(3) What can be learned about airborne hazards and open burn pits exposures and health outcomes through the analysis of registry data?

An examination of multiple indices of exposure to burn pit emissions and other hazards associated with deployment showed that these indices had strong and consistent relationships with essentially all the health outcomes considered. That is, registry participants who reported more exposures of all types also tended to report more health problems of all types. Exposure potential variables used in the multivariable analysis (burn pits, dust, construction, diesel/exhaust/fuel, combat) were strongly and uniformly associated with increased odds of each of the respiratory and cardiovascular health outcomes. The magnitude of the statistical associations was substantial, there were generally monotonic gradients across quartiles of exposure, and all associations were statistically significant. Registry participants who report being closer to burn pits, having more hours of smoke exposure, and having burn pit duties also report a greater prevalence of respiratory and cardiovascular disease.

The committee's exposure potential variables had strong and consistent associations with self-reported asthma; emphysema, chronic bronchitis, or COPD; any respiratory symptom; functional limitations due to a lung or breathing problem; cardiovascular disease; and hypertension. This was observed for the three indicators of burn pit exposure and for the composite exposure potential measure as well. However, diseases and conditions that are not known to be related to these exposures, such as sleep apnea, showed similar positive associations. In addition, analyses found that several other exposures and diseases that were not plausibly related to one another were also statistically associated, such as asbestos exposure and neurologic conditions or immune problems.

Such outcomes strongly suggest that the results of analyses of registry data cannot be taken at face value and that the identified associations may be an artifact of the population's selection and the limitations of the self-reported exposure and disease data. The registry analyses are not generalizable and can only describe what exposures and conditions the population of registry respondents are reporting. Registry data cannot be used to determine cause or estimate prevalence in the total eligible population of service members or veterans.

Given the implausible consistency of the findings of statistical associations between exposures and outcomes—including the example of asbestos and neurologic conditions that are not well supported by the scientific literature—the committee concludes that the associations are likely to reflect biases arising from selective participation in the registry, problems with self-reported exposure and disease data, or other factors unrelated to exposure–disease associations. The registry data by themselves have, at best, limited utility in exposure and health studies but may be more useful when combined with other sources of information, including deployment data and VA and DoD medical records.

The committee wishes to emphasize that it would have reached the same determination regarding the weaknesses of the registry data had the analyses found no associations or weak associations between the exposures and health outcomes. The strong conclusion that can be drawn from the results of the committee's analyses is that a more rigorous and appropriate study design is needed to examine the relationship between the exposures encountered during deployment to the Southwest Asia theater of operations and health outcomes.

While the registry provides a forum for collecting and recording information on those who were deployed and are motivated to participate, it cannot answer such questions. Thus, **the committee recommends that other means for evaluating the potential health effects associated with airborne hazards and open burn pit exposures be developed, such as a well-designed epidemiologic study.** The 2011 Institute of Medicine report *Long-Term Health Consequences of Exposure to Burn Pits in Iraq and Afghanistan* contains advice and recommendations on how such a study might be conducted.

Given this and the committee's other findings regarding the registry, **the committee recommends that VA's messaging be explicit about the limitations on the ability of the registry to generate valid information that can be used to improve VA health and benefits programs or inform the treatment of individuals potentially exposed to burn pits or other airborne hazards in theater in order to ensure that participants and others do not form unrealistic expectations about the value of participation or capabilities of the registry.**

(4) How can VA health care data be used constructively in conjunction with registry data to examine exposures and health outcomes?

As reported in Chapter 4, respondents were more likely to use VA health care than the eligible population (69% versus 46%), and these same respondents were also more likely to report disease. Examination of VA user status and self-reported health outcomes among registry participants showed higher prevalences of most health conditions among VA users compared with nonusers (Gasper and Katawa, 2015). Sensitivity was moderate or low for most conditions, suggesting either that persons diagnosed with these conditions are not reporting them on the questionnaire or people who report having the condition do not have it documented in their VA medical records. High specificity (greater than 90%) for most conditions indicates that respondents who do not have the conditions (according to their VA medical records) also do not report having the conditions on the questionnaire.

Although information is only available for a subset of the population that uses VA health care, there is potential value in linking the registry data to health care utilization and conducting analyses for this subset of registry participants—something that was not possible with the data available to the committee. Self-reported diseases follow expected patterns in relation to known predictors and are corroborated to some extent by VA health care data. The committee therefore concludes that the registry's self-report disease information has at least modest validity. A periodic reassessment of the comparisons between self-reported information collected by the questionnaire and diagnoses in VA medical records for respondents who use VA health care would provide further evidence concerning the level of validity of self-reported health outcomes in the population of respondents.

REFERENCES

Brenner, H., D. A. Savitz, and O. Gefeller. 1993. The effects of joint misclassification of exposure and disease on epidemiologic measures of association. *Journal of Clinical Epidemiology* 46(10):1195–1202.

CDC (Centers for Disease Control and Prevention). 2014a. Current cigarette smoking among adults in the United States. https://www.cdc.gov/tobacco/data_statistics/fact_sheets/adult_data/cig_smoking/index.htm#national (accessed October 15, 2016).

CDC. 2014b. National Health Interview Survey summary health statistics. http://ftp.cdc.gov/pub/Health_Statistics/NCHS/NHIS/SHS/2014_SHS_Table_A-2.pdf (accessed September 23, 2016).

Gasper, J. and J. Kawata. 2015. *An evaluation of selection bias and data quality: Open Burn Pit and Airborne Hazards Registry*. Prepared for the Department of Veterans Affairs Office of Public Health, Post-Deployment Health Group by Westat. September.

Hill, A. B. 1965. The environment and disease: Association or causation? *Proceedings of the Royal Society of Medicine* 58(5):295–300.

IOM (Institute of Medicine). 2011. *Long-term health consequences of exposure to burn pits in Iraq and Afghanistan*. Washington, DC: The National Academies Press.

King, M. S., R. Eisenberg, J. H. Newman, J. J. Tolle, F. E. Harrell, H. Nian, M. Ninan, E. S. Lambright, J. R. Sheller, and J. E. Johnson. 2011. Constrictive bronchiolitis in soldiers returning from Iraq and Afghanistan. *New England Journal of Medicine* 365:222–230.

Liu, J., N. Lezama, J. Gasper, J. Kawata, S. Morley, D. Helmer, and P. Ciminera. 2016. Burn pit emissions exposure and respiratory and cardiovascular conditions among Airborne Hazards and Open Burn Pit Registry participants. *Journal of Occupational and Environmental Medicine* 58(7):e249–e255.

Mozaffarian, D., E. J. Benjamin, A. S. Go, D. K. Arnett, M. J. Blaha, M. Cushman, S. de Ferranti, J. P. Despres, H. J. Fullerton, V. J. Howard, M. D. Huffman, S. E. Judd, B. M. Kissela, D. T. Lackland, J. H. Lichtman, L. D. Lisabeth, S. Liu, R. H. Mackey, D. B. Matchar, D. K. McGuire, E. R. Mohler, C. S. Moy, P. Muntner, M. E. Mussolino, K. Nasir, R. W. Neumar, G. Nichol, L. Palaniappan, D. K. Pandey, M. J. Reeves, C. J. Rodriguez, P. D. Sorlie, J. Stein, A. Towfighi, T. N. Turan, S. S. Virani, J. Z. Willey, D. Woo, R. W. Yeh, M. B. Turner, and the American Heart Association Statistics Committee and Stroke Statistics Subcommitee. 2015. Heart disease and stroke statistics—2015 update: A report from the American Heart Association. *Circulation* 131:e29–e322.

Yoon, S. S., C. D. Fayar, and M. D. Carroll. 2015. *Hypertension prevalence and control among adults: United States, 2011–2014*. National Center for Health Statistics Data Brief No. 220. Hyattsville, MD: National Center for Health Statistics.

7

Findings, Conclusions, and Recommendations

This chapter builds on the foundation laid in Chapters 1–6 to draw out the overarching themes of the report and present the committee's findings, conclusions, and recommendations related to its statement of task. It focuses on the major issues that the Department of Veterans Affairs (VA) will need to address as it chooses how to administer and make best use of the registry in the future. Details of and the scientific backing for the findings, conclusions, and recommendations offered may be found in those earlier chapters.

OPENING OBSERVATIONS

VA was presented with a challenge when it was directed by Congress to design, test, and implement an environmental health registry for "individuals who may have been exposed to toxic airborne chemicals and fumes caused by open burn pits" in 12 months. It set a goal to create a Web-only instrument that would cover military personnel who served in the Southwest Asia theater of operations from the beginning of the Gulf War conflict in 1990 through the present (2016) and include data from other VA and from Department of Defense (DoD) databases to lessen the burden on participants.

The National Academies of Sciences, Engineering, and Medicine,[1] which was tasked with organizing an expert committee to analyze the initial months of data gathered by the registry and offering observations regarding its operation and the use of the information it generates, appreciates the difficulty of the work undertaken and the obvious effort VA put into accomplishing it. The committee's responsibility, though, was to conduct a critical inquiry. It would like to make clear that the comments it offers are made in the spirit of helping VA to make the best use of registry they were directed to establish.

The committee wishes to acknowledge the contributions made by the participants in its 2015 workshop. Their expertise, insights, and personal stories, summarized in Chapter 1, greatly aided the committee's understanding of airborne hazards, open burn pits, and service members' health issues.

[1] The work was done through the operational unit formerly referred to as the Institute of Medicine (IOM); as of March 2016, the Health and Medicine Division continues the consensus studies and convening activities previously undertaken by the IOM.

THE AIRBORNE HAZARDS AND OPEN BURN PIT REGISTRY QUESTIONNAIRE

While registries that rely on voluntary participation and self-reported information are a common means of collecting data on large populations, they are an intrinsically poor source of information on exposures, health outcomes, and possible associations among these events. Even under the best of circumstances, there are substantial limits to the accuracy of the data and—when the respondents constitute only a small, unrepresentative fraction of the eligible population—the generalizability of analyses made with them as well.

These weaknesses are apparent in the Airborne Hazards and Open Burn Pit (AH&OBP) Registry questionnaire and in the data collected in the registry's first 13 months. The weaknesses have been exacerbated by a series of flaws in the structure and operation of the questionnaire and in the questions that are asked and the way they are asked. The AH&OBP Registry questionnaire is flawed in that it

- inappropriately uses questions that were validated for and meant to be administered by other survey means such as a face-to-face or computer-assisted phone interview;
- asks questions that may be confusing for respondents because they are ambiguous or otherwise poorly written;
- elicits information on topics such as hobbies and places of childhood residence that do not yield information that could be productively used in any analysis that would be appropriate to undertake using registry data;
- fails to ask questions (regarding non-burn-pit trash burning, for example) that could yield information related to relevant exposures;
- does not take full advantage of its Web-based format to streamline and focus questions based on previous responses;
- does not permit answers to be supplemented or updated later in time; and
- requires respondents to complete a sometimes lengthy set of repetitive questions regarding deployments before addressing core issues such as health, increasing the possibility of response fatigue.

Examples of these weaknesses are offered in Chapter 2. Their cumulative effect is evidenced by the high percentage of respondents who initiated but did not complete the instrument and the number of questions that had large nonresponse rates.

The issue of how to improve the questionnaire depends critically on the registry's intended purpose(s) going forward. VA has articulated several different purposes in various documents: to help monitor health conditions affecting eligible veterans and service members; to improve VA programs to help veterans and service members with deployment exposure concerns; to generate potential hypotheses about exposure–response relationships; to improve programs in the Veterans Health Administration; and to provide outreach to veterans who may have experienced adverse health outcomes as a result of their exposures. However, with the exception of compiling a list of persons who may benefit from future outreach efforts and the possible exception of hypothesis generation, it does not appear that the registry as currently configured is fit for the articulated purposes.

Given the inherent weaknesses of registries that rely on voluntary participation and self-reports of crucial information, the committee concludes that the best ways to make use of the registry are to

1. make it a means for the eligible population to document their concerns over health problems that may have resulted from their service, bring those concerns to the attention of VA and their health care providers, and supply VA with a list of persons who are interested in burn pit exposure issues; and
2. generate data on the prevalence of health problems in the respondents that might possibly be used to stimulate research using more sophisticated analysis means.

If VA chooses to use the registry for these purposes, then the questionnaire may be simplified as follows:

The committee recommends that VA eliminate the questionnaire sections addressing locations of previous residences (Section 4), non-military work history (5) and home environment, community, or hobbies (6), which collect data that might only be useful in epidemiologic studies of the population.

Eliminating these categories would make the questionnaire easier and faster to complete and would better focus it on the needs of the eligible population.

More generally, the AH&OBP Registry's data collection, administration, and management efforts would be improved by taking these steps:

The committee recommends that once VA clarifies the intent and purpose of the registry, it develop a specific plan for more seamlessly integrating relevant VA and DoD data sources with the registry's data, with the goals of reducing future participant burden, increasing data quality by restructuring questions to minimize recall and other biases, and improving the usefulness of the registry database as an information source for health care professionals and researchers.

The committee recommends that alternative means of completing the questionnaire such as a mail-in form or via a computer-assisted phone interview be offered in order to ensure that the subset of eligible persons who do not use or are not facile with the Internet have the opportunity to participate in the registry.

The committee recommends that VA involve external survey experts experienced in Web-based instruments in any restructuring of the registry questionnaire.

ANALYSIS METHODS AND DESCRIPTIVE STATISTICS FOR THE REGISTRY DATA

VA made data from the first 13 months of the operation of the AH&OBP Registry (June 2014–July 2015) available for analysis by the committee's contractor. Not all of the fields requested were provided because none of the data that the committee were allowed to access could contain information that VA deemed to be personally identifiable. This restricted the type and level of detail of the analyses that could be conducted on the data, and it prevented the committee from carrying out some of the work specified in its statement of task. These constraints also affected the confidence with which the committee can draw conclusions regarding the process of data acquisition and the validity of the information reported on exposure and health outcomes.

Another major limitation is that questionnaire and other data were made available only for those who finished and submitted the questionnaire. A VA report (2015) indicated that nearly 40% of those who initiated an AH&OBP Registry questionnaire did not complete it; this is an outcome that should be followed up.

The committee recommends that VA evaluate whether and how registrants who did not complete the questionnaire differ from those who did, analyze the determinants of non-completion, and use this information to formulate strategies to encourage registrants to finish and submit their responses and improve the completion rate for future participants.

The committee supplemented the information made available for its analysis with reports and a peer-reviewed paper generated under the direction of VA. The resulting analyses of registry data had access to more—and more detailed—information than was available to the committee, and while they covered different time periods than the dataset used for the committee's analyses, they provided some additional insights.

Over the registry's first 13 months, approximately 47,000 people completed the questionnaire, representing 1.0% of eligible Gulf War veterans and 1.7% of eligible post-9/11 veterans. Approximately 7.5% of registry respondents served in the 1990–1991 Gulf War only. Compared to post-9/11 respondents, they are more racially diverse, had less education, deployed fewer times, and were older, more likely to be enlisted, to have served in the Army, and to have been active-duty. Post-9/11 respondents comprises the majority—more than 85%. Analyses were adjusted for demographic and military characteristics, but factors such as the older age of Gulf War veterans might be more salient when examining associations with respiratory and cardiovascular diseases, which are more likely to become more prevalent as the population ages.

Among the most notable of the observations that can be drawn from these data are that nearly all respondents reported one or more airborne hazards encountered in theater: 96% of all respondents reported being exposed to a burn pit on at least one deployment, and 85.6% of Gulf War era respondents reported exposure to smoke from oil-well fires, while 85.2% of all respondents reported exposure to dust storms. The lack of data on those who were deployed and do not believe they were exposed to burn pits precludes using the registry to compare exposed to unexposed individuals. Therefore, the only means available for evaluating burn pit exposure is to examine gradations of exposure among the respondents.

Several other variables have high rates of consistent responses (showing little variability), making them ultimately of little use for analyses, and a number of questions had nonresponse rates of greater than 15%. These findings lend additional supporting evidence that many of the questions are poorly worded or otherwise problematic.

Analyses of demographic data indicate that neither the Gulf War nor the post-9/11 era registry respondents can be considered representative of their respective eligible non-respondent populations. This means that findings made using the registry data—which represent the experience of a small, non-random, self-selected sample—are not generalizable to the broader, eligible population and cannot be used for making inferences concerning it.

ANALYSIS METHODS AND INTERPRETATION OF REGISTRY EXPOSURE DATA

The committee identified several problems with the way the registry's exposure data were collected that were compounded by the inherent limitations of self-reported information. One problem, already mentioned, was found with deployment-related exposure questions, which required respondents to recall specific information for each of the locations they were assigned to. Another was that the questions do not provide information on the intensity of exposure. A high fraction of registry participants reported potential exposures to both burn pit emissions and dust, and there was a tendency for individuals reporting exposures to one type of source to report exposures to other sources as well. This raises concerns about the representativeness of the data and its usefulness in evaluating associations between exposures and health outcomes.

Given the charge—and a concern for overinterpreting the data at hand—the committee developed a reduced set of metrics to categorize exposure potential for the purpose of analysis. Because there were many sources of airborne emissions that contributed to a service member's exposures to particulate matter and chemical exposures and insufficient data by which to determine which sources contributed the most or posed the most harm, the committee chose to weigh each potential exposure equally and focus on the totality of exposures.

On the basis of its evaluation, the committee concludes that the exposure data are of insufficient quality or reliability to make them useful in anything other than the most general assessments of exposure potential. Given this limitation, the committee believes that there may be some circumstances where supplementing these data with information from on-site environmental monitoring or with meteorological, satellite, or other relevant measurements or observations might yield results that would suggest that some individuals or groups experienced greater or lesser exposures to specific constituents that might stimulate more detailed assessments of health outcomes in particular populations.

ANALYSIS METHODS AND INTERPRETATION OF REGISTRY HEALTH OUTCOMES DATA

The committee took an approach analogous to that used for exposure data in order to characterize the health outcomes data for analysis purposes—specifically, generating variables using multiple grouped indicators of these outcomes. While the limitations of the AH&OBP Registry questionnaire and the data collected by it are too great to allow any firm conclusions to be drawn from its analysis, the health outcomes data related to the symptoms, conditions, and diseases associated with the respiratory and the cardiovascular systems are the best candidates for study because these constitute the most plausible and well documented potential health effects of the exposures of concern.

Generally speaking, the committee found that the observed prevalences of respiratory and cardiovascular outcomes appear consistent with what would be expected in a population that is predominantly male, aged 25–60, and for whom about one-third report a current or former history of smoking. It concluded that the health data may be of sufficient quality to justify internal comparisons in which subsets of registry participants with varying levels of potential exposure are compared with one another. An examination of multiple indices of exposure to burn pit emissions and other hazards associated with deployment showed that registry participants who reported more exposures of all types also tended to report more health problems of all types.

The committee's exposure potential variables had strong and consistent associations with self-reported asthma; any respiratory symptom; emphysema, chronic bronchitis, or chronic obstructive pulmonary disease (COPD); functional limitations due to lung or breathing problems; cardiovascular disease; and hypertension. Importantly, though, the analyses also uncovered some unexpected findings that are not consistent with currently understood scientific mechanisms of exposure and outcome, such as a statistically significant association between higher self-reported levels of asbestos exposure and a higher prevalence of neurologic, immune, or liver conditions. Such outcomes strongly suggest that the results of analyses of the registry data cannot be taken at face value and that the identified associations may be an artifact of the population's selection and the limitations of the self-reported exposure and disease data.

Again, the bottom line is that registry analyses are not generalizable and can only describe what exposures and conditions the population of registry respondents are reporting: registry data cannot be used to determine cause or to estimate prevalence in the total eligible population of service members or veterans. The committee wishes to emphasize that it would have made this same determination had the analyses found no associations or weak associations between the exposures and health outcomes.

The strong conclusion that can be drawn is that a more rigorous and appropriate study design is needed to examine the relationship between the exposures encountered during deployment to the Southwest Asia theater of operations and health outcomes. While the registry provides a forum for collecting and recording information on those who were deployed and who are motivated to participate, it cannot answer such questions.

The committee recommends that other means for evaluating the potential health effects associated with airborne hazards and open burn pit exposures be developed, such as a well designed epidemiologic study.

The 2011 Institute of Medicine (IOM) report *Long-Term Health Consequences of Exposure to Burn Pits in Iraq and Afghanistan* contains advice and recommendations on how such a study might be conducted.

The National Research Council (NRC) *Review of the Department of Defense Enhanced Particulate Matter Surveillance Program Report* recommended that a "complete inventory of all major sources of ambient pollutants and potential emissions in the theater should be constructed before assessment of health effects to ensure that all relevant pollutants are monitored" (NRC, 2010, p. 73), and recent advances in low-cost air pollution monitoring devices (Manikonda et al., 2016; Wang et al., 2015) make such data collection efforts more feasible than ever before.

The committee believes that, while medically verified health outcomes information exists only for the subset of the population that uses VA health care,[2] there is potential value in linking the registry data to health care use and conducting analyses on registry participants. Comparisons between self-reported information collected by the questionnaire and diagnoses in VA medical records for respondents who use VA health care would provide further information concerning the level of validity of self-reported health outcomes in the population of respondents.

[2] These data were not available to the committee but are contained in the Veterans Health Administration records.

Given this and the committee's other findings regarding the registry:

The committee recommends that VA's messaging be explicit about the limitations on the ability of the AH&OBP Registry to generate valid information that can be used to improve VA health and benefits programs or inform treatment of individuals potentially exposed to burn pits or other airborne hazards in theater in order to ensure that participants and others do not form unrealistic expectations about the value of participation or capabilities of the registry.

The AH&OBP Registry has many flaws, but even a well-designed and executed registry would have little value as a scientific tool for health-effects research compared to a well-designed epidemiologic study. Addressing the issues identified by the committee would, though, improve the registry's utility as a means of

- generating a roster of concerned individuals that VA can use for targeted outreach, surveillance, and health-risk communication;
- creating, via the completed questionnaire, a record of potential exposures and health concerns that is recorded in the participant's VA electronic health record; and
- allowing VA users and nonusers who take part in the optional clinical exam to articulate concerns they may have to a health care provider and, if warranted, undergo appropriate diagnostic testing or referral, and begin treatment to improve symptoms.

OTHER FINDINGS, CONCLUSIONS, AND RECOMMENDATIONS

The committee was also asked to offer observations on some additional issues surrounding the registry and the actions being taken by DoD and VA to address airborne hazards and open burn pit questions. Specifically, the legislation that directed VA to establish the registry called for

- an assessment of the effectiveness of actions taken by the [Department of Veterans Affairs and the Department of Defense] to collect and maintain information on the health effects of exposure to toxic airborne chemicals and fumes caused by open burn pits;
- recommendations to improve the collection and maintenance of such information; and
- using established and previously published epidemiological studies, recommendations regarding the most effective and prudent means of addressing the medical needs of eligible individuals with respect to conditions that are likely to result from exposure to open burn pits (Public Law 112-260 § 201(b)(1)(A)(i–iii)).

To date, other than the AH&OBP Registry and the airborne exposures and health information collected as part of such efforts as the Gulf War Registry and Millennium Cohort Study, there are no systematic data collection or maintenance efforts focused on the effects of burn pit emissions.[3] Very limited in-theater air pollution data gathering efforts have generated information that would aid in studies of those who served in the same place and at the same time as measurements were made. Two previous reports have offered recommendations on how more rigorous and useful data could be collected: *Review of the Department of Defense Enhanced Particulate Matter Surveillance Program Report* (NRC, 2010) and *Long-Term Health Consequences of Exposure to Burn Pits in Iraq and Afghanistan* (IOM, 2011). This committee concludes that the recommendations these reports offer regarding, respectively, environmental sampling in a combat theater and the conduct of a prospective study of the long-term health effects of exposure to burn-pit emissions are still salient and, if implemented, would materially improve the knowledge base on the health effects of past, present, and future in-theater exposures.

The committee's assessment of "established and previously published epidemiological studies" requested by

[3] DoD and VA collect and analyze data on all medical conditions in the populations that participate in their health care programs, but these are not specific to airborne hazards or burn pit emissions.

the registry's enabling legislation[4] found very few that addressed service members and veterans exposed to open burn pits. The results of these studies do not suggest any general course of action for addressing the medical needs of this population beyond the steps that health care providers should already be carrying out: taking a thorough history, including all occupational exposures, and listening carefully to each patient to determine whether personalized diagnostic testing or treatment is indicated. The health care provider instructions for AH&OBP Registry clinical examinations published by VA (2016) are sound guidance on this.

The information developed by the registry has limited value for improving individual patient care. However, while these data may be inappropriate for evaluating the association between exposures and health outcomes, there are other ways in which they may be useful. As has already been mentioned, the committee believes that the registry's primary utility is that it provides a means for veterans and service members to document their concerns about wartime exposures and the health problems that might have resulted from them and to bring these to the attention of both VA and its health care providers. The self-reported signs, symptoms, and diseases identified by registrants constitute a record that can alert providers to concerns and problems that may be forgotten or missed during clinical encounters.

The registry questionnaire collects a number of pieces of information that would facilitate conversations between a patient and a health care provider, without regard to whether the information might be relevant to AH&OBP exposures. For example, someone who reported difficulty walking long distances or climbing stairs might be experiencing joint pain, respiratory problems, atherosclerotic vascular disease, congestive heart failure, obesity, or even anxiety. Similarly, a complaint of chest pain can have multiple causes in addition to angina and coronary artery disease, including gastroesophageal reflux disease, chest wall pain or costochondritis, and anxiety. And often these symptoms can be multifactorial in origin. Registry questionnaire responses are already accessible to VA health care providers as part of a veteran's electronic health record, and a complete set of responses may be downloaded and printed for a respondent to take to a clinical visit with a provider outside of the VA system.

The committee recommends that VA enhance the utility of the AH&OBP Registry by developing a concise version of participant's questionnaire responses focused on information that would be most useful in a routine clinical encounter and make it available for download.

Providers often have little time to get histories and patients do not always do a good job of raising concerns so a succinct summary would greatly benefit both.

The data the registry generates on the number of respondents who report particular health problems may also be useful to VA. For example, several thousand individuals have indicated that they have diagnosed or self-reported cardiopulmonary symptoms. If these persons subsequently present for evaluation or treatment at rates that would not otherwise have been anticipated by VA, it would indicate that the registry could be used as tool for anticipating future demand for particular provider services. However, it remains to be seen whether this would be the case, and the number of individuals who have thus far completed the questionnaire is only a tiny fraction of the overall population eligible for VA care.

Given the demonstrated concerns of respondents regarding the health effects of exposure to airborne hazards and open burn pit emissions, it is unclear why so few have yet to arrange for the optional in-person clinical evaluation by a VA provider that is made available as part of the registry.

The committee recommends that VA continue its efforts to make it easier for participants to schedule and get the optional health examination offered as part of the AH&OBP Registry—such as through targeted follow-up of respondents who indicate interest—and that it investigate the reasons why such a small percentage of respondents who indicate interest in an exam (~2.5%, to date) request one.

[4] Presented in Chapter 1 of the report.

Adding a means of scheduling an exam as part of the questionnaire—a capability that the committee understands is being implemented—is a useful first step.

CLOSING OBSERVATIONS

The committee recognizes the great interest that active duty military personnel and veterans who served in Iraq, Afghanistan, and the greater Southwest Asia theater of operations have in understanding potential threats to their health from airborne hazards and open burn pit exposures. As its analysis has made clear, though, there are inherent features of registries that rely on voluntary participation and self-reported information that make them fundamentally unsuitable for addressing the question of whether these exposures have, in fact, caused health problems. All parties—service members, veterans, and their families; VA; Congress; and other concerned people—would benefit from having a realistic understanding of the strengths and limitations of registry data so that they can make best use of them and, if desired, conduct the kind of investigations that might yield salient health information and enhance health care for those affected.

REFERENCES

IOM (Institute of Medicine). 2011. *Long-term health consequences of exposure to burn pits in Iraq and Afghanistan.* Washington, DC: The National Academies Press.

Manikonda, A., N. Zíková, P. K. Hopke, and A. R. Ferro. 2016. Laboratory assessment of low-cost PM monitors. *Journal of Aerosol Science* 102:29–40.

NRC (National Research Council). 2010. *Review of the Department of Defense Enhanced Particulate Matter Surveillance Program report.* Washington, DC: The National Academies Press.

VA (Department of Veterans Affairs). 2015. *Report on data from the Airborne Hazards and Open Burn Pit (AH&OBP) Registry, June 2015.* http://www.publichealth.va.gov/docs/exposures/va-ahobp-registry-data-report-june2015.pdf (accessed September 10, 2016).

VA. 2016. War Related Illness and Injury Study Center. Airborne Hazards & Open Burn Pit Registry: Introduction to airborne hazards for providers. http://www.warrelatedillness.va.gov/education/factsheets/airborne-hazards-and-open-burn-pit-registry-for-providers.asp (accessed October 18, 2016).

Wang, Y., J. Li, H. Jing, Q. Zhang, J. Jiang, and P. Biswas. 2015. Laboratory evaluation and calibration of three low-cost particle sensors for particulate matter measurement. *Aerosol Science and Technology* 49(11):1063–1077.

Appendix A

Public Law 112-260

Public Law 112-260
112th Congress

An Act

To amend title 38, United States Code, to ensure that deceased veterans with no known next of kin can receive a dignified burial, and for other purposes.

Jan. 10, 2013
[S. 3202]

Be it enacted by the Senate and House of Representatives of the United States of America in Congress assembled,

SECTION 1. SHORT TITLE; TABLE OF CONTENTS.

Dignified Burial and Other Veterans' Benefits Improvement Act of 2012. 38 USC 101 note.

(a) SHORT TITLE.—This Act may be cited as the "Dignified Burial and Other Veterans' Benefits Improvement Act of 2012".

(b) TABLE OF CONTENTS.—The table of contents for this Act is as follows:

...

38USC527note.

SEC. 201. ESTABLISHMENT OF OPEN BURN PIT REGISTRY.

(a) ESTABLISHMENT OF REGISTRY.—

(1) IN GENERAL.—Not later than one year after the date of the enactment of this Act, the Secretary of Veterans Affairs shall—

(A) establish and maintain an open burn pit registry for eligible individuals who may have been exposed to toxic airborne chemicals and fumes caused by open burn pits;

(B) include any information in such registry that the Secretary of Veterans Affairs determines necessary to ascertain and monitor the health effects of the exposure of members of the Armed Forces to toxic airborne chemicals and fumes caused by open burn pits;

(C) develop a public information campaign to inform eligible individuals about the open burn pit registry, including how to register and the benefits of registering; and

(D) periodically notify eligible individuals of significant developments in the study and treatment of conditions associated with exposure to toxic airborne chemicals and fumes caused by open burn pits.

(2) COORDINATION.—The Secretary of Veterans Affairs shall coordinate with the Secretary of Defense in carrying out paragraph (1).

(b) REPORT TO CONGRESS.—

(1) REPORTS BY INDEPENDENT SCIENTIFIC ORGANIZATION.—The Secretary of Veterans Affairs shall enter into an agreement with an independent scientific organization to prepare reports as follows:

(A) Not later than two years after the date on which the registry under subsection (a) is established, an initial report containing the following:

(i) An assessment of the effectiveness of actions taken by the Secretaries to collect and maintain information on the health effects of exposure to toxic airborne chemicals and fumes caused by open burn pits.

information.

(ii) Using established and previously published epidemiological studies, recommendations regarding the most effective and prudent means of addressing the medical needs of eligible individuals with respect to conditions that are likely to result from exposure to open burn pits.

(B) Not later than five years after completing the initial report described in subparagraph (A), a follow-up report containing the following:

(i) An update to the initial report described in subparagraph (A).

(ii) An assessment of whether and to what degree the content of the registry established under subsection (a) is current and scientifically up-to-date.

(2) SUBMITTAL TO CONGRESS.—

(A) INITIAL REPORT.—Not later than two years after the date on which the registry under subsection (a) is established, the Secretary of Veterans Affairs shall submit to Congress the initial report prepared under paragraph (1)(A).

(B) FOLLOW-UP REPORT.—Not later than five years after submitting the report under subparagraph (A), the Secretary of Veterans Affairs shall submit to Congress the follow-up report prepared under paragraph (1)(B).

(b) DEFINITIONS.—In this section:

(1) ELIGIBLE INDIVIDUAL.—The term "eligible individual" means any individual who, on or after September 11, 2001—

(A) was deployed in support of a contingency operation while serving in the Armed Forces; and

(B) during such deployment, was based or stationed at a location where an open burn pit was used.

(2) OPEN BURN PIT.—The term "open burn pit" means an area of land located in Afghanistan or Iraq that—

(A) is designated by the Secretary of Defense to be used for disposing solid waste by burning in the outdoor air; and

(B) does not contain a commercially manufactured incinerator or other equipment specifically designed and manufactured for the burning of solid waste.

LEGISLATIVE HISTORY—S. 3202:

CONGRESSIONAL RECORD, Vol. 158 (2012):

Dec. 19, considered and passed Senate. Dec. 30, considered and passed House.

Appendix B

Workshop Agenda

May 1, 2015
Keck Center of the National Academies of Sciences, Engineering, and Medicine, Washington, DC

10:00 a.m. **Welcome | Notes on the Conduct of the Open Session | Introduction of Participants**
David Savitz, Ph.D.
Committee Chair

10:15 a.m. **Ambient Air Monitoring at Deployment Locations in the Middle East**
John E. Kolivosky, PE
Lead, Deployment Occupational and Environmental Health Surveillance, Army Institute of Public Health

10:45 a.m. **Balad Air Base Burn Pit Study Observations (Boots on the Ground)**
Maj Charlie Toth, USAF, PE
Defense Fellow—Office of Senator Patty Murray, U.S. Senate

11:15 a.m. **Retrospective Geospacial Modeling of PM10 Exposures from Open Burning at Joint Base Balad, Iraq**
John Rinker, CIH
Industrial Hygienist, Kirk U.S. Army Health Clinic

11:45 a.m. **General Discussion of Open Burn Pit Emissions Data and Modeling Issues**
David Savitz, Ph.D.
Moderator

12:00 p.m. Lunch Break—Keck Center Atrium

1:00 p.m. **Lessons Learned from Other Registries and the Millennium Cohort Study**
Gary Gackstetter, D.V.M., M.P.H., Ph.D.
Associate Professor, Uniformed Services University of the Health Sciences

Tomoko I. Hooper, M.D., M.P.H.
Professor, Department of Preventive Medicine and Biostatistics, Uniformed Services University of the Health Sciences

1:40 p.m. **Analysis of VA Burn Pits Registry: Testimony at Workshop**
Anthony M. Szema, M.D.
Adjunct Professor, Department of Technology & Society, College of Engineering & Applied Science, Stony Brook University; Clinical Assistant Professor, Department of Occupational Medicine, Epidemiology and Prevention, Hofstra Northwell School of Medicine at Hofstra University [via teleconference]

2:20 p.m. Break

2:30 p.m. **Roundtable—Comments and Perspectives on the Airborne Hazards and Open Burn Pit Registry**
Adrian Atizado, Disabled American Veterans [via teleconference]
Thomas Berger, Ph.D., Vietnam Veterans of America
Carlos Fuentes, Veterans of Foreign Wars
Daniel Sullivan, The Thomas Joseph Sergeant Sullivan Center
CPT (Ret.) Le Roy Torres, U.S. Army Reserve, Veteran, U.S. Army-Iraq War Campaign and Founder, BURNPITS 360° [via Web conference]
Rosie Torres, BURNPITS 360°
Rick Weidman, Vietnam Veterans of America

4:00 p.m. **General Discussion of Issues Raised in the Workshop**
David Savitz, Ph.D.
Moderator

4:30 p.m. Workshop Adjourns

Appendix C

Airborne Hazards and Open Burn Pit Registry Self-Assessment Questionnaire

DRAFT – Working Document 15 December 2014

Airborne Hazards And Open Burn Pit Registry Self-Assessment Questionnaire

OMB 2900- XXXX

VA Form 10-10066

This information is collected in accordance with section 3507 of the Paperwork Reduction Act of 1995. Accordingly, VA may not conduct or sponsor and you are not required to respond to a collection of information unless it displays a valid OMB number. We anticipate that the time expended by all individuals who complete this questionnaire will average 40 minutes. This includes the time it will take to read instructions, gather the necessary facts and fill out the form. The results of this questionnaire will lead to improvement in the quality of service delivery by helping to shape the direction and focus of specific programs and services. Submission of this form is voluntary and failure to respond will have no impact on benefits to which you may be entitled.

DRAFT – Working Document 15 December 2014

Airborne Hazards and Open Burn Pit Registry Self-Assessment Questionnaire

Contents

Note: items in square parenthesis, "[]", and item selection number are not displayed to the user.

1. Deployment History

1.1. Deployment Data from the VA Defense Information Repository (VADIR) and DMDC

[Note: Deployment and demographic data will be obtained from DoD data sources (VADIR) after the user's personal identifier is authenticated and stored in the registry database. All deployments from DoD data are displayed to the user. The user will then indicate if the deployment dates are valid or not, add missing deployments, and select which base names they were at while deployed. Guidance will be provided to facilitate direct contact with the appropriate DoD service to correct entries in the official system of record for the DoD deployment data.]

Report Section	Report Field	Note
Deployment Periods	Service	**User Validates**
GWVIS Note	Note	**Indicates Operation Desert Storm and Operation Desert Shield service**

1.2. Location Specific Deployment Exposures

"Tell us about potential exposures while you were deployed."

[Note: Section 1.2 questions are asked for each deployment or deployment segment in the VADIR data]

"During **this** deployment or portion of your deployment:"

A. [if deployment dates within 1990 – 1992, e.g. VADIR GWVIS indicator set], Were you exposed to soot, ash, smoke, or fumes from the Gulf War oil well fires?

1. Yes, 2. No, 3. I do not wish to answer, 4. Don't know

B. Where did you spend **most** of your time during these dates?
[if deployment dates **not** within 1990 – 1992, e.g. VADIR GWVIS indicator not set:list base names, see Appendix A]
[Select from list], Other (text entry), I do not wish to answer, Don't know

C. If you were at more than one base, where did you spend the **second most** amount of time during these dates?

[if deployment dates **not** within 1990 – 1992, e.g. VADIR GWVIS indicator not set: list base names, see Appendix A]

[Select from list], Other (text entry), I was not at any other bases, I do not wish to answer, Don't know

D. Were you near a burn pit during these dates (on the base or close enough to the base for you to see the smoke)?
1. Yes, 2. No, 3. I do not wish to answer, 4. Don't know

E. [If 'D' = yes], Who ran this burn pit (circle all that apply)?
1. U.S. forces or Contractor, 2. Coalition forces, 3. Host nation, 4. I do not wish to answer, 5. Don't know

F. [If 'D' = yes] Did your duties during these dates include the burn pit (examples include trash burning, hauling trash to the burn pit, burn pit security, trash sorting at the burn pit)?
1. Yes, 2. No, 3. I do not wish to answer, 4. Don't know

G. [If 'D' = yes] On a typical day, how many hours did smoke or fumes from the burn pit enter your **work site or housing**?
1. Never, 2. Enter {1, 2, 3, ... 24} hours, 3. I do not wish to answer, 4. Don't know

H. On a typical day, how many hours were you outside or in an open tent or shelter (for example a single wall tent with open seams or drafty "B" hut)?
1. Never, 2. Enter {1, 2, 3, ... 24} hours, 3. I do not wish to answer, 4. Don't know

I. On a **typical day**, how many hours were you near (for example you could smell or see it) sewage ponds?
1. Never, 2. Enter {1, 2, 3, ... 24} hours, 3. I do not wish to answer, 4. Don't know

1.3. General Military Occupational Exposures

During **any** of your deployments:

A. Were you ever close enough to feel the blast from an IED (improvised explosive device) or other explosive device?
   1. Yes, 2. No, 3. I do not wish to answer, 4. Don't know

B. In a typical month, how many days were you near heavy smoke from weapons, signal smoke, markers, or other combat items?
   1. Never, 2. Enter {1, 2, 3 ... 31} days, 3. I do not wish to answer, 4. Don't know

C. In a typical month, how many days were you in convoy or other vehicle operations?
   1. Never, 2. Enter {1, 2, 3 ... 31} days, 3. I do not wish to answer, 4. Don't know

D. In a typical month, how many days did you perform refueling operations?
   1. Never, 2. Enter {1, 2, 3 ... 31} days, 3. I do not wish to answer, 4. Don't know

E. In a typical month, how many days did you perform aircraft, generator, or other large engine maintenance?
   1. Never, 2. Enter {1, 2, 3 ... 31} days, 3. I do not wish to answer, 4. Don't know

F. In a typical month, how many days did you perform construction duties?
   1. Never, 2. Enter {1, 2, 3 ... 31} days, 3. I do not wish to answer, 4. Don't know

G. In a typical month, how many days did you perform pesticide duties for your unit?
   1. Never, 2. Enter {1, 2, 3 ... 31} days, 3. I do not wish to answer, 4. Don't know

1.4. Environmental Exposures, Regional Air Pollution

A. Did you do anything differently during your deployment(s), when you thought or were informed air quality was bad (for example during dust storms or heavy pollution days)?
   1. Yes, 2. No, 3. Never thought of this, 4. I was not informed or aware of bad air quality, 5. I do not wish to answer, 6. Don't know

B. [A=yes], What did you do differently (select all that apply)?
   1. Wore a mask, cravat, or bandana over your mouth or nose
   2. Spent less time outdoors
   3. Did less strenuous activities (i.e. avoided physical training (PT))
   4. Took medication

5. Closed windows of your sleeping quarters
6. Spent less time in convoy
7. Canceled outdoor activities
8. Exercised indoors instead of outdoors
9. Used or changed air filter/air cleaner
10. Other
11. I did not (or could not) do anything differently
12. I do not wish to answer

C. In a typical month during your deployment(s), how many days did you experience dust storms?
 1. Never, 2. Enter {1, 2, 3 ... 31} days, 3. I do not wish to answer, 4. Don't know

D. During your deployment(s), did you experience wheezing, difficulty breathing, an itchy or irritated nose, eyes or throat that you thought was the result of poor air quality?
 1. Yes, 2. No, 3. I do not wish to answer, 4. Don't know

E. [If 'D'=yes], How many days in an average month did you experience wheezing, difficulty breathing, an itchy or irritated eyes, nose or throat that you thought was the result of poor air quality?
 1. Enter {1, 2, 3 ... 31} days, 2. Never, 3. I do not wish to answer, 4. Don't know

F. During your deployment(s), did you seek medical care for wheezing, difficulty breathing, itchy or irritated nose, eyes or throat that you thought was the result of poor air quality?
 1. Yes, 2. No, 3. I do not wish to answer, 4. Don't know

2. Symptoms and Medical History

"Tell us your health history. Please list all conditions even if you don't think they're related to a deployment exposure."

2.1. Functional Limitations and Reported Cause

[Source: NHIS Adult Health Status & Limitations starting with AHS.091_01.000]

A. How difficult is it to run or jog one mile on a level surface?

1. Not at all difficult, 2. only a little difficult, 3. somewhat difficult, 4. very difficult, 5. can't do it at all, 6. do not do this activity, 7. I do not wish to answer, 8. Don't know

B. How difficult is it to walk on a level surface for one mile?

1. Not at all difficult, 2. only a little difficult, 3. somewhat difficult, 4. very difficult, 5. can't do it at all, 6. do not do this activity, 7. I do not wish to answer, 8. Don't know

C. How difficult is it to walk a ¼ of a mile – about 3 city blocks?

1. Not at all difficult, 2. only a little difficult, 3. somewhat difficult, 4. very difficult, 5. can't do it at all, 6. do not do this activity, 7. I do not wish to answer, 8. Don't know

D. How difficult is it to walk up a hill or incline?

1. Not at all difficult, 2. only a little difficult, 3. somewhat difficult, 4. very difficult, 5. can't do it at all, 6. do not do this activity, 7. I do not wish to answer, 8. Don't know

E. How difficult is it to walk up 10 steps or climb a flight of stairs?

1. Not at all difficult, 2. only a little difficult, 3. somewhat difficult, 4. very difficult, 5. can't do it at all, 6. do not do this activity, 7. I do not wish to answer, 8. Don't know

[Source: NHIS: Adult Health Status & Limitations AHS.200_00.000, selection 14 modified]

F. [If any question A-E = "difficult"] What condition or health problem causes you to have difficulty with these activities? (Check all that apply.)

01 Arthritis/rheumatism
02 Back or neck problem
03 Benign Tumors, Cysts
04 Birth defect
05 Brain injury (for example, Traumatic Brain Injury/TBI, Intellectual disability)
06 Cancer
07 Circulation problems (including blood clots)
08 Depression/anxiety/emotional problem
9 Diabetes
10 Epilepsy, seizures
11 Fibromyalgia, lupus
12 Fracture, bone/joint injury
13 Hearing problem
14 Heart problem
15 Hernia

16 Hypertension/high blood pressure
17 Kidney, bladder or renal problems
18 Knee problems (not arthritis, not joint injury)
19 Lung/breathing problem (for example, asthma and emphysema)
20 Migraine headaches (not just headaches)
21 Missing limbs (fingers, toes or digits), amputee
22 Multiple Sclerosis (MS), Muscular Dystrophy (MD)
23 Other developmental problem (for example, cerebral palsy)
24 Other injury
25 Other nerve damage, including carpal tunnel syndrome
26 Osteoporosis, tendinitis
27 Parkinson's disease, other tremors
28 Polio(myelitis), paralysis, para/quadriplegia
29 Senility
30 Stroke problem
31 Thyroid problems, Grave's disease, gout
32 Ulcer
33 Varicose veins, hemorrhoids
34 Vision/problem seeing
35 Weight problem
36 Other impairment/problem (Specify one)
37 I do not wish to answer
38 Don't know/Not sure

2.2. Health Conditions

2.2.1. Respiratory Conditions

[Source: NHIS Adult Conditions ACN.031 series]

A. Have you **ever** been told by a doctor or other health professional that you had Hay fever or allergies to pollen, dust, or animals?
   1. Yes, 2. No, 3. I do not wish to answer, 4. Don't know

[Source: NHIS Adult Conditions ACN.080_00.000]

B. Have you **ever** been told by a doctor or other health care professional that you had asthma?
   1. Yes, 2. No, 3. I do not wish to answer, 4. Don't know

DRAFT – Working Document 15 December 2014

C. Have you **ever** been told by a doctor or other health care professional that you had emphysema?

1. Yes, 2. No, 3. I do not wish to answer, 4. Don't know

D. Have you **ever** been told by a doctor or other health care professional that you had chronic bronchitis?

1. Yes, 2. No, 3. I do not wish to answer, 4. Don't know

[Source: NHIS Adult Conditions ACN.035 series]

E. Have you **ever** been told by a doctor or other health care professional that you had chronic obstructive pulmonary disease also called COPD?

1. Yes, 2. No, 3. I do not wish to answer, 4. Don't know

F. Have you **ever** been told by a doctor or other health care professional that you had some lung disease or condition other than asthma, emphysema, chronic bronchitis or COPD?

1. Yes, 2. No, 3. I do not wish to answer, 4. Don't know

G. [if F="Yes"] Have you **ever** been told by a doctor or other health care professional that you had constrictive bronchiolitis (CB)?

1. Yes, 2. No, 3. I do not wish to answer, 4. Don't know

H. [if F="Yes"] Have you **ever** been told by a doctor or other health care professional that you had pulmonary fibrosis or idiopathic pulmonary fibrosis (IPF)?

1. Yes, 2. No, 3. I do not wish to answer, 4. Don't know

I. [if B-F = yes], When you were told you had asthma, emphysema, chronic bronchitis, COPD or some other lung disease by a doctor or other health care professional, were you told before, during, or after deployment? (check all that apply.)

1. Before deployment, 2. During deployment, 3. After deployment, 4. I do not wish to answer, 5. Don't know

J. [if I = Before], Did this lung disease get better, worse, or about the same during deployment?

1. Better, 2. Worse, 3. About the Same, 4. Not applicable, 5. I do not wish to answer, 6. Don't know

K. Do you currently have any of the following symptoms? (Check all that apply.)

1. Cough for more than 3 weeks
2. Sputum or phlegm production for more than 3 weeks
3. Wheezing or whistling in the chest

4. Shortness of breath; breathlessness
5. Decreased ability to exercise
6. Hay fever or other respiratory allergy
7. Sore throat, hoarseness, or change in voice
8. Chest pain, chest discomfort or chest tightness
9. Chronic sinus infection/sinusitis
10. I do not wish to answer
11. I do not have these symptoms

L. In the past 12 months did you have any of the following symptoms? (Check all that apply.)

1. Cough for more than 3 weeks
2. Sputum or phlegm production for more than 3 weeks
3. Wheezing or whistling in the chest
4. Shortness of breath; breathlessness
5. Decreased ability to exercise
6. Hay fever or other respiratory allergy
7. Sore throat, hoarseness, or change in voice
8. Chest pain, chest discomfort or chest tightness
9. Chronic sinus infection/sinusitis
10. I do not wish to answer
11. I do not have these symptoms

[Source: Medical Research Chronic (MRC) Breathlessness scale]

M. [IF ANSWER TO "L" Current Health symptoms = 04] How would you rate your shortness of breath or breathlessness? (Check the description/grade that applies to you.) I'm:

1. Not troubled by breathlessness except on strenuous exercise
2. Short of breath when hurrying on the level or walking up a slight hill
3. Walking slower than most people on level ground, stop after one mile, or stop after 15 minutes walking at my own pace
4. Stopping for breath after walking about 100 yards or after a few minutes on level ground
5. Too breathless to leave the house, or breathless when dressing or undressing
6. I do not wish to answer

2.2.2. Cardiovascular Conditions

A. Have you **ever** been told by a doctor or other health care professional that you had hypertension, also called high blood pressure?
    1. Yes, 2. No, 3. I do not wish to answer, 4. Don't know

B. Have you **ever** been told by a doctor or other health care professional that you had coronary artery disease?
    1. Yes, 2. No, 3. I do not wish to answer, 4. Don't know

C. Have you **ever** been told by a doctor or other health care professional that you had angina pectoris?
    1. Yes, 2. No, 3. I do not wish to answer, 4. Don't know

D. Have you **ever** been told by a doctor or other health care professional that you had a heart attack, also called myocardial infarction?
    1. Yes, 2. No, 3. I do not wish to answer, 4. Don't know

E. Have you **ever** been told by a doctor or other health care professional that you had a heart condition other than coronary artery disease or angina or myocardial infarction?
    1. Yes, 2. No, 3. I do not wish to answer, 4. Don't know

F. [if any A-E = yes], When you were told you had hypertension, coronary artery disease, angina pectoris, a heart attack, or some other heart condition by a doctor or other health care professional, were you told before, during, or after deployment? (check all that apply.)
    1. Before deployment, 2. During deployment, 3. After deployment, 4. I do not wish to answer, 5. Don't know

2.2.3. Other Conditions

[Source NHIS ACN.125_00.250]

A. During the **past 12 months**, have you regularly had insomnia or trouble sleeping?
    1. Yes, 2. No, 3. I do not wish to answer, 4. Don't know

[Source modified from NHIS ACN.125_00.130]

B. During the **past 12 months**, have you had Neurological problems? (Some examples of neurological problems may include numbness, tingling, or weakness in your arms or legs or difficulties with thinking or memory.)

1. Yes, 2. No, 3. I do not wish to answer, 4. Don't know

[Source modified from NHIS ACN.125_00.100]

C. During the **past 12 months**, have you had problems of the immune system?

1. Yes, 2. No, 3. I do not wish to answer, 4. Don't know

[Source NHIS ACN.201_05.000]

D. During the **past 12 months**, have you been told by a doctor or other health professional that you had any kind of liver condition?

1. Yes, 2. No, 3. I do not wish to answer, 4. Don't know

E. During the **past 12 months**, have you been told by a doctor or other health professional that you had any a chronic multi-symptom illness (examples include irritable bowel syndrome, chronic fatigue syndrome, and fibromyalgia)?

1. Yes, 2. No, 3. I do not wish to answer, 4. Don't know

F. [if B-E = yes], Did your, neurological or immune problems, chronic multi-symptom illness, or liver condition **first** occur before, during, or after deployment? (check all that apply.)

1. Before deployment, 2. During deployment, 3. After deployment, 4. I do not wish to answer, Don't know

G. On average, how many hours of sleep do you get in a 24-hour period? (Round up 30 minutes or more to the next whole hour.)

1 Enter {1,2,3 ... 24} hours
2 I do not wish to answer
3 Don't know

"Questions H and I are about snoring and breathing during sleep. To answer these questions, please consider both what others have told you **and** what you know about yourself."

H. How often do you snore?

1 Never
2 Rarely - less than one night a week
3 Sometimes - 1 or 2 nights a week
4 Frequently - 3 to 5 nights a week

5 Always or almost always - 6 or 7 nights a week
6 I do not wish to answer
7 Don't know

I. How often do you have times when you stop breathing during your sleep?

1 Never
2 Rarely - less than one night a week
3 Sometimes - 1 or 2 nights a week
4 Frequently - 3 to 5 nights a week
5 Always or almost always - 6 or 7 nights a week
6 I do not wish to answer
7 Don't know

2.3. Height and Weight

A. How tall are you without shoes?
1 Enter (x feet, y inches), 2 I do not wish to answer, 3 Don't know

B. How much do you weigh without shoes?
1 Enter X pounds, 2 I do not wish to answer, 3 Don't know

2.4. Cancer History

[Source NHIS ACN.130_00.000]

A. Have you **ever** been told by a doctor or other health professional that you had Cancer or a malignancy (tumor) of any kind?
1. Yes, 2. No, 3. I do not wish to answer, 4. Don't know

[if 'A' <> "Yes" skip to section 2.5]

B. What kind of cancer was it?
01 Bladder
2 Blood
3 Bone
4 Brain
5 Breast
6 Cervix

7 Colon
8 Esophagus
9 Gallbladder
10 Kidney
11 Larynx-windpipe
12 Leukemia
13 Liver
14 Lung
15 Lymphoma
16 Melanoma
17 Mouth/tongue/lip
18 Ovary
19 Pancreas
20 Prostate
21 Rectum
22 Skin (non-melanoma)
23 Skin (Don't Know what kind)
24 Soft tissue (muscle or fat)
25 Stomach
26 Testis
27 Throat - pharynx
28 Thyroid
29 Uterus
30 Other
31 None
32 I do not wish to answer
33 Don't know

C. [if 'B' < 30] How old were you when this cancer was first diagnosed?
1. Enter {00-99} Years
2. I do not wish to answer, 3. Don't know

D. [if 'B' < 30] If you were diagnosed with a second cancer, what kind of cancer was it?
01 Bladder
2 Blood
3 Bone
4 Brain
5 Breast
6 Cervix
7 Colon
8 Esophagus
9 Gallbladder
10 Kidney
11 Larynx-windpipe

12 Leukemia
13 Liver
14 Lung
15 Lymphoma
16 Melanoma
17 Mouth/tongue/lip
18 Ovary
19 Pancreas
20 Prostate
21 Rectum
22 Skin (non-melanoma)
23 Skin (Don't Know what kind)
24 Soft tissue (muscle or fat)
25 Stomach
26 Testis
27 Throat - pharynx
28 Thyroid
29 Uterus
30 Other
31 None
32 I do not wish to answer
33 Don't know

E. [if 'D' < 30] How old were you when this cancer was first diagnosed?
1. Enter {00-99} Years
2. I do not wish to answer, 3. Don't know

F. [if 'D' < 30] If you were diagnosed with a third cancer, what kind of cancer was it?
01 Bladder
2 Blood
3 Bone
4 Brain
5 Breast
6 Cervix
7 Colon
8 Esophagus
9 Gallbladder
10 Kidney
11 Larynx-windpipe
12 Leukemia
13 Liver
14 Lung
15 Lymphoma
16 Melanoma

17 Mouth/tongue/lip
18 Ovary
19 Pancreas
20 Prostate
21 Rectum
22 Skin (non-melanoma)
23 Skin (Don't Know what kind)
24 Soft tissue (muscle or fat)
25 Stomach
26 Testis
27 Throat - pharynx
28 Thyroid
29 Uterus
30 Other
32 I do not wish to answer
33 Don't know

G. [if 'F' < 30] How old were you when this cancer was first diagnosed?
1. Enter {00-99} Years
2. I do not wish to answer, 3. Don't know

2.5. Tobacco Exposure

[Source: NHIS Adult Health Behaviors: AHB.010_00.000]

A. Have you smoked at least 100 cigarettes in your entire life?
   1. Yes
   2. No
   3. I do not wish to answer
   4. Don't know

[if A=Yes continue to 'B' else skip to 'F']

B. How old were you when you first started to smoke fairly regularly?
   1. Enter X (age in years),
   2. Never smoked regularly
   3. I do not wish to answer
   4. Don't know

[if B=age continue to 'C' else skip to 'F']

C. Do you now smoke cigarettes every day, some days or not at all?
   1. Every day
   2. Some days
   3. Not at all
   4. I do not wish to answer
   5. Don't know

D. [if 'C'=not at all], How long has it been since you quit smoking cigarettes?
   1. Enter {00-99} (Years since quit)
   2. I do not wish to answer
   3. Don't know

E. [if 'C'=some days or every day], On the average, how many cigarettes do you now smoke a day?
   1. Enter {00-99} (Number of cigarettes per day)
   2. I do not wish to answer
   3. Don't know

F. Have you ever smoked tobacco products other than cigarettes even one time? (Such as cigars, pipes, water pipes or hookahs, small cigars that look like cigarettes, bidis, cigarillos, marijuana?)
   1. Yes

2. No
3. I do not wish to answer
4. Don't know

[if F=Yes continue to 'G' else skip to 'H']

G. Do you now smoke tobacco products other than cigarettes every day, some days, rarely, or not at all?

1. Every day
2. Some days
3. Rarely
4. Not at all
5. I do not wish to answer
6. Don't know

H. Have you ever used smokeless tobacco products even one time? (Such as chewing tobacco, snuff, dip, snus, or dissolvable tobacco.)

1. Yes
2. No
3. I do not wish to answer
4. Don't know

[if H=Yes continue to 'I' else skip to 'J']

I. Do you **now** use smokeless tobacco products every day, some days, rarely, or not at all?

1. Every day
2. Some days
3. Rarely
4. Not at all
5. I do not wish to answer
6. Don't know

J. Are you exposed to second-hand smoke or environmental tobacco smoke every day, some days, rarely, or not at all?

1. Every day
2. Some days
3. Rarely
4. Not at all
5. I do not wish to answer
6. Don't know

2.6. Deployment Smoking History

[Source: modified from DoD USAPHC DARE H2-5c]

A. [if 2.5.A = yes], Did you start smoking for the first time while being deployed?

1. Yes, 2. No, 3. I do not wish to answer, 4. Don't know

B. [if 2.6.A = No], How did deployment(s) change how much you smoked?

1. No change, 2. I smoked more while deployed, 3. I smoked less while deployed, 4. I do not wish to answer, 5. Don't know

2.7. 12 Month Alcohol Use

A. In the PAST YEAR, how often did you ever drink any type of alcoholic beverage (Included are liquor such as whiskey or gin, beer, wine, wine coolers, and any other type of alcoholic beverage)? "On average, how many days per week did you drink?"

1. Never, 2. Less than one, 3. 1-7 days per week, 4. I do not wish to answer, 5. Don't know

DRAFT – Working Document 15 December 2014

3. Health Concerns

"Help us focus our efforts on health issues you care about."

A. Compared to pre-deployment, would you say your overall health is better, worse, or about the same?

  1. Better, 2. Worse, 3. About the same, 4. I do not wish to answer, 5. Don't know

B. During your deployment(s), do you believe you were sick because of something you breathed?

  1. Yes, 2. No, 3. I do not wish to answer, 4. Don't know

C. Do you **currently** have a sickness or condition you think began or got worse because of something you breathed during deployment(s)?

  1. Yes, 2. No, 3. I do not wish to answer, 4. Don't know

D. [If 'C'=yes], When did the problem start?

  1. Before deployment
  2. During Deployment
  3. 6 months or less after deployment
  4. More than 6 months later after deployment
  5. Not sure
  6. I do not wish to answer

E. Please rate your concern that something you breathed during deployment **has already affected** your health.

  1. Not at all concerned, 2. a little concerned, 3. very concerned, 4. I do not wish to answer

[If 'E'=very or little concerned continue to F, else skip to H]

F. Please identify your biggest health concern that something you breathed during deployment **has already affected** your health.

  1. Lung/Respiratory/Breathing problem
  2. Heart problem
  3. Skin problem
  4. Eye problem
  5. Gastrointestinal (GI) problem
  6. Neurological problem
  7. Immune problem
  8. Effect on children or ability to have children
  9. Cancer

10. Other problem
11. I do not wish to answer

G. Have you discussed this concern with your health care provider, medical professional or team?

1. Yes, 2. No, 3. Not yet but I would like to talk with a medical professional

H. Are you concerned that **in the future** that your health will be affected by something you breathed during deployment(s)

1. Yes, 2. No, 3. I do not wish to answer, 4. Don't know

I. [If 'H'=yes], Please rate your concern that something you breathed during deployment will affect your **future health**.

1. Not at all concerned, 2. a little concerned, 3. very concerned, 4. I do not wish to answer

J. [If 'I'=very or little concerned], Please identify your biggest health concern that something you breathed during deployment will affect your **future health**.

1. Lung/Respiratory/Breathing
2. Heart
3. Skin
4. Eyes
5. Effect on children or ability to have children
6. Cancer
7. Other
8. I do not wish to answer

K. [If 'E' or 'I'=very or little concerned], Which exposure do you think has the **biggest** overall effect on your health?

1. **Off** base air pollution during deployment (factories, cars, burning trash, dust)
2. **On** base air pollution during deployment (burning fuel, burn pits)
3. Hobbies and non-military jobs
4. Military jobs while I'm not deployed
5. Smoking (by you or those near you)
6. I do not wish to answer
7. Don't know

4. Places You've Lived

"Poor air quality in places where you've lived may impact how deployment exposures affect you."

[System displays current address]

A. What is your current address (complete here if not shown above [from VADIR and VA BIRLS sources], if correct skip to "5")? Please include the city, state, zip code, and country.

1. Country _

   [If country <> "USA" then skip to 5]

2. City Name_
3. State_ (two letter code)
4. Zip code (if known):_ (5 digit number)
5. How many years have you lived at your current address (listed above)? _ years
6. Do you live nine or more months of the year at the address listed above? Yes, No
7. If not, indicate the other residence.
   a. Other city name_
   b. Other state_ (two letter code)
   c. Other zip code (if known):_ (5 digit number)
   d. Other country _

B. Where have you lived the longest? Please include the city, state, zip code, and country.

1. The address where I lived the longest is the same as my current address.
   Yes (if yes go to next question), No
2. Country _

   [If country <> "USA" then skip to 6]

3. City Name_
4. State_ (two letter code)
5. Zip code (if known):_ (5 digit number)
6. Indicate the approximate year you moved to this address: _
7. Indicate the approximate year you moved out of this address: _

C. Please provide the address where you lived the longest before age 13. Please include the city, state, zip code, and country.

1. Country _

   [If country <> "USA" then skip to 5]

2. City Name_
3. State_ (two letter code)
4. Zip code (if known):_ (5 digit number)
5. Indicate the approximate age you moved to this address. _ years (Enter "0" if you lived there before age 1)
6. Indicate the approximate age you moved out of this address: _ years

5. Work History

"Exposures in your non-military jobs may impact how deployment exposures affect you."

5.1. Current Occupational Status

A. Which of the following were you doing last week?
   1 Working for pay at a job or business
   2 With a job or business but not at work (e.g. a volunteer)
   3 Looking for work
   4 Working, but not for pay, at a family-owned job or business
   5 Not working at a job or business and not looking for work
   6 I do not wish to answer
   7 Don't know

B. [if A=3 or 5] What is the main reason you did not [3 or 5 text: work last week/have a job or business last week]?
   1 Taking care of house or family
   2 Going to school
   3 Retired
   4 On a planned vacation from work
   5 On family or maternity leave
   6 Temporarily unable to work for health reasons

7 Have job/contract and off-season
8 On layoff/laid-off from a job
9 Disabled
10 Other
11 I do not wish to answer
12 Don't know

5.2. Main Occupation

A. Select the occupational category that best describes your main occupation (the civilian job you've held the longest). **Do not include your occupation during military service.** If your occupation is not included, select "other occupation":

1. Agricultural and fishing/hunting workers	2. Automotive, aircraft and marine mechanics and service technicians	3. Construction trade workers, helpers and other construction related workers
4. Driver/sales workers and truck drivers	5. Extraction workers (e.g. mining or drilling)	6. Firefighters
7. Food processing	8. Forest conservation and logging workers	9. Police and sheriff's patrol officers
10. Welding, soldering and brazing	11. Other occupation [text entry]	12. I do not wish to answer

B. Total years in this non-military job {0...99} years (enter 0 if less than one year).

1. Enter {00-99} years
2. I do not wish to answer, 3. Don't know

5.3. Dust Exposures

A. Have you ever worked for a year or more in any dusty job **outside the military**?

1. Yes, 2. No, 3. I do not wish to answer, 4. Don't know

[if A = yes, continue, else jump to 5.4]

B. For the job with the biggest dust exposure:

1. Select the occupational category that best describes the job with the longest dust exposure. If your occupation is not included, select "other occupation":

1. Agricultural and fishing/hunting workers	2. Automotive, aircraft and marine mechanics and service technicians	3. Construction trade workers, helpers and other construction related

		workers
4. Driver/sales workers and truck drivers	5. Extraction workers (e.g. mining or drilling)	6. Firefighters
7. Food processing	8. Forest conservation and logging workers	9. Police and sheriff's patrol officers
10. Welding, soldering and brazing	11. Other occupation [text entry]	12. I do not wish to answer

2. In this job, what were the most common kinds of dust to which you were exposed (select all that apply)?

1. Animal dander	2. Wood or sawdust	3. Metal (aluminum, copper, iron, steel, or other types)
4. Cotton, wool, or other cloth or textile	5. Asbestos	6. Plaster
7. Flour	8. Cement	9. Sand or silica
10. Grain	11. Coal	12. Talc
13. Hay	14. Fiberglass	15. Lime
16. Paper or cardboard	17. Granite or other rock	18. Plastic or rubber
19. Soil or dirt	20. Other dust [text entry]	21. I do not wish to answer

3. Total years in this job {0...99} years (enter 0 if less than one year).
   1. Enter {00-99} years
   2. I do not wish to answer, 3. Don't know

4. Are you working in this dusty job now?

1. Yes, 2. No, 3. I do not wish to answer, 4. Don't know

5.4. Gas, Smoke, Vapors or Fumes Exposures

A. Have you ever been exposed to gas, smoke, chemical vapors or fumes in your work **outside the military**?

1. Yes, 2. No, 3. I do not wish to answer, 4. Don't know

[if A = yes, continue, else jump to 5.5]

B. For the job with the biggest gas, smoke, vapor or fume exposure:

DRAFT – Working Document 15 December 2014

1. Select the occupational category that best describes the job with the longest gas, smoke, chemical vapor, or fume exposures. If your occupation is not included, select "other occupation":

1. Agricultural and fishing/hunting workers	2. Automotive, aircraft and marine mechanics and service technicians	3. Construction trade workers, helpers and other construction related workers
4. Driver/sales workers and truck drivers	5. Extraction workers (e.g. mining or drilling)	6. Firefighters
7. Food processing	8. Forest conservation and logging workers	9. Police and sheriff's patrol officers
10. Welding, soldering and brazing	11. Other occupation [text entry]	12. I do not wish to answer

2. In this job, what were the most common kinds of gas, smoke, or chemical vapors or fumes to which you were exposed (select all that apply)?

1. Cutting oils or mists	2. Exhaust: primarily diesel engine	3. Exhaust: primarily gasoline engine
4. Exhaust: both diesel and gasoline engine	5. Exhaust: primarily another kind	6. Fumes from chemicals
7. Gasoline or other fuel fumes	8. Paint or lacquers	9. Pesticides or insecticides
10. Smoke from burning buildings, fuel oil, refuse, or wood	11. Solvents	12. Welding
13. Other gas, smoke, or chemical vapor or fume (indicate kind)____________	14. I do not wish to answer	15. Don't know

3. Total years in this job {0...99} years (enter 0 if less than one year).
   1. Enter {00-99} years
   2. I do not wish to answer, 3. Don't know

4. Are you working in this job with gas, smoke, or chemical vapors or fumes now?
   1. Yes, 2. No, 3. I do not wish to answer, 4. Don't know

5.5. Asbestos Exposure

A. Have you ever worked in a job with asbestos exposure, including military service?
   1. Yes, 2. No, 3. I do not wish to answer, 4. Don't know

   [if A = yes, continue, else jump to 5.6]

B. Circle the type(s) of asbestos exposure that describe(s) how you were exposed to.
   1. I did not handle asbestos directly, but asbestos was present on overhead pipes or ceilings, flooring, brakes, or other materials.
   2. I did not handle asbestos directly, but I worked in area where asbestos dust was created by others.
   3. I handled asbestos or asbestos containing products directly and created asbestos dust.
   4. I do not wish to answer
   5. Don't know

C. How many years did you work in a job with asbestos exposure (enter 0 if less than one year)?
   1. Enter {00-99} years
   2. I do not wish to answer, 3. Don't know

D. Are you working in a job with asbestos exposure now?
   1. Yes, 2. No, 3. I do not wish to answer, 4. Don't know

6. Home Environment and Hobbies

"Exposures in your home environment or hobbies may impact how deployment exposures affect you."

A. Are there any traditional farm animals that live on your land or that you visit on a regular basis?
   1. Yes, 2. No, 3. I do not wish to answer, 4. Don't know

B. Have you ever removed mold in your home because of its effect on your health?
   1. Yes, 2. No, 3. I do not wish to answer, 4. Don't know

C. Have you ever lived in a home that had elevated radon levels?
   1. Yes, 2. No, 3. I do not wish to answer, 4. Don't know

D. Please select from the list below any hobbies you participate in. [Source: DoD USAPHC DARE questionnaire, page 14, section G].

1. Woodworking, including sanding	2. Welding, brazing or soldering	3. Metal working, including machining, grinding
4. Stained glass work	5. Hobbies utilizing epoxy resin adhesives	6. Pottery work, including glazing
7. Indoor swimming and/or indoor ice-skating	8. None	9. I do not wish to answer,

E. [if item selected in 'D'] How many total hours a week, on average, do you participate in all the above hobbies combined?
   1. Enter: 1,2, 3, 4, 5, 6, 7, 8, 9, 10 or more, 2. I do not wish to answer,
   3. Don't know

7. Health Care Utilization

[Source NHIS: Adult Access to Health Care & Utilization, AAU.305_00.000]

A. About how long has it been since you last saw or talked to a doctor or other health care professional about your own health? Include doctors seen while a patient in a hospital.
   1. Never
   2. 6 months or less
   3. More than 6 months, but not more than 1 yr ago
   4. At least 1 year, but not more than 2 yrs ago
   5. At least 2 years, but not more than 5 yrs ago
   6. At least 5 years ago
   7. I do not wish to answer
   8. Don't know

B. Do you wish to see a DoD or VA health care provider to discuss your health concerns related to airborne hazards during deployment?
   1. Yes, 2. No, 3. Don't know

8. Contact Preferences

"Help us communicate in ways that are most effective. VA will review these responses to determine the best ways to conduct outreach."

A. How do you prefer to receive updated information on burn pits and other airborne exposures?
   1. Email from VA
   2. VA Web site
   3. Through my health care provider
   4. VA social media (e.g. Twitter, Facebook)
   5. Letter/U.S. Mail
   6. Through the Department of Defense
   7. Through a Veterans Service Organization
   8. I do not wish to receive any updated information

B. Do you use the Internet?
   1. Yes
   2. No
   3. I do not wish to answer
   4. Don't know

C. Do you send or receive emails?
   1. Yes
   2. No
   3. I do not wish to answer
   4. Don't know

Appendix D

Data Requested

Data elements to be provided to the National Academies of Sciences, Engineering, and Medicine for use in review of initial information collected by the Airborne Hazards and Open Burn Pit Registry.

TABLE D-1 Data Sources and Items

Requested Item	Description	De-Identified Files to Be Included in CD-ROM File Transfer NI = Not Included ✓ = Included
Dataset: SAQ_Main.sas7bdat	***Current File Date: July 2015***	
Random ID	New variable needed: Random identifier that allows linking of respondents across files	✓ RANDOM_ID
Q1_3_A	Were you ever close enough to feel the blast from an IED (improvised explosive device) or other explosive device?	✓
Q1_3_B	In a typical month, how many days were you near heavy smoke from weapons, signal smoke, markers or other combat items?	✓
Q1_3_C	In a typical month, how many days were you in convoy or other vehicle operations?	✓
Q1_3_D	In a typical month, how many days did you perform refueling operations?	✓
Q1_3_E	In a typical month, how many days did you perform aircraft, generator, or other large engine maintenance?	✓
Q1_3_F	In a typical month, how many days did you perform construction duties?	✓
Q1_3_G	In a typical month, how many days did you perform pesticide duties for your unit?	✓
Q1_4_A	Did you do anything differently during your deployment(s), when you thought or were informed air quality was bad (for example, during dust storms or heavy pollution days)?	✓
Q1_4_B_Mask	What did you do differently? Wore a mask, cravat, or bandana over your mouth or nose	✓
Q1_4_B_LessOut	What did you do differently? Spent less time outdoors	✓
Q1_4_B_LessAct	What did you do differently? Did less strenuous activities (i.e., avoided physical training (PT))	✓
Q1_4_B_Meds	What did you do differently? Took medication	✓
Q1_4_B_Windows	What did you do differently? Closed windows of your sleeping quarters	✓
Q1_4_B_Convoy	What did you do differently? Spent less time in convoy	✓
Q1_4_B_CancelOut	What did you do differently? Canceled outdoor activities	✓
Q1_4_B_Indoors	What did you do differently? Exercised indoors instead of outdoors	✓
Q1_4_B_Filter	What did you do differently? Use or change air filter/air cleaner	✓
Q1_4_B_Other	What did you do differently? Other	✓

TABLE D-1 Continued

Requested Item	Description	De-Identified Files to Be Included in CD-ROM File Transfer NI = Not Included ✓ = Included
Q1_4_B_Nothing	What did you do differently? I did not (or could not) do anything differently	✓
Q1_4_B_RF	What did you do differently? I do not wish to answer	✓
Q1_4_C	In a typical month during your deployment(s), how many days did you experience dust storms?	✓
Q1_4_D	During your deployment(s), did you experience wheezing, difficulty breathing, an itchy or irritated nose, eyes, or throat that you thought was the result of poor air quality?	✓
Q1_4_E	How many days in an average month did you experience wheezing, difficulty breathing, an itchy or irritated eyes, nose, or throat that you thought was the result of poor air quality?	✓
Q1_4_F	During your deployment(s), did you seek medical care for wheezing, difficulty breathing, an itchy or irritated nose, eyes, or throat that you thought was the result of poor air quality?	✓
Q2_1_A	How difficult is it to run or jog 1 mile on a level surface?	✓
Q2_1_B	How difficult is it to walk on a level surface for 1 mile?	✓
Q2_1_C	How difficult is it to walk a 1/4 of a mile—about 3 city blocks?	✓
Q2_1_D	How difficult is it to walk up a hill or incline?	✓
Q2_1_E	How difficult is it to walk up 10 steps or climb a flight of stairs?	✓
Q2_1_F_Count	How many conditions or health problems caused you to have difficulty with these activities? (Derived)	✓
Q2_1_F_heart	What condition or health problem causes you to have difficulty with these activities? (Check all that apply.) — Heart problem [derived: 1 indicates condition selected, 0 indicates condition not selected]	✓
Q2_1_F_htn	What condition or health problem causes you to have difficulty with these activities? — Hypertension/high blood pressure [derived: 1 indicates condition selected, 0 indicates condition not selected]	✓
Q2_1_F_lung	What condition or health problem causes you to have difficulty with these activities? (Check all that apply.) — Lung/breathing problem (for example, asthma and emphysema) [derived: 1 indicates condition selected, 0 indicates condition not selected]	✓
Q2_1_F_other	What condition or health problem causes you to have difficulty with these activities? (Check all that apply.) — Other impairment/problem [derived: 1 indicates condition selected, 0 indicates condition not selected]	✓
Q2_1_F_other_desc	What condition or health problem causes you to have difficulty with these activities? (Check all that apply.) — Other impairment/problem (Participant entered free text)	✓
Q2_1_F_ref	What condition or health problem causes you to have difficulty with these activities? (Check all that apply.) — I do not wish to answer [derived: 1 indicates condition selected, 0 indicates condition not selected]	✓
Q2_1_F_dk	What condition or health problem causes you to have difficulty with these activities? (Check all that apply.) — Don't know/Not sure [derived: 1 indicates condition selected, 0 indicates condition not selected]	✓

continued

TABLE D-1 Continued

Requested Item	Description	De-Identified Files to Be Included in CD-ROM File Transfer NI = Not Included ✓ = Included
Q2_2_1_A	Have you been told by a doctor or other health professional that you had hay fever or allergies to pollen, dust, or animals?	✓
Q2_2_1_B	Have you ever been told by a doctor or other health care professional that you had asthma?	✓
Q2_2_1_C	Have you ever been told by a doctor or other health care professional that you had emphysema?	✓
Q2_2_1_D	Have you ever been told by a doctor or other health care professional that you had chronic bronchitis?	✓
Q2_2_1_E	Have you ever been told by a doctor or other health care professional that you had chronic obstructive pulmonary disease also called COPD?	✓
Q2_2_1_F	Have you ever been told by a doctor or other health care professional that you had some lung disease or condition other than asthma, emphysema, chronic bronchitis, or COPD?	✓
Q2_2_1_G	Have you ever been told by a doctor or other health care professional that you had constrictive bronchiolitis (CB)?	✓
Q2_2_1_H	Have you ever been told by a doctor or other health care professional that you had idiopathic pulmonary fibrosis (IPF)?	✓
Q2_2_1_I_Before	When you were told you had asthma, emphysema, chronic bronchitis, COPD, or some other lung disease by a doctor or other health care professional, were you told ... Before deployment	✓
Q2_2_1_I_During	When you were told you had asthma, emphysema, chronic bronchitis, COPD, or some other lung disease by a doctor or other health care professional, were you told ... During deployment	✓
Q2_2_1_I_After	When you were told you had asthma, emphysema, chronic bronchitis, COPD, or some other lung disease by a doctor or other health care professional, were you told ... After deployment	✓
Q2_2_1_I_DK	When you were told you had asthma, emphysema, chronic bronchitis, COPD, or some other lung disease by a doctor or other health care professional, were you told ... Don't know	✓
Q2_2_1_I_RF	When you were told you had asthma, emphysema, chronic bronchitis, COPD, or some other lung disease by a doctor or other health care professional, were you told ... I do not wish to answer	✓
Q2_2_1_J	Did this lung disease get better, worse, or about the same during deployment?	✓
Q2_2_1_K_Cough	Do you currently have any of the following symptoms? Cough for more than 3 weeks	✓
Q2_2_1_K_Sputum	Do you currently have any of the following symptoms? Sputum or phlegm production for more than 3 weeks	✓

TABLE D-1 Continued

Requested Item	Description	De-Identified Files to Be Included in CD-ROM File Transfer NI = Not Included ✓ = Included
Q2_2_1_K_Wheeze	Do you currently have any of the following symptoms? Wheezing or whistling in the chest	✓
Q2_2_1_K_Breath	Do you currently have any of the following symptoms? Shortness of breath; breathlessness	✓
Q2_2_1_K_Exercise	Do you currently have any of the following symptoms? Decreased ability to exercise	✓
Q2_2_1_K_Hayfever	Do you currently have any of the following symptoms? Hay fever or other respiratory allergy	✓
Q2_2_1_K_Throat	Do you currently have any of the following symptoms? Sore throat, hoarseness, change in voice	✓
Q2_2_1_K_Chest	Do you currently have any of the following symptoms? Chest pain, chest discomfort, or chest tightness	✓
Q2_2_1_K_Sinus	Do you currently have any of the following symptoms? Chronic sinus infection/sinusitis	✓
Q2_2_1_K_RF	Do you currently have any of the following symptoms? I do not wish to answer	✓
Q2_2_1_K_None	Do you currently have any of the following symptoms? I do not have these symptoms	✓
Q2_2_1_L_Cough	In the past 12 months did you have any of the following symptoms? Cough for more than 3 weeks	✓
Q2_2_1_L_Sputum	In the past 12 months did you have any of the following symptoms? Sputum or phlegm production for more than 3 weeks	✓
Q2_2_1_L_Wheeze	In the past 12 months did you have any of the following symptoms? Wheezing or whistling in the chest	✓
Q2_2_1_L_Breath	In the past 12 months did you have any of the following symptoms? Shortness of breath; breathlessness	✓
Q2_2_1_L_Exercise	In the past 12 months did you have any of the following symptoms? Decreased ability to exercise	✓
Q2_2_1_L_Hayfever	In the past 12 months did you have any of the following symptoms? Hay fever or other respiratory allergy	✓
Q2_2_1_L_Throat	In the past 12 months did you have any of the following symptoms? Sore throat, hoarseness, change in voice	✓
Q2_2_1_L_Chest	In the past 12 months did you have any of the following symptoms? Chest pain, chest discomfort, or chest tightness	✓
Q2_2_1_L_Sinus	In the past 12 months did you have any of the following symptoms? Chronic sinus infection/sinusitis	✓
Q2_2_1_L_RF	In the past 12 months did you have any of the following symptoms? I do not wish to answer	✓
Q2_2_1_L_None	In the past 12 months did you have any of the following symptoms? I do not have these symptoms	✓
Q2_2_1_M	How would you rate your shortness of breath or breathlessness? (check the description/grade that applies to you.)	✓

continued

TABLE D-1 Continued

Requested Item	Description	De-Identified Files to Be Included in CD-ROM File Transfer NI = Not Included ✓ = Included
Q2_2_2_A	Have you ever been told by a doctor or other health care professional that you had hypertension, also called high blood pressure?	✓
Q2_2_2_B	Have you ever been told by a doctor or other health care professional that you had coronary artery disease?	✓
Q2_2_2_C	Have you ever been told by a doctor or other health care professional that you had angina pectoris?	✓
Q2_2_2_D	Have you ever been told by a doctor or other health care professional that you had a heart attack, also called myocardial infarction?	✓
Q2_2_2_E	Have you ever been told by a doctor or other health care professional that you had a heart condition other than coronary artery disease or angina or myocardial infarction?	✓
Q2_2_2_F_Before	When you were told you had hypertension, coronary artery disease, angina pectoris, a heart attack, or some other heart condition by a doctor or health care professional, were you told ... Before deployment	✓
Q2_2_2_F_During	When you were told you had hypertension, coronary artery disease, angina pectoris, a heart attack, or some other heart condition by a doctor or health care professional, were you told ... During deployment	✓
Q2_2_2_F_After	When you were told you had hypertension, coronary artery disease, angina pectoris, a heart attack, or some other heart condition by a doctor or health care professional, were you told ... After deployment	✓
Q2_2_2_F_DK	When you were told you had hypertension, coronary artery disease, angina pectoris, a heart attack, or some other heart condition by a doctor or health care professional, were you told ... Don't know	✓
Q2_2_2_F_RF	When you were told you had hypertension, coronary artery disease, angina pectoris, a heart attack, or some other heart condition by a doctor or health care professional, were you told ... I do not wish to answer	✓
Q2_2_3_A	During the past 12 months, have you regularly had insomnia or trouble sleeping?	✓
Q2_2_3_B	During the past 12 months, have you had neurological problems? (Some examples of neurological problems may include numbness, tingling, or weakness in your arms or legs or difficulties with thinking or memory.)	✓
Q2_2_3_C	During the past 12 months, have you had problems of the immune system?	✓
Q2_2_3_D	During the past 12 months, have you been told by a doctor or other health professional that you had any kind of liver condition?	✓
Q2_2_3_E	During the past 12 months, have you been told by a doctor or other health professional that you had any chronic multisymptom illness (examples include irritable bowel syndrome, chronic fatigue syndrome, and fibromyalgia)?	✓
Q2_2_3_F_Before	Did your, neurological or immune problems, chronic multisymptom illness, or liver condition first occur ... Before deployment	✓
Q2_2_3_F_During	Did your, neurological or immune problems, chronic multisymptom illness, or liver condition first occur ... During deployment	✓

TABLE D-1 Continued

Requested Item	Description	De-Identified Files to Be Included in CD-ROM File Transfer NI = Not Included ✓ = Included
Q2_2_3_F_After	Did your, neurological or immune problems, chronic multisymptom illness, or liver condition first occur ... After deployment	✓
Q2_2_3_F_DK	Did your, neurological or immune problems, chronic multisymptom illness, or liver condition first occur ... Dont know	✓
Q2_2_3_F_RF	Did your, neurological or immune problems, chronic multisymptom illness, or liver condition first occur ... I do not wish to answer	✓
Q2_2_3_G	On average, how many hours of sleep do you get in a 24-hour period (round up 30 minutes or more to the next whole hour)?	✓
Q2_2_3_H	Questions H and I are about snoring and breathing during sleep. To answer these questions, please consider both what others have told you and what you know about yourself. How often do you snore?	✓
Q2_2_3_I	How often do you have times when you stop breathing during your sleep?	✓
Q2_3_A	How tall are you without shoes?	✓
Q2_3_A_specify	How tall are you without shoes?	✓
Q2_3_B	How much do you weigh without shoes? (Specify)	✓
Q2_4_A	Have you ever been told by a doctor or other health professional that you had cancer or a malignancy (tumor) of any kind?	✓
Q2_4_B	What kind of cancer was it?	✓
Q2_4_C	How old were you when this cancer was first diagnosed?	✓
Q2_4_D	If you were diagnosed with a second cancer, what kind of cancer was it?	✓
Q2_4_E	How old were you when this cancer was first diagnosed?	✓
Q2_4_F	If you were diagnosed with a third cancer, what kind of cancer was it?	✓
Q2_4_G	How old were you when this cancer was first diagnosed?	✓
Q2_5_A	Have you smoked at least 100 cigarettes in your entire life?	✓
Q2_5_B	How old were you when you first started to smoke fairly regularly?	✓
Q2_5_C	Do you now smoke cigarettes every day, some days or not at all?	✓
Q2_5_D	How long has it been since you quit smoking?	✓
Q2_5_E	On the average, how many cigarettes do you now smoke a day?	✓
Q2_5_F	Have you ever smoked tobacco products other than cigarettes even one time? (Such as cigars, pipes, water pipes or hookahs, small cigars that look like cigarettes, bidis, cigarillos, marijuana)?	✓
Q2_5_G	Do you now smoke tobacco products other than cigarettes every day, some days, rarely, or not at all?	✓
Q2_5_H	Have you ever used smokeless tobacco products even one time? (Such as chewing tobacco, snuff, dip, snus, or dissolvable tobacco)?	✓
Q2_5_I	Do you now use smokeless tobacco products every day, some days, rarely, or not at all?	✓
Q2_5_J	Are you exposed to secondhand smoke or environmental tobacco smoke every day, some days, rarely, or not at all?	✓

continued

TABLE D-1 Continued

Requested Item	Description	De-Identified Files to Be Included in CD-ROM File Transfer NI = Not Included ✓ = Included
Q2_6_A	Did you start smoking for the first time while being deployed?	✓
Q2_6_B	How did deployment(s) change how much you smoked?	✓
Q2_7_A	In the PAST YEAR, how often did you drink any type of alcoholic beverage. (Included are liquor such as whiskey or gin, beer, wine, wine coolers, and any other type of alcoholic beverage)? On average, how many days per week did you drink?	✓
Q3_1_A	Compared to pre-deployment, would you say your overall health is better, worse, or about the same?	✓
Q3_1_B	During your deployment(s), do you believe you were sick because of something you breathed?	✓
Q3_1_C	Do you currently have a sickness or condition you think began or got worse because of something you breathed during deployment(s)?	✓
Q3_1_D	When did the problem start?	✓
Q3_1_E	Please rate your concern that something you breathed during deployment has already affected your health?	✓
Q3_1_F	Please identify your biggest health concern that something you breathed during deployment has already affected your health.	✓
Q3_1_G	Have you discussed this concern with your health care provider, medical professional, or team?	✓
Q3_1_H	Are you concerned that in the future that your health will be affected by something you breathed during deployment(s)?	✓
Q3_1_I	Please rate your concern that something you breathed during deployment will affect your future health.	✓
Q3_1_J	Please identify your biggest health concern that something you breathed during deployment will affect your future health.	✓
Q3_1_K	Which exposure do you think has the biggest overall effect on your health?	✓
Q5_1_A	Which of the following were you doing last week?	✓
Q5_1_B	What is the main reason you did not work last week/have a job or business last week?	✓
Q5_2_A	Select the occupational category that best describes your main occupation (the civilian job you've held the longest). Do not include your occupation during military service. If your occupation is not included, select other occupation:	✓
Q5_2_A_specify	Select the occupational category that best describes your main occupation (the civilian job you've held the longest). Do not include your occupation during military service. If your occupation is not included, select other occupation: (Specify)	✓
Q5_2_B	Total years in this non-military job {0 . . . 99} years (enter 0 if less than 1 year).	✓
Q5_3_A	Have you ever worked for a year or more in any dusty job outside the military?	✓

TABLE D-1 Continued

Requested Item	Description	De-Identified Files to Be Included in CD-ROM File Transfer NI = Not Included ✓ = Included
Q5_3_B_1	For the job with the biggest dust exposure: Select the occupation category that best describes the job with the longest dust exposure. If your occupation is not included, select other occupation:	✓
Q5_3_B_1_specify	For the job with the biggest dust exposure: Select the occupation category that best describes the job with the longest dust exposure. If your occupation is not included, select other occupation: (Specify)	✓
Q5_3_B_2_Dander	In this job, what were the most common kinds of gas, smoke, or chemical vapors or fumes to which you were exposed? Animal dander	✓
Q5_3_B_2_Dust	In this job, what were the most common kinds of gas, smoke, or chemical vapors or fumes to which you were exposed? Wood or sawdust	✓
Q5_3_B_2_Metal	In this job, what were the most common kinds of gas, smoke, or chemical vapors or fumes to which you were exposed? Metal (aluminum, copper, iron, steel, or other types)	✓
Q5_3_B_2_Textile	In this job, what were the most common kinds of gas, smoke, or chemical vapors or fumes to which you were exposed? Cotton, wool, or other cloth or textile	✓
Q5_3_B_2_Asbestos	In this job, what were the most common kinds of gas, smoke, or chemical vapors or fumes to which you were exposed? Asbestos	✓
Q5_3_B_2_Plaster	In this job, what were the most common kinds of gas, smoke, or chemical vapors or fumes to which you were exposed? Plaster	✓
Q5_3_B_2_Flour	In this job, what were the most common kinds of gas, smoke, or chemical vapors or fumes to which you were exposed? Flour	✓
Q5_3_B_2_Cement	In this job, what were the most common kinds of gas, smoke, or chemical vapors or fumes to which you were exposed? Cement	✓
Q5_3_B_2_Sand	In this job, what were the most common kinds of gas, smoke, or chemical vapors or fumes to which you were exposed? Sand or silica	✓
Q5_3_B_2_Grain	In this job, what were the most common kinds of gas, smoke, or chemical vapors or fumes to which you were exposed? Grain	✓
Q5_3_B_2_Coal	In this job, what were the most common kinds of gas, smoke, or chemical vapors or fumes to which you were exposed? Coal	✓
Q5_3_B_2_Talc	In this job, what were the most common kinds of gas, smoke, or chemical vapors or fumes to which you were exposed? Talc	✓
Q5_3_B_2_Hay	In this job, what were the most common kinds of gas, smoke, or chemical vapors or fumes to which you were exposed? Hay	✓
Q5_3_B_2_Fiberglass	In this job, what were the most common kinds of gas, smoke, or chemical vapors or fumes to which you were exposed? Fiberglass	✓
Q5_3_B_2_Lime	In this job, what were the most common kinds of gas, smoke, or chemical vapors or fumes to which you were exposed? Lime	✓
Q5_3_B_2_Paper	In this job, what were the most common kinds of gas, smoke, or chemical vapors or fumes to which you were exposed? Paper or cardboard	✓
Q5_3_B_2_Granite	In this job, what were the most common kinds of gas, smoke, or chemical vapors or fumes to which you were exposed? Granite or other rock	✓

continued

TABLE D-1 Continued

Requested Item	Description	De-Identified Files to Be Included in CD-ROM File Transfer NI = Not Included ✓ = Included
Q5_3_B_2_Plastic	In this job, what were the most common kinds of gas, smoke, or chemical vapors or fumes to which you were exposed? Plastic or rubber	✓
Q5_3_B_2_Soil	In this job, what were the most common kinds of gas, smoke, or chemical vapors or fumes to which you were exposed? Soil or dirt	✓
Q5_3_B_2_Other	In this job, what were the most common kinds of gas, smoke, or chemical vapors or fumes to which you were exposed? Other dust	✓
Q5_3_B_2_RF	In this job, what were the most common kinds of gas, smoke, or chemical vapors or fumes to which you were exposed? I do not wish to answer	✓
Q5_3_B_3	Total years in this job {0 ... 99} years (enter 0 if less than 1 year).	✓
Q5_3_B_4	Are you working in this dusty job now?	✓
Q5_4_A	Have you ever been exposed to gas, smoke, chemical vapors, or fumes in your work outside the military?	✓
Q5_4_B_1	For the job with the biggest gas, smoke, vapor, or fume exposure: Select the occupational category that best describes the job with the longest gas, smoke, chemical vapor, or fume exposures. If your occupation is not included, select other occupation:	✓
Q5_4_B_1_specify	For the job with the biggest gas, smoke, vapor or fume exposure: Select the occupational category that best describes the job with the longest gas, smoke, chemical vapor, or fume exposures. If your occupation is not included, select other occupation: (Specify)	✓
Q5_4_B_2_Oils	In this job, what were the most common kinds of gas, smoke, or chemical vapors or fumes to which you were exposed? Cutting oils or mists	✓
Q5_4_B_2_DieselExhaust	In this job, what were the most common kinds of gas, smoke, or chemical vapors or fumes to which you were exposed? Exhaust: primarily diesel engine	✓
Q5_4_B_2_GasExhaust	In this job, what were the most common kinds of gas, smoke, or chemical vapors or fumes to which you were exposed? Exhaust: primarily gasoline engine	✓
Q5_4_B_2_BothExhaust	In this job, what were the most common kinds of gas, smoke, or chemical vapors or fumes to which you were exposed? Exhaust: both diesel and gasoline engine	✓
Q5_4_B_2_AnotherExhaust	In this job, what were the most common kinds of gas, smoke, or chemical vapors or fumes to which you were exposed? Exhaust: primarily another kind	✓
Q5_4_B_2_Fumes	In this job, what were the most common kinds of gas, smoke, or chemical vapors or fumes to which you were exposed? Fumes from chemicals	✓
Q5_4_B_2_Gas	In this job, what were the most common kinds of gas, smoke, or chemical vapors or fumes to which you were exposed? Gasoline or other fuel fumes	✓
Q5_4_B_2_Paint	In this job, what were the most common kinds of gas, smoke, or chemical vapors or fumes to which you were exposed? Paint or lacquers	✓
Q5_4_B_2_Pesticides	In this job, what were the most common kinds of gas, smoke, or chemical vapors or fumes to which you were exposed? Pesticides or insecticides	✓

TABLE D-1 Continued

Requested Item	Description	De-Identified Files to Be Included in CD-ROM File Transfer NI = Not Included ✓ = Included
Q5_4_B_2_Smoke	In this job, what were the most common kinds of gas, smoke, or chemical vapors or fumes to which you were exposed? Smoke from burning buildings, fuel oil, refuse, or wood	✓
Q5_4_B_2_Solvents	In this job, what were the most common kinds of gas, smoke, or chemical vapors or fumes to which you were exposed? Solvents	✓
Q5_4_B_2_Welding	In this job, what were the most common kinds of gas, smoke, or chemical vapors or fumes to which you were exposed? Welding	✓
Q5_4_B_2_Other	In this job, what were the most common kinds of gas, smoke, or chemical vapors or fumes to which you were exposed? Other gas, smoke, or chemical vapor or fume (indicate kind)	✓
Q5_4_B_2_DK	In this job, what were the most common kinds of gas, smoke, or chemical vapors or fumes to which you were exposed? Don't know	✓
Q5_4_B_2_RF	In this job, what were the most common kinds of gas, smoke, or chemical vapors or fumes to which you were exposed? I do not wish to answer	✓
Q5_4_B_3	Total years in this job {0 ... 99} years (enter 0 if less than 1 year).	✓
Q5_4_B_4	Are you working in this job with gas, smoke, chemical vapors, or fumes now?	✓
Q5_5_A	Have you ever worked in a job with asbestos exposure, including military service?	✓
Q5_5_B	Select the type(s) of asbestos exposure that describe(s) how you were exposed	✓
Q5_5_C	How many years did you work in a job with asbestos exposure? (enter 0 if less than 1 year)	✓
Q5_5_D	Are you working in a job with asbestos exposure now?	✓
Q6_1_A	Are there any traditional farm animals that live on your land or that you visit on a regular basis?	✓
Q6_1_B	Have you ever removed mold in your home because of its effect on your health?	✓
Q6_1_C	Have you ever lived in a home that had elevated radon levels?	✓
Q6_1_D_Wood	Please select from the list below any hobbies you participate in. Woodworking, including sanding?	✓
Q6_1_D_Weld	Please select from the list below any hobbies you participate in. Welding, brazing, or soldering?	✓
Q6_1_D_Metal	Please select from the list below any hobbies you participate in. Metal working, including machining, grinding?	✓
Q6_1_D_Glass	Please select from the list below any hobbies you participate in. Stained glass work?	✓
Q6_1_D_Epoxy	Please select from the list below any hobbies you participate in. Hobbies utilizing epoxy resin adhesives?	✓
Q6_1_D_Pottery	Please select from the list below any hobbies you participate in. Pottery work, including glazing?	✓

continued

TABLE D-1 Continued

Requested Item	Description	De-Identified Files to Be Included in CD-ROM File Transfer NI = Not Included ✓ = Included
Q6_1_D_Indoor	Please select from the list below any hobbies you participate in. Indoor swimming and/or indoor ice-skating?	✓
Q6_1_D_None	Please select from the list below any hobbies you participate in. None?	✓
Q6_1_D_RF	Please select from the list below any hobbies you participate in. I do not wish to answer?	✓
Q6_1_E	How many total hours a week, on average, do you participate in all the above hobbies combined?	✓
Q7_1_A	About how long has it been since you last saw or talked to a doctor or other health care professional about your own health? Include doctors seen while a patient in a hospital.	✓
Q7_1_B	Do you wish to see a DoD or VA health care provider to discuss your health concerns related to airborne hazards during deployment?	✓
Q8_1_A	How do you prefer to receive updated information on burn pits and other airborne exposures?	✓
Q8_1_B	Do you use the internet?	✓
Q8_1_C	Do you send or receive emails?	✓
SAQ Deployment	***Current File Date: July 2015***	
Random ID	New variable needed: Random identifier that allows linking of respondents across files	✓ RANDOM_ID
deploymentId	Deployment segment ID	✓
Q1_2_A	Were you exposed to soot, ash, smoke, or fumes from the Gulf War oil fires?	✓
Q1_2_B	Where did you spend most of your time during these dates?	✓
Q1_2_B_SPECIFY	Where did you spend most of your time during these dates? (Specify from list or user)	NI (base name)
Q1_2_C	If you were at more than one base, where did you spend the second most amount of time during these dates?	✓
Q1_2_C_SPECIFY	If you were at more than one base, where did you spend the second most amount of time during these dates? (Specify from list or user)	NI (base name)
Q1_2_D	Were you near a burn pit during these dates (on the base or close enough to the base for you to see the smoke)?	✓
Q1_2_E_CF	Who ran this burn pit? Coalition forces	✓
Q1_2_E_Host	Who ran this burn pit? Host nation	✓
Q1_2_E_US	Who ran this burn pit? U.S. forces or contractor	✓
Q1_2_E_DK	Who ran this burn pit? Dont know	✓
Q1_2_E_RF	Who ran this burn pit? I do not wish to answer	✓
Q1_2_F	Did your duties during these dates include the burn pit (examples include trash burning, hauling trash to the burn pit, burn pit security, trash sorting at the burn pit)?	✓
Q1_2_G	On a typical day, how many hours did smoke or fumes from the burn pit enter your work site or housing?	✓

TABLE D-1 Continued

Requested Item	Description	De-Identified Files to Be Included in CD-ROM File Transfer NI = Not Included ✓ = Included
Q1_2_H	On a typical day, how many hours were you outside or in an open tent or shelter? (for example, a single wall tent with open seams or drafty B hut)	✓
Q1_2_I	On a typical day, how many hours were you near (for, example you could smell or see it) sewage ponds?	✓
Branch	Branch of service	✓
deploymentBase	Deployment base	NI (base name)
deploymentCountry	Deployment country	✓
deploymentEndDate	Deployment segment end date	Year ONLY deploymentEndDate_yr
deploymentStartDate	Deployment segment start date	Year ONLY deploymentStartDate_yr
userEntered	Deployment segment entered by user	✓
userVerified	Deployment segment verified by user as correct	✓
NumDays	Number of days deployed in segment (derived)	✓
SAQ Participant Data ***Current File Date: July 2015***		
Random ID	New variable needed: Random identifier that allows linking of respondents across files	✓ RANDOM_ID
bpr_form_id	Version of the web survey	✓
questionnaireStartedDate	DateTime started questionnaire	✓
questionnaireCompletedDate	DateTime completed questionnaire	✓
CompletedDate	Date completed	✓
Age	Age at questionnaire start (derived)	✓
AgeCO	Age at questionnaire completion (derived)	✓
YearDOB	Year of birth	✓
serviceStatus	ServiceStatus from BPR_USER table (told not to use in analyses)	✓
CTS Extract (Provided by DMDC) ***Current File Date: April 2015***		
Random ID	New variable needed: Random identifier that allows linking of respondents across files	✓ RANDOM_ID
AGE	**New variable needed: Current age in years**	(Age as of 4/30/2015) AGE_20150430 ✓
BIRYR	**New variable needed: Year of birth**	✓
VITAL STATUS (death flag; date)	DTHFLAG New variable needed: Death date	NI *Element Not Available*

continued

TABLE D-1 Continued

Requested Item	Description	De-Identified Files to Be Included in CD-ROM File Transfer NI = Not Included ✓ = Included
SEX	GENDER	✓ PN_SEX_CD Provided at time of segment, as available
MARITAL	MARITAL STATUS (latest)	MRTL_STAT_CD Provided at time of segment, as available
EDUC	EDUCATION LEVEL (latest)	EDU_LVL_CD ✓
RACE	RACE	RACE_CD ✓
ETHNIC	ETHNICITY	ETH_AFF_CD NI
SERVICE	BRANCH OF SERVICE (Current)	*Element Not Available* NI
SERVICE	BRANCH OF SERVICE (Upon first login)	*Element Not Available* NI
SERVICE	BRANCH OF SERVICE (For every deployment segment)	SVC NI
COMPON	UNIT COMPONENT (current)	*Element Not Available* NI
COMPON	UNIT COMPONENT (upon first login)	*Element Not Available* NI
COMPON	UNIT COMPONENT (For every deployment segment)	✓ COMP
SEGBEGIN SEGEND	ALL DEPLOYMENT SEGMENTS (start date, end date)	Year ONLY DEP_BGN_CDT_yr DEP_END_CDT_yr LOC_BEGIN_DATE_yr LOC_END_DATE_yr ✓
Country	Deployment country (for each deployment segment)	LOC_CTRY_CD
DUTYMOS	MOS (during each deployment segment)	DTY_DOD_OCC_CD ✓
RANK (E1–E9; O1–O10; W1–W4)	RANK (during each deployment segment)	GRADE NI
ISCTYPE	SEPARATION TYPE--ISC (most recent date) New variable needed: Separation Date	*Element Not Available* NI

TABLE D-1 Continued

Requested Item	Description	De-Identified Files to Be Included in CD-ROM File Transfer NI = Not Included ✓ = Included
ISCTYPE	SEPARATION TYPE--ISC (upon first login) New variable needed: Separation Date	*Element Not Available* NI
	Gulf War Oil Well Fire Smoke Registry File	
Random ID	New variable needed: Random identifier that allows linking of respondents across files	✓ RANDOM_ID
AAE	Age at Entry	✓
AAS	Age at As-of-File Date	✓
ADSTD	Active Duty Start Date Day	NI
ADSTM	Active Duty Start Date Month	NI
ADSTY	Active Duty Start Date Year	✓
AFQT	AFQT Percentile	✓
BASDD	Basic Active Service Date Day	NI
BASDM	Basic Active Service Date Month	NI
BASDY	Basic Active Service Date Year	✓
BRANCH	Branch of the Military (Code)	✓
COMMIS	Source of Commission	✓
COMP		✓
COUNTRY_NAME	Name of Country Veteran was located; some located in water	✓
CSVC	Character of Service	✓
DATE_IN_JUL_DAY	Julian Day of Date In of Theatre	NI
DATE_OUT_JUL_DAY	Julian Day of Date Out of Theatre	NI
DATE_IN_CAL_DAY	Calendar Day of Date In of Theatre	✓ (Year Only)
DATE_OUT_CAL_DAY	Calendar Day of Date Out of Theatre	✓ (Year Only)
DDOC	Duty DoD Occupation Code	✓
DEPENDENTS	Number of Dependents	✓
DLEY	Date of Latest Enlistment Year	✓
DLEM	Date of Latest Enlistment Month	NI
DLOC	Duty Location State/Country	✓
DMOS	Duty Service Occupation	✓
DOGY	Date of Grade Year	✓
DOGM	Date of Grade Month	NI
DOSY	Date of Separation Year	✓
DOSM	Date of Separation Month	NI
DOSD	Date of Separation Day	NI

continued

TABLE D-1 Continued

Requested Item	Description	De-Identified Files to Be Included in CD-ROM File Transfer NI = Not Included ✓ = Included
DPOC	Primary DoD Occupation Code	✓
EDCERT	Education Certification	✓
EDUC	Education Level	✓
ETHNIC	Ethnicity (Code)	✓
ETSY	ETS Date Year	✓
ETSM	ETS Date Month	NI
HMRD	Home of Record State/Country	✓
HYE	Highest Year of Education Completed	✓
INENY	In Theater End Date Year	✓
INENM	In Theater End Date Month	NI
INEND	In Theater End Date Day	NI
INSTY	In Theater Start Date Year	✓
INSTM	In Theater Start Date Month	NI
INSTD	In Theater Start Date Day	NI
ISC	Interservice Separation Code	✓
MARITAL	Marital Status (Code)	✓
MCAT	Mental Category at Entry	✓
MIG	Months in Grade	✓
PEBDY	Pay Entry Base Date Year	✓
PEBDM	Pay Entry Base Date Month	NI
PEBDD	Pay Entry Base Date Day	NI
PMOS	Primary MOS (Code)	✓
PYGR	Pay Grade	✓
RACE	Race (Code)	✓
RACEETH	Race Ethnic Code	✓
RE	Reenlistment Eligibility	✓
SDCC	Secondary DoD Occupation	✓
SERV_COMP_CODE	Component	✓
SEX	Sex/Gender (Code	✓
SPD	Separation Program Designator	✓
TAFMS	Total Active Federal Military Service Months	✓
UIC_DATE_IN	Unit Date In of Theatre (in Julian Day)	NI
UIC_DATE_OUT	Unit Date Out of Theatre (in Julian Day)	NI
YCS	Year of Commissioned Service	✓
YOS	Years of Service	✓

TABLE D-1 Continued

Requested Item	Description	De-Identified Files to Be Included in CD-ROM File Transfer NI = Not Included ✓ = Included
	OEF/OIF/OND Roster file—Unavailable	
Random ID	New variable needed: Random identifier that allows linking of respondents across files	NI
ACTTYPE	Processing variable indicating the file from which the record was extracted. This can change from month to month; COMPON is later set to equal this value (ACT or RES)	NI
BASD	For AD personnel only, this is the date that they first reported for service (base date)	NI
BEGDATE	Original variable; first date of a deployment	NI
BRANCH	Branch of Service	NI
CNTRY	Country—At time of military discharge	NI
COMPON	Unit component	NI
COUNTRY	Deployment country	NI
BIRYR	New variable needed: Year of birth	NI
DTHFLAG	Death flag — from any source, DoD, Birls, SSA	NI
DUTYMOS	Duty military occupational specialty	NI
EDLEVEL	Education level	NI
EDUC	Education code	NI
EDUCR	Education code recoded	NI
ENDDATE	Original variable; last date of a deployment	NI
ETHNIC	Ethnicity	NI
INDATE	[OPH] Processing variable; first date of a deployment	NI
ISCTYPE	Separation type	NI
LOSSDATE	Date of separation — associated to Type of separation	NI
MARITAL	Marital status	NI
OUTDATE	[OPH] Processing variable; last date of a deployment	NI
PRIMOS	Primary military occupational specialty	NI
RACE	Race code	NI
RACER	Race code recoded	NI
RANK	Rank	NI
RANKD	Detailed listing of rank	NI
RESTYPE	[OPH] Processing variable indicating the file from which the record was extracted. This can change from month to month; COMPON is later set to equal this value (ACT or RES)	NI
SEGBEGIN	First date of a segment of a deployment	NI
SEGEND	End date of a segment of a deployment	NI
SERVICE	Service branch	NI

continued

TABLE D-1 Continued

Requested Item	Description	De-Identified Files to Be Included in CD-ROM File Transfer NI = Not Included ✓ = Included
SEX	Sex code	NI
STATE	State — At time of military discharge	NI
UNITTYPE	Unit type	NI

Appendix E

Multivariate Model Results

TABLE E-1 Multivariate Model Results of Burn Pit Exposure Variables and Respiratory and Cardiovascular Health Outcomes

	Asthma (N = 36,446)				Emphysema, Chronic Bronchitis, or COPD (N = 35,383)				Any Respiratory Symptoms (N = 39,271)				Functional Limitation Due to Lung Problem (N = 37,903)			
	OR	LCL	UCL	P	OR	LCL	UCL	P	OR	LCL	UCL	P	OR	LCL	UCL	P
Near Burn Pit																
Q 2	1.1	1.0	1.2	0.24	1.0	0.9	1.2	0.53	1.2	1.1	1.3	*	1.1	1.0	1.2	*
Q 3	1.2	1.1	1.4	*	1.3	1.1	1.4	*	1.3	1.2	1.5	*	1.2	1.1	1.4	*
Q 4	1.3	1.2	1.6	*	1.5	1.3	1.8	*	1.8	1.6	2.0	*	1.4	1.3	1.6	*
Burn Pit Duty																
Q 2	1.3	1.1	1.4	*	1.5	1.3	1.6	*	1.4	1.3	1.5	*	1.2	1.1	1.3	*
Q 3	1.2	1.1	1.3	*	1.3	1.2	1.4	*	1.4	1.3	1.5	*	1.3	1.2	1.4	*
Q 4	1.4	1.2	1.5	*	1.6	1.4	1.7	*	1.8	1.7	2.0	*	1.4	1.3	1.5	*
Smoke Hours																
Q 2	1.1	1.0	1.2	0.07	1.3	1.2	1.5	*	1.3	1.2	1.4	*	1.2	1.1	1.3	*
Q 3	1.2	1.1	1.3	*	1.6	1.5	1.8	*	1.7	1.6	1.8	*	1.5	1.4	1.6	*
Q 4	1.5	1.3	1.6	*	2.0	1.8	2.2	*	2.3	2.1	2.4	*	1.9	1.8	2.1	*

TABLE E-1 Continued

	Hypertension (N = 36,770)				CAD, MI, or Angina Pectoris (N = 35,740)			
	OR	LCL	UCL	P	OR	LCL	UCL	P
Near Burn Pit								
Q 2	1.0	0.9	1.1	0.74	1.1	0.9	1.4	0.40
Q 3	1.2	1.1	1.3	*	1.3	1.0	1.7	0.04
Q 4	1.2	1.1	1.4	*	1.6	1.2	2.2	*
Burn Pit Duty								
Q 2	1.2	1.1	1.3	*	1.3	1.0	1.6	0.02
Q 3	1.1	1.0	1.2	*	1.0	0.8	1.2	1.00
Q 4	1.3	1.2	1.4	*	1.4	1.1	1.7	*
Smoke Hours								
Q 2	1.1	1.1	1.2	*	1.3	1.1	1.6	*
Q 3	1.3	1.2	1.4	*	1.4	1.1	1.7	*
Q 4	1.4	1.3	1.5	*	1.8	1.5	2.2	*

NOTES: The first quartile is the reference group. CAD = coronary artery disease; COPD = chronic obstructive pulmonary disease; LCL = lower confidence level; MI = myocardial infarction; OR = odds ratio; P = p-value; Q2 = second quartile; Q3 = third quartile; Q4 = fourth quartile; UCL = upper confidence level.
* Denotes p-value <0.01.

TABLE E-2 Model Estimates of Plausible and Implausible Levels of Burn Pit Exposure on Respiratory and Cardiovascular Health Outcomes

	Asthma (N = 36,446)				Emphysema, Chronic Bronchitis, or COPD (N = 35,383)				Any Respiratory Symptoms (N = 39,271)				Functional Limitation Due to Lung Problems (N = 37,903)			
	OR	LCL	UCL	P	OR	LCL	UCL	P	OR	LCL	UCL	P	OR	LCL	UCL	P
Plausible Near Burn Pit																
Q 2	1.0	0.9	1.1	0.68	0.9	0.8	1.0	0.20	1.1	1.0	1.2	*	1.0	1.0	1.1	0.26
Q 3	1.0	0.9	1.1	0.64	1.0	0.9	1.1	0.93	1.2	1.1	1.2	*	1.1	1.0	1.2	0.06
Q 4	1.0	0.9	1.1	0.65	0.9	0.8	1.0	0.01	1.3	1.2	1.4	*	1.1	1.0	1.2	0.28
Implausible Near Burn Pit																
Q 2	1.0	0.9	1.1	0.51	1.0	1.0	1.1	0.35	1.0	1.0	1.1	0.51	1.1	1.0	1.2	*
Q 3	1.0	1.0	1.1	0.31	1.2	1.1	1.3	*	1.0	0.9	1.0	0.51	1.1	1.0	1.1	0.04
Q 4	1.0	0.9	1.1	0.88	1.1	1.0	1.2	0.18	0.8	0.7	0.9	*	0.9	0.9	1.0	0.17
Plausible Burn Pit Duty																
Q 2	1.4	1.2	1.5	*	1.5	1.3	1.6	*	1.4	1.3	1.5	*	1.2	1.1	1.4	*
Q 3	1.2	1.1	1.3	*	1.2	1.1	1.3	*	1.4	1.3	1.5	*	1.2	1.2	1.3	*
Q 4	1.3	1.1	1.4	*	1.4	1.3	1.5	*	1.7	1.6	1.8	*	1.4	1.3	1.5	*
Implausible Burn Pit Duty																
Q 2	0.9	0.8	1.0	0.07	1.0	0.9	1.1	0.92	0.9	0.8	0.9	*	0.9	0.9	1.0	0.04
Q 3	0.8	0.8	0.9	*	0.9	0.8	1.0	*	0.7	0.7	0.8	*	0.8	0.7	0.8	*
Q 4	0.8	0.7	0.9	*	0.7	0.6	0.8	*	0.5	0.5	0.6	*	0.7	0.6	0.8	*
Plausible Smoke Hours																
Q 2	1.1	1.0	1.2	0.19	1.3	1.2	1.4	*	1.3	1.2	1.4	*	1.2	1.1	1.3	*
Q 3	1.2	1.1	1.3	*	1.6	1.4	1.7	*	1.6	1.5	1.7	*	1.5	1.4	1.6	*
Q 4	1.3	1.2	1.4	*	1.7	1.5	1.9	*	2.1	1.9	2.2	*	1.8	1.6	1.9	*

TABLE E-2 Continued

	Hypertension (N = 36,770)				CAD, MI, or Angina Pectoris (N = 35,740)			
	OR	LCL	UCL	P	OR	LCL	UCL	P
Plausible Near Burn Pit								
Q 2	0.8	0.8	0.9	*	0.7	0.6	0.9	*
Q 3	0.9	0.9	1.0	0.13	0.7	0.6	0.9	*
Q 4	0.8	0.8	0.9	*	0.6	0.5	0.8	*
Implausible Near Burn Pit								
Q 2	1.0	1.0	1.1	0.37	1.0	0.8	1.2	0.98
Q 3	1.1	1.1	1.2	*	1.3	1.0	1.5	0.02
Q 4	1.2	1.1	1.3	*	1.3	1.1	1.6	0.01
Plausible Burn Pit Duty								
Q 2	1.2	1.1	1.3	*	1.3	1.1	1.6	0.01
Q 3	1.0	1.0	1.1	0.31	0.9	0.8	1.1	0.35
Q 4	1.1	1.1	1.2	*	1.1	0.9	1.4	0.21
Implausible Burn Pit Duty								
Q 2	1.0	1.0	1.1	0.47	1.0	0.8	1.2	0.94
Q 3	1.0	0.9	1.0	0.28	1.0	0.8	1.2	0.76
Q 4	1.0	0.9	1.0	0.28	0.9	0.7	1.1	0.19
Plausible Smoke Hours								
Q 2	1.1	1.0	1.2	0.01	1.3	1.1	1.5	0.01
Q 3	1.2	1.1	1.3	*	1.3	1.1	1.6	*
Q 4	1.3	1.2	1.4	*	1.5	1.2	1.8	*

NOTES: The first quartile is the reference group. CAD = coronary artery disease; COPD = chronic obstructive pulmonary disease; LCL = lower confidence level; MI = myocardial infarction; OR = odds ratio; P = p-value; Q2 = second quartile; Q3 = third quartile; Q4 = fourth quartile; UCL = upper confidence level.
* Denotes p-value <0.01.

TABLE E-3 Multivariate Model Results of Respiratory and Cardiovascular Health Outcomes by Exposure Category and Level

	Asthma				Emphysema, Chronic Bronchitis, or COPD				Any Respiratory Symptoms				Functional Limitation Due to Lung Problem			
	OR	LCL	UCI	P	OR	LCL	UCL	P	OR	LCL	UCL	P	OR	LCL	UCL	P
Burn Pits																
1	1.1	0.9	1.3	0.31	1.2	1.0	1.5	0.03	1.1	1.0	1.2	0.05	1.1	1.0	1.3	0.07
2	1.3	1.1	1.5	*	1.5	1.2	1.7	*	1.3	1.2	1.4	*	1.4	1.2	1.6	*
3	1.5	1.3	1.7	*	1.6	1.3	1.9	*	1.5	1.4	1.7	*	1.6	1.4	1.8	*
4	1.4	1.2	1.6	*	1.7	1.4	2.0	*	1.5	1.4	1.7	*	1.4	1.3	1.6	*
5	1.5	1.3	1.8	*	2.1	1.8	2.4	*	2.0	1.8	2.2	*	1.8	1.6	2.0	*
6	1.8	1.6	2.1	*	2.5	2.1	2.9	*	2.8	2.5	3.0	*	2.3	2.0	2.5	*
Diesel/Exhaust/Fuel																
1	1.2	1.1	1.4	*	1.2	1.1	1.4	*	1.2	1.1	1.3	*	1.1	1.0	1.3	0.01
2	1.3	1.2	1.5	*	1.5	1.3	1.7	*	1.4	1.3	1.5	*	1.2	1.1	1.3	*
3	1.4	1.2	1.6	*	1.5	1.3	1.7	*	1.7	1.5	1.8	*	1.4	1.3	1.6	*
4	1.5	1.3	1.7	*	1.6	1.4	1.8	*	1.8	1.6	2.0	*	1.5	1.4	1.7	*
5	1.6	1.4	1.9	*	2.0	1.7	2.3	*	2.2	2.0	2.5	*	1.7	1.5	1.9	*
6	1.6	1.4	1.9	*	2.1	1.8	2.5	*	2.4	2.1	2.8	*	1.8	1.6	2.0	*

Construction																
3	1.1	1.0	1.2	0.05	1.3	1.2	1.4	*	1.4	1.3	1.5	*	1.2	1.2	1.3	*
6	1.3	1.2	1.4	*	1.5	1.4	1.6	*	1.7	1.6	1.8	*	1.4	1.3	1.5	*
Dust																
1	1.1	0.9	1.3	0.64	0.9	0.8	1.1	0.51	1.1	1.0	1.3	0.09	1.1	0.9	1.2	0.48
2	1.1	0.9	1.4	0.19	1.1	0.9	1.4	0.19	1.3	1.2	1.5	*	1.2	1.0	1.4	0.01
3	1.3	1.1	1.6	*	1.4	1.2	1.7	*	1.7	1.5	1.9	*	1.4	1.2	1.7	*
4	1.4	1.2	1.7	*	1.7	1.4	2.0	*	2.0	1.8	2.3	*	1.6	1.4	1.9	*
5	1.6	1.3	2.0	*	1.9	1.5	2.3	*	2.6	2.3	3.0	*	1.9	1.6	2.2	*
6	1.8	1.4	2.2	*	2.5	2.0	3.1	*	3.2	2.7	3.8	*	2.3	1.9	2.7	*
Combat																
1	1.1	0.9	1.4	0.27	1.2	0.9	1.4	0.20	1.0	0.8	1.1	0.49	1.1	0.9	1.3	0.22
2	1.4	1.2	1.6	*	1.7	1.5	2.0	*	1.5	1.3	1.6	*	1.4	1.2	1.5	*
3	1.4	1.2	1.6	*	1.6	1.4	1.9	*	1.5	1.4	1.7	*	1.4	1.3	1.6	*
4	1.6	1.4	1.8	*	2.0	1.8	2.4	*	1.8	1.7	2.0	*	1.5	1.4	1.7	*
5	1.7	1.5	2.0	*	2.4	2.0	2.7	*	2.1	1.9	2.3	*	1.8	1.7	2.0	*
	1.9	1.7	2.2	*	2.9	2.5	3.3	*	2.8	2.5	3.1	*	2.1	1.9	2.4	*

continued

TABLE E-3 Continued

	Hypertension				CAD, MI, or Angina Pectoris			
	OR	LCL	UCL	P	OR	LCL	UCL	P
Burn Pits								
1	1.1	1.0	1.2	0.25	1.0	0.7	1.5	0.89
2	1.2	1.1	1.4	*	1.1	0.8	1.6	0.48
3	1.2	1.1	1.4	*	1.0	0.7	1.4	1.00
4	1.3	1.1	1.4	*	1.4	1.0	1.9	0.07
5	1.4	1.2	1.5	*	1.4	1.0	1.9	0.05
6	1.6	1.4	1.8	*	1.8	1.3	2.4	*
Diesel/Exhaust/Fuel								
1	1.1	1.0	1.2	0.01	1.5	1.1	2.0	*
2	1.2	1.1	1.3	*	1.7	1.3	2.3	*
3	1.3	1.2	1.4	*	1.9	1.4	2.5	*
4	1.2	1.1	1.3	*	1.4	1.0	1.9	0.03
5	1.4	1.3	1.6	*	2.1	1.5	2.9	*
6	1.6	1.4	1.8	*	2.5	1.7	3.5	*

Construction								
3	1.2	1.0	1.4	0.04	1.2	1.0	1.4	0.04
6	1.6	1.3	1.9	*	1.6	1.3	1.9	*
Dust								
1	0.9	0.8	1.1	0.31	1.0	0.6	1.5	0.92
2	1.1	1.0	1.2	0.21	1.2	0.8	1.8	0.42
3	1.2	1.1	1.4	*	1.4	0.9	2.2	0.11
4	1.3	1.2	1.5	*	1.5	1.0	2.3	0.05
5	1.4	1.2	1.7	*	2.0	1.3	3.1	*
6	1.7	1.4	2.0	*	2.3	1.5	3.7	*
Combat								
1	1.1	1.0	1.3	0.17	1.4	0.9	2.2	0.10
2	1.3	1.2	1.4	*	1.3	0.9	1.7	0.11
3	1.3	1.1	1.4	*	1.6	1.1	2.3	0.01
4	1.5	1.3	1.6	*	1.8	1.3	2.4	*
5	1.6	1.5	1.8	*	2.0	1.4	2.6	*
6	1.7	1.6	1.9	*	2.1	1.5	2.8	*

NOTES: The unexposed group is the reference group. CAD = coronary artery disease; COPD = chronic obstructive pulmonary disease; LCL = lower confidence limit; MI = myocardial infarction; OR = odds ratio; P = p-value; UCL = upper confidence limit.

* Denotes p-value <0.01.

TABLE E-4 Multivariate Model Results of Composite Exposure Potential Measure and Respiratory and Cardiovascular Health Outcomes

	Asthma (N = 36,446)				Emphysema, Chronic Bronchitis or COPD (N = 35,383)				Any Respiratory Symptoms (N = 39,271)				Functional Limitation Due to Lung Problem (N = 37,903)			
Composite Exposure	OR	LCL	UCL	P	OR	LCL	UCL	P	OR	LCL	UCL	P	OR	LCL	UCL	P
Q 2	1.2	1.1	1.3	*	1.4	1.2	1.5	*	1.4	1.3	1.5	*	1.3	1.2	1.4	*
Q 3	1.5	1.4	1.6	*	1.7	1.6	1.9	*	1.9	1.7	2.0	*	1.5	1.4	1.7	*
Q 4	1.7	1.5	1.8	*	2.3	2.1	2.6	*	2.7	2.5	2.9	*	2.0	1.8	2.1	*

TABLE E-4 Continued

	Hypertension (N = 36,770)				CAD, MI, or Angina Pectoris (N = 35,740)			
Composite Exposure	OR	LCL	UCL	P	OR	LCL	UCL	P
Q 2	1.2	1.1	1.3	*	1.2	1.0	1.6	0.07
Q 3	1.4	1.3	1.5	*	1.5	1.3	1.9	*
Q 4	1.7	1.6	1.8	*	2.1	1.7	2.6	*

NOTES: The first quartile is the reference group. CAD = coronary artery disease; COPD = chronic obstructive pulmonary disease; LCL = lower confidence level; MI = myocardial infarction; OR = odds ratio; P = p-value; Q2 = second quartile; Q3 = third quartile; Q4 = fourth quartile; UCL = upper confidence level.
* Denotes p-value <0.01.

Appendix F

Biographical Sketches of Committee Members and Staff

COMMITTEE

David A. Savitz, Ph.D. (*Chair*), is the vice president for research at Brown University. He is a professor of epidemiology in the Brown University School of Public Health, with a joint appointment in obstetrics and gynecology in the Alpert Medical School. He came to Brown in 2010 from Mount Sinai School of Medicine, where he had served as the Charles W. Bluhdorn Professor of Community and Preventive Medicine and the director of the Disease Prevention and Public Health Institute since 2006. Earlier, he taught and conducted research at the University of North Carolina School of Public Health and at the Department of Preventive Medicine and Biometrics at the University of Colorado School of Medicine. Dr. Savitz received his undergraduate training in psychology at Brandeis University, a master's degree in preventive medicine at The Ohio State University, and a Ph.D. in epidemiology from the University of Pittsburgh Graduate School of Public Health. His epidemiological research has addressed a wide range of public health issues including hazards in the workplace, the environmental effects of energy development, childhood obesity, pesticides and breast cancer, pregnancy health risks from environmental exposures, drinking water safety, and ethnicity and birth outcomes. Dr. Savitz has directed 30 doctoral dissertations and 15 master's theses. He is the author of nearly 350 papers in professional journals and the editor or author of three books on environmental epidemiology. He has served as editor at the *American Journal of Epidemiology* and *Epidemiology*, and as a member of the Epidemiology and Disease Control-1 study section of the National Institutes of Health. He was President of the Society for Epidemiologic Research and the Society for Pediatric and Perinatal Epidemiologic Research and North American Regional Councilor for the International Epidemiological Association. Dr. Savitz is a member of the National Academy of Medicine.

Vinícius Antão, M.D., M.Sc., Ph.D., is the director of patient registries at the Hospital for Special Surgery in New York, New York. Prior to this position, he was the lead in the registries team for the Centers for Disease Control and Prevention (CDC) where he was the principal and co-investigator of numerous research projects, including analysis of large databases from 2008 to 2015. Before this he was the senior manager for GlaxoSmithKline Biologicals, where he designed and implemented epidemiological studies in many countries in Europe, Asia, and Latin America. He has received numerous awards including Excellence in Surveillance and Health Monitoring from the CDC, Director's Award for Innovation (National Amyotrophic Lateral Sclerosis Registry) from the CDC, Occupational Health and Safety Scientific Research Award from the American Public Health Association, Excellence in Program Delivery Award (National Amyotrophic Lateral Sclerosis Registry) from the CDC, and the Collaborative

Success Award (Respiratory Outbreak Working Group) from the CDC. Dr. Antão received his M.D. from Petropolis Medical School in Rio de Janeiro, Brazil, his M.Sc. in respiratory medicine from Fluminense Federal University in Rio de Janeiro, Brazil, and his Ph.D. in respiratory medicine at Sao Paulo University.

Jane E. Clougherty, Sc.D., is an assistant professor of environmental and occupational health and the director of exposure science at the Graduate School of Public Health at the University of Pittsburgh. Before joining the faculty, she worked with the City of New York Department of Health and Mental Hygiene to collect and analyze year-round measures of fine particles and metals constituents, elemental carbon, nitrogen oxides, sulfur dioxide, and ozone at 150 sites throughout New York City, developing predictive models to explore exposure variability across a large population. As an interdisciplinary environmental health scientist, Dr. Clougherty's training and experience lie predominantly in air pollution exposure assessment and environmental epidemiology, but her interests also include occupational health, social epidemiology, community-based research, and toxicology. She serves on the editorial boards of the *Journal of Exposure Science & Environmental Epidemiology* and *Environmental Health Perspectives* and as a reviewer for several other publications in the field of environmental health. Dr. Clougherty received her undergraduate degree in economics and environmental studies from the University of Chicago, her masters of science in geography (environmental health sciences) from McMaster University, and her Sc.D. in exposure, epidemiology, and risk program from Harvard University School of Public Health.

Montserrat Fuentes, Ph.D., became the Dean of the College of Humanities and Sciences at Virginia Commonwealth University in Richmond, Virginia, in 2016. Prior to that position, she served as the head and a professor of statistics at North Carolina State University (NCSU). She is the principal investigator and director of the Research Network for Statistical Methods for Atmospheric and Oceanic Sciences, has authored more than 75 scientific publications, and served as principal investigator (or co-principal investigator) on more than 20 research grants. Dr. Fuentes was named an American Statistics Association Fellow (2008) for outstanding contributions to research in spatial statistics, for excellence in the development and application of statistical methodology in atmospheric sciences, air pollution and oceanography; and for service to the profession. She is the editor of the *Journal of Agricultural, Biological, and Environmental Statistics of the International Biometrics Society*. Dr. Fuentes is a member of the Science Advisory Board Integrated Human Exposure Committee of the Environmental Protection Agency, and the U.S. representative in the board of directors of the International Environmetrics Society. She was a member of the biostatistical methods and research design study section of the National Institutes for Health, and she is currently a member of the scientific review committee of Health Canada. She was also a member of a committee of the National Research Council working on the impact of ozone on mortality. Dr. Fuentes was awarded the NCSU 2013 Equity of Women award, for major contributions to the equity and well-being of women at NCSU. She received her B.S. in mathematics and music (piano) from the University of Valladolid (Spain) and her Ph.D. in statistics from the University of Chicago.

Richard (Dick) A. Kulka, Ph.D., is an independent consultant in statistical, survey and social research. He has served on the staff of four major research organizations: the Survey Research Center at the University of Michigan, the National Opinion Research Center (NORC) at the University of Chicago, RTI International (formerly Research Triangle Institute), and Abt Associates Inc. (also holding senior management positions at RTI, NORC, and Abt). He has been involved in the design, conduct, and analysis of numerous statistical surveys on health, mental health, and other social policy issues for more than 30 years, while also conducting a broad range of applied research on survey research methods. Dr. Kulka has authored or co-authored numerous papers, articles, and chapters based on this work as well as three research monographs. He has also provided ongoing methodological design, analysis, oversight, and consultation on a broad range of federal statistical surveys as well as advice to several other statistical survey organizations, including the Census Bureau and ongoing longitudinal surveys conducted at the University of Michigan. Also active in several professional organizations, he is a member and fellow of the American Statistical Association (including service on the Census Advisory Committee and Committee on Privacy and Confidentiality) and a member of the American Association for Public Opinion Research since 1980, where he has served the organization in a number of capacities, including as president (2008–2009). Other activities include the board of directors for the Council of Professional Associations on Federal Statistics, the National Institute of Statistical Sciences, and the triennial Conferences on Health Survey Research Methods. He has also served at the National Academies of Sciences, Engineering, and Medicine on several committees and expert panels and as a report coordinator for the Committee on National Statistics, the National Research Council, and the Institute of Medicine. Dr. Kulka earned his B.A. from Tulane University and M.A. and Ph.D. from the University of Michigan.

Frances Murphy, M.D., M.P.H., is the president and chief executive officer of Sigma Health Consulting, LLC, a woman-owned, veteran-owned small business. Dr. Murphy is a health care executive with extensive experience in managing, operating, and transforming large health care organizations. She serves as a consultant on health information technology, leadership development, health care management, neuroscience and mental health, women's health, quality and safety, study design and research management, and veterans' and military health. Dr. Murphy had a 20-year career working in the Department of Veterans Affairs (VA) at medical centers and in the VA Central Office. She served as the deputy under secretary for health (DUSH) for health policy coordination from 2002 to 2006 and as the principle DUSH from 1999 to 2002. She served in the U.S. Air Force as the staff neurologist at Andrews Air Force Base, Maryland, from 1983 to 1987. She is board certified in neurology and earned her M.D. at Georgetown University School of Medicine and her M.P.H. from the Uniformed Services University of the Health Sciences.

Cecile S. Rose, M.D., M.P.H., is a professor of medicine in the Division of Environmental and Occupational Health Sciences at National Jewish Health (NJH). She has academic appointments in the Division of Pulmonary Sciences and Critical Care Medicine at the University of Colorado and in the Department of Environmental and Occupational Health in the Colorado School of Public Health. Dr. Rose has long-standing research and clinical interests in occupational and environmental lung diseases. She is the principal investigator on a collaborative research project in deployment-related lung diseases and is medical director of the clinical program in deployment lung disease at NJH. Her other research interests focus on noninfectious granulomatous lung diseases and mining-related cardiopulmonary diseases. Dr. Rose earned her B.A. from Northwestern University. She received a masters degree in public health and an M.D. from the University of Illinois at Chicago.

Armistead (Ted) G. Russell, Ph.D., is the Howard T. Tellepsen Chair and Regents Professor at the Georgia Institute of Technology. Dr. Russell arrived at Georgia Tech in 1996, from Carnegie Mellon University, and has expertise in air quality engineering, with a particular emphasis in air quality modeling, air quality monitoring and analysis. He has been a member of a number of the National Academies committees, including chairing the Committee to Review EPA's Mobile Model and the Committee on Carbon Monoxide Episodes in Meteorological and Topographical Problem Areas, and he served on the committee on Tropospheric Ozone Formation and Measurement, the committee on the ozone-forming potential of reformulated fuels and the committee on Risk Assessment of Hazardous Air Pollutants. Dr. Russell was a member of the Environmental Protection Agency's (EPA's) Clean Air Scientific Advisory Committee and the subcommittee on Air Quality Modelling Subcommittee of the Advisory Council on Clean Air Compliance Analysis. He was a member of the Health Effects Institute Research Review Committee, the EPA Federal Advisory Committee Act Subcommittee on Ozone, Particulate Matter and Regional Haze, the North American Research Strategy for Tropospheric Ozone, and California's Reactivity Science Advisory Committee. Dr. Russell earned his M.S. and Ph.D. degrees in mechanical engineering at the California Institute of Technology. His B.S. is from Washington State University.

David H. Trump, M.D., M.P.H., M.P.A., is the former chief deputy commissioner for public health and preparedness for the Virginia Department of Health. Dr. Trump oversaw the Office of Epidemiology, Office of Drinking Water, Office of Radiological Health, Office of Emergency Services, Office of Licensure and Certification, Office of Chief Medical Examiner, Office of Emergency Preparedness, and Office of Risk Communication and Education. From March 2012 to May 2014 he was the director, Office of Epidemiology, and the state epidemiologist. The Office of Epidemiology oversees and is responsible for disease reporting, investigation, surveillance, and control programs statewide for communicable diseases; HIV and other sexually transmitted infections; vaccine preventable diseases; and waterborne, vectorborne, and zoonotic diseases in humans. From 2005 to 2012 he was the director of the Peninsula Health District, which provides public health services for the 340,000 residents of the cities of Newport News, Williamsburg, and Poquoson and the counties of York and James City in southeastern Virginia. Prior to joining the Virginia Department of Health, he completed a 24-year career as a Navy medical officer specializing in public health and preventive medicine. His career included service as officer in charge of Navy Environmental and Preventive Medicine Unit No. 7 in Naples, Italy, and senior public health leadership assignments with the Navy Surgeon General's office and at the Pentagon. He retired at the rank of Captain from his last assignment as an associate professor of preventive medicine at the Uniformed Services University of the Health Sciences in Bethesda, Maryland. Dr. Trump received his M.D. from Jefferson Medical College in Philadelphia, an M.P.H. from the School of Hygiene and Public Health at the Johns Hopkins University, and an M.P.A. from George Mason University. He is a fellow of the American College of Preventive Medicine and is board certified in family medicine and general preventive medicine.

Joyce S. Tsuji, Ph.D., DABT, is a board-certified toxicologist and a fellow of the Academy of Toxicological Sciences. She specializes in assessing exposure and risks associated with chemicals and in the communication of scientific issues. Dr. Tsuji has worked on projects in the United States and internationally for industry, trade associations, the Environmental Protection Agency (EPA) and state agencies, the U.S. Department of Justice, the Australian EPA, municipalities, and private citizens. Her experience includes human health and environmental toxicology related to a wide variety of chemicals in the environment, consumer products, and medical devices. She has designed and directed dietary and environmental exposure studies and community programs involving health education and biomonitoring for populations potentially exposed to chemicals in the environment, including soil, water, and food-chain exposures. Dr. Tsuji has also assessed exposure and health risks associated with chemical exposures from air, foods, medical devices, and a variety of consumer products (e.g., cleaners, air fresheners, cosmetics, personal care products, paints and coatings, carpets, glues, wood preservatives, building materials, and children's toys and play equipment), including those containing nanotechnology or nanomaterials. She has served on expert panels on toxicology and health risks issues for the National Academy of Sciences/National Research Council (including the Board on Environmental Studies and Toxicology), Institute of Medicine, and federal and state agencies. Dr. Tsuji earned her B.S. in biological sciences from Stanford University and a Ph.D. focused in environmental physiology from the Department of Zoology, University of Washington.

Mark J. Utell, M.D., is a professor of medicine and environmental medicine, the director of occupational and environmental medicine, and the former director of pulmonary and critical care medicine at the University of Rochester Medical Center. His research interests have centered on the effects of environmental toxicants on the human respiratory tract. Dr. Utell has published extensively on the health effects of inhaled gases, particles, and fibers in the workplace and indoor and outdoor environments. He was the co-principal investigator of an Environmental Protection Agency (EPA) Particulate Matter Center and chair of the Health Effects Institute's Research Committee. He has served as chair of EPA's Environmental Health Committee and on the executive committee of the EPA's science advisory board. He is a former recipient of the National Institute of Environmental Health Sciences Academic Award in Environmental and Occupational Medicine. Dr. Utell currently chairs the National Academies Committee to Review EPA's "Science to Achieve Results" Research Grant Program. He previously served as chair of the Institute of Medicine Committee to Review the Department of Labor's Site Exposure Matrix (SEM); as chair of the National Research Council (NRC) Committee to Review the Department of Defense Enhanced Particulate Matter Surveillance Program Report; and as chair of the NRC Committee to Review the National Institute for Occupational Safety and Health Respiratory Disease Research Program. He was a member of the Board on Environmental Studies and Toxicology (BEST) and the Committee on Research Priorities for Airborne Particulate Matter. He received his M.D. from the Tufts University School of Medicine.

STAFF

David A. Butler, Ph.D., is a scholar in and the director of the Office of Military and Veterans Health in the Health and Medicine Division of the National Academies of Sciences, Engineering, and Medicine. He earned his B.S. and M.S. degrees in engineering from the University of Rochester and his doctoral degree in public policy analysis from Carnegie Mellon University. Before joining the National Academies, Dr. Butler served as an analyst for the U.S. Congress Office of Technology Assessment, was a research associate in the Department of Environmental Health of the Harvard School of Public Health, and conducted research at Harvard's Kennedy School of Government. He has directed several National Academies studies on military and veterans health, environmental health, and risk assessment topics, including ones that produced *Research on the Health Effects of Low-Level Ionizing Radiation—Opportunities for the Armed Forces Radiobiology Research Institute*; *Future Uses of the DoD Joint Pathology Center Biorepository*; *Provision of Mental Health Counseling Services Under TRICARE*; *PTSD Compensation and Military Service*; *Veterans and Agent Orange: Update 1998*, and *Update 2000*; *Disposition of the Air Force Health Study*; and the report series *Characterizing the Exposure of Veterans to Agent Orange and Other Herbicides Used in Vietnam*. Dr. Butler was also a co-editor of *Systems Engineering to Improve Traumatic Brain Injury Care in the Military Health System*.

Anne N. Styka, M.P.H., is a program officer in the Health and Medicine Division at the National Academies. Over her tenure she has worked on a broad range of topics related to the health of military and veteran populations, including mental health treatment offered in the Department of Defense and the Department of Veterans Affairs (VA), epidemiological research using VA data, and directing a research program of fostering new research studies using data and biospecimens collected as part of the 20-year Air Force Health Study. Before coming to the Academies, Ms. Styka spent several years working as an epidemiologist for the New Mexico Department of Health and the Albuquerque Area Southwest Tribal Epidemiology Center and even spent several months in Zambia as the epidemiologist on a study of silicosis and other nonmalignant respiratory diseases among copper miners. She has several peer-reviewed publications and has contributed to numerous state and national reports. She received her B.S. in cell and tissue bioengineering from the University of Illinois at Chicago and has an M.P.H. in epidemiology from the University of Michigan. Ms. Styka was the 2015 recipient of the Institute of Medicine and National Academy of Medicine Multitasker Award and a member of the 2011 National Academies' Distinguished Group Award.

Cary Haver, M.P.H., is a program officer in the Health and Medicine, and has worked with the Committee on Gulf War Health: Treatment of Chronic Multisymptom Illness; the Committee to Review the Department of Labor Site Exposure Matrix; and the Committee on Long-Term Health Consequences of Exposure to Burn Pits in Iraq and Afghanistan. Before joining the National Academies, Ms. Haver worked for Tetra Tech Sciences, a consulting group dedicated to a variety of environmental and occupational epidemiology and toxicology projects, including exposures to asbestos, chromium, and mercury. She earned her M.P.H. with a concentration in epidemiology from George Washington University.

Pamela Ramey-McCray is an administrative assistant in the Health and Medicine Division. She has worked to support numerous studies on military and veterans health, malaria research, and studies on U.S. veteran twins since coming to the National Academies in 1993. Ms. Ramey-McCray is a recipient of the Institute of Medicine's 2009 Veteran Award. She earned her bachelor's degree in human relations at Trinity Washington University in Washington, DC. Before coming to the National Academies, Ms. Ramey-McCray worked for the American Psychological Society and the Consumer Product Safety Commission.

Sulvia Doja, M.S.H.C.P.M., was a research associate at the National Academies through August 2016, and worked on studies addressing occupational and environmental exposures, health diagnostic criteria, and treatment guidelines. She earned her masters of science degree in health care policy and management from Carnegie Mellon University's H. John Heinz III College and her undergraduate degree in biology with a minor in chemistry from Chatham College. Before coming to the National Academies, she researched the effect of national health care reform on American Indian and Alaskan Native populations and pioneered a social-media marketing outlet to increase customer visibility and enhance communication with that population.

Nicole Freid was a senior program assistant at the National Academies until October 2016. She worked to support several studies addressing occupational and environmental exposures among the veteran population. She earned her bachelor's degree in economics and political science at American University in Washington, DC. Before coming to the National Academies, she worked at Avalere Health, a health care consulting firm, where she incorporated health policy analysis in final deliverables for pharmaceutical and medical device manufacturers.